T0182147

The Anthropocene: Politik—Economics—Society—Science

Volume 26

Series editor

Hans Günter Brauch, Mosbach, Germany

More information about this series at http://www.springer.com/series/15232
http://www.afes-press-books.de/html/APESS.htm
http://www.afes-press-books.de/html/APESS_26.htm

Amar K. J. R. Nayak
Editor

Transition Strategies for Sustainable Community Systems

Design and Systems Perspectives

 Springer

Editor
Amar K. J. R. Nayak
Professor of Strategic Management
Xavier Institute of Management
Xavier University Bhubaneswar
Bhubaneswar, India

More on this book is at: http://www.afes-press-books.de/html/APESS_26.htm

ISSN 2367-4024 ISSN 2367-4032 (electronic)
The Anthropocene: Politik—Economics—Society—Science
ISBN 978-3-030-00355-5 ISBN 978-3-030-00356-2 (eBook)
https://doi.org/10.1007/978-3-030-00356-2

Library of Congress Control Number: 2018963843

Series Editor: PD Dr. Hans Günter Brauch, AFES-PRESS e.V., Mosbach, Germany
English Language Editor: Dr. Vanessa Greatorex, Chester, UK

The image in the cover page is based on the TSSCS Framework, Bhubaneswar, India. The image has been developed by the Editor and permission has been received for it to be used here

This Springer imprint is published by the registered company Springer Nature Switzerland AG
The registered company address is: Gewerbestrasse 11, 6330 Cham, Switzerland

Foreword

This book, on Transition Strategies for Sustainable Community Systems (TSSCS) and the TSSCS Symposium it amplifies to a larger audience, represents a powerful and robust spinning together, by hundreds of people, of five vibrant strands (called here dimensions or themes) of thinking and action, the better to explore and achieve Sustainable Community Systems in rural India. We have all been so greatly enriched by participating in this spinning, publicly, for the past four years, that I am humbled and honoured to offer a brief Foreword at the invitation of the National Apex Bank for Agricultural and Rural Development (NABARD) Chair Unit at Xavier University Bhubaneswar, Odisha, India, and grateful to its academic and organising partners convened by Prof. Amar K. J. R. Nayak and his teammates.

A thirty-three-year career in the UN Food and Agriculture Organisation (FAO), during each year of which I worked in agricultural fields and communities (mostly rice paddies and their villages) in India, embolden me to offer spotlights on a very few comparative international examples that could benefit from the thread spun together in this collective work. During these four years, most of TSSCS' collaborative work drew out strands from sustainable agricultural systems, organisations, governance and institutions, each of which could reflect more or less autonomous units in FAO, including my earlier home Division: Plant Production and Protection. It has been refreshing that the TSSCS' work style and values have vigorously and relentlessly punctured the intellectual and managerial/bureaucratic walls isolating these strands or dimensions, enabling all of us to listen better to each other. For these four strands, and the sections of the book they comprise, this active demolition of conventional professional silos or strongholds has strengthened and deepened the discussions and reflective analyses the readers will join.

After hearing, reading and reflecting on nearly all the papers, however, I choose to focus this Foreword on the fifth dimension or strand, found at the core of the "mandala". The TSSCS' initiative uniquely sets and keeps at the centre this strand: *Relationships*. The exemplary attractive variable in this dimension is *trust*. Four decades of international experience of village-level organisations, institutions and governance convince me that trust is always necessary for a successful transition to sustained community systems. Only trust **within** the community will protect new institutions and organisations. Trust is certainly a truism of the literature from management "gurus" to (not so) "new" institutional economics. But it is even more essential in the life or death of local community systems. It is one of the very few necessary variables to confront and disassemble inequality, which otherwise grows without limit in the present time. Trust is necessary for rural households to keep together the vital relationships among families, across castes, classes, genders and ages, even when under pressure from "natural" disasters, wars or civil unrest, and "boom and bust" price cycles and environmental overshoots driven by market forces. I humbly submit that a major methodological key to the growing success of the TSSCS' approach can best be seen in how the collaborators keep trust—and its fellow relationship neighbours like faith, morals, ethics, belief, interdependence and identifying oneself with others—alive and present in so many formal discussions and informal conversations ostensibly started about farmer producer organisations, Panchayati Raj institutions, women's self-help groups, or soils and moisture.

Moral Rice, Dharma Garden Temple and Buddhist Eco-spirituality in Yasothon Province, North-East Thailand.
This exploration demonstrates how stronger and healthier relationships can increase benefits among rural Asian villagers growing the same crops for the same range of local to national to global consumers in the same production ecosystems. Yasothon Province is part of the Isan region in north-east Thailand on the Lao–Thai–Cambodian border. The critical comparative study (Kaufman and Mock 2014) demonstrated that a Buddhist temple-based group of organic rice farmers held deeper Buddhist eco-spiritual values than a non-profit, but non-temple-based group of organic farmers in the same villages who retained a more commercial world view. Both groups had access to government-sponsored subsidies that encouraged them to plant organic rice crops, including organically certifiable fertiliser.

While yields and incomes did not differ significantly, the Buddhist temple group had a statistically significantly higher percentage of members, however, who *eliminated their debts* after becoming more spiritual as well as adopting more organic methods of farming. A significantly higher proportion of the Buddhist temple group also perceived that their physical health improved as they continued to practise organic agriculture; the more debt farmers had to carry, the less healthy they felt. The Buddhist temple group members eliminated the use of agrochemicals, used organic fertilisers, ate mostly or exclusively organic food and found increased biodiversity on their farms. When the global price of (non-organic) rice spiked in 2009, a larger proportion of non-temple-based organic rice farmers abandoned their organic crops and simply grew a second conventional crop of rice to take advantage of that price increase.

The context for this comparison grew out of Thailand's religious traditions, starting with local animist relationships with sacred groves or trees, and wild animals such as propitiating the Rice Soul, the Rice Mother, the Earth Mother and the River Mother, all reinforcing the role of rice as a live-giving force. Theravada Buddhism spread from Sri Lanka and has been practised in Thailand since the sixth to ninth centuries CE, i.e. over for 1,200–1,500 years. The core principles emphasised in Theravada are (I) the five precepts of the Pali canon: 1—not to kill; 2—not to steal; 3—not to engage in sexual misconduct; 4—not to lie; and 5—not to consume intoxicants; (II) the Eightfold Path; (III) Dependent Origination (*Paticcasamuppada*) and (IV) the Law of Karma (connected with III) (Kaufman 2012).

The Green Revolution arrived in the 1970s; most Central Thailand rice farmers adopted seeds, inorganic fertiliser and agrochemicals. Private corporate discounts and bonuses together with government subsidies increased the rates of adoption of the Green Revolution along with abandoning traditional practices based on local ecological and technical knowledge. Small enclaves of farmers, including those affected by amplified price swings, drew upon the local Buddhist and syncretic animist traditions and explored market opportunities for organic rice and other products. When Thailand was severely hit by the exchange-rate crisis of 1998, the government, which had until then supported Green Revolution practices, began to encourage organic production.

Many farmers in the Isan region, already at lower income levels than Central Thai rice bowl farmers, explored organic production to reduce price fluctuation-driven risks. The King, HMK Bhumibol or HMK Rama IX, encouraged a national sufficiency economy to reduce the costs of production, and reinforced traditional veneration of rice spirits and local animist traditions. Monks in temple or monastery communities emphasised the ecological dimensions of Theravada Buddhist spirituality. Charismatic individual monks and priests built followings that included laypersons. As in many economically poorer rural villages, men tended to seek employment in larger cities. Women, young people and older men became the largest proportion of farm operators. These farmers joined and became more active in village-level organisations, such as the Dharma Garden Temple's "Moral Rice" network. Farmers who did not share as deep a commitment to Buddhist eco-spirituality often joined marketing collectives that provided cheaper inputs but without the high level of solidarity of the temple-based group. Both types of group member had access to local, national and international markets, as Thailand is traditionally among the highest rice-exporting countries, and people around the world are familiar with Thai rice. This is how the comparison between demographically similar groups who showed higher and lower degrees of Buddhist devotion emerged. The comparison revealed one of the clearest examples of how twenty-first-century family farmers in a global economy become more resilient in economic and health terms through stronger relationships provided by a religious organisation.

My second exploration starts from Chap. 8 in the Relationships section, by Prof. Praveen Kumar, on *Paticcasamuppada*, or Dependent Origination, already mentioned as a core belief of Buddhism. The strong connections between more than two actions or thoughts, or economic institutions, mean that people have the power to change their situations, their position, to a new and different outcome, arising from their actions, because they acted differently, in deeper understanding of the multiple forces acting on each other. Professor P. Kumar gives an especially vivid metaphor of the origination of greed—as a passion for eating the earth that grows out of control until it destroys relationships and, ultimately, the environment.

This metaphor echoes a famous paragraph written over 60 years ago by the economic historian Prof. Karl Polanyi. It is found in *The Great Transformation*, Polanyi's masterpiece on the "political and economic origins of our time". This subtitle clearly echoes his exploration in a direction paralleling *Paticcasamuppada*, to understand multiple interdependent originations. The paragraph examines the impact of the exploding scope and power of markets on natural resources (in this case, soil) and people.

> ... leaving the fate of soil and people to the market would be tantamount to annihilating them. Accordingly, the countermove consisted in checking the action of the market in respect to the factors of production, labor, and land. This was the main function of intervention (Polanyi 1957, p. 131).

Two more recent scholars of Buddhism, who each went on to have long academic careers and published a number of books and hundreds of articles, chose to write their Ph.D. dissertations (each of which then became a book) on philosophical interpretations of causality in Buddhism, concentrating on *Paticcasamuppada*. Professor David Kalupahana was from Sri Lanka, studied there and at the University of London, then became the Chair of the Philosophy Department and Senior Professor of Buddhism at the University of Hawaii. He focused, including with grammatical analysis, on the earliest Buddhist texts, mostly in Pali, and sought to demystify the concept of incarnations carrying the sins of earlier lives, replacing it with interdependent origination during the current life. In later publications, he pointed out numerous parallels between *Paticcasamuppada* in Buddhist causality with the position of American pragmatism, especially of William James, C. S. Peirce and later John Dewey. The position of that group of philosophers was that people are fully part of the environment around them, so that the perceptions and responses by a person to the world are created together with all other people who interact with that person.

Professor Joanna Macy, who is still publishing as emerita professor, also looked into early Pali writings, but then focused her dissertation, which was also about Buddhist causality, on parallels with the General Systems Theory she had studied with Professor Ervin Laszlo. General Systems Theory evolved in close association with biology, both physiology and ecology. Professor Macy went on to publish extensively on systems approaches to mindful action in the modern, including twenty-first century, world from Buddhist and deep ecology perspectives.

Both these scholars provided paths for us to explore applications of *Paticcasamuppada*.

A final exploration and inspiration comes from the work of Dr. B. R. Ambedkar. Dr. Ambedkar applied *Paticcasamuppada* to underpin his argument against the existence of an omnipotent God as part of the Buddha's Dharma (Book 3, Part IV, Section 2, paragraphs 50–58, of *The Buddha and His Dharma*). *Paticcasamuppada* provides a coherent logic and applicable tools to analyse causes in order to lead to a path of action. Among many other instances, Dr. Ambedkar applied and acknowledged the American pragmatist philosopher, Prof. John Dewey of Columbia University, in his classic (and undelivered) lecture on *Annihilation of Caste* (1936, reprinted 2016; Chap. 25, Sect. 25.4, p. 313):

> Secondly, the Hindus must consider whether they should conserve the whole of their social heritage, or select what is helpful and transmit to future generations only that much and no more. Prof John Dewey, who was my teacher and to whom I owe so much, has said:

> 'Every society gets encumbered with what is trivial, with dead wood from the past, and with what is positively perverse…As a society becomes more enlightened, it realizes that it is responsible not to conserve and transmit the whole of its existing achievements, but only such as make for a better future society.'

I hope these explorations indicate the potential range and power that the TSSCS approach, centred around relationships but interrogating and applying the other four strands, contribute deeply to investigating practical questions on transitions to sustainable community systems.

Peter Kenmore
FAO-UN, California, USA

References

Ambedkar, B.R. (Bhimrao Ramji). 2016. *The Annihilation of Caste: The annotated critical edition*/edited and annotated by S. Anand. London and New York: Verso.

Ambedkar, B.R. (Bhimrao Ramji). 2006. *The Buddha and His Dharma*. Delhi: Siddarth Books.

Falvey, John L. 2000. *Thai Agriculture: Golden Cradle of Millennia*. Bangkhen: Thailand. Kaesesart University Press. https://www.researchgate.net/publication/236694126_Thai_Agriculture_Golden_Cradle_of_Millennia. (30 November 2016).

Kalupahana, David J. 1989. Toward a middle path of survival, in Callicott, J. Baird and Roger T. Ames, eds. 1989. *Nature in Asian Traditions of Thought: Essays in Environmental Philosophy*. Albany: State University of New York Press pp. 247–256.

Kalupahana, David J. 1975. *Causality: The Central Philosophy of Buddhism*. Honolulu: University Press of Hawaii.

Kaufman, Alexander H. 2012. Organic farmers' connectedness with nature: Exploring Thailand's alternative agriculture network. *Worldviews: Global Religions, Culture, and Ecology*, Volume 16, Issue 2, pp. 154–178.

Kaufman, Alexander H.; Jeremiah Mock 2014. Cultivating greater well-being: The benefits Thai organic farmers experience from adopting Buddhist eco-spirituality. *Journal of Agricultural and Environmental Ethics*. December 2014, Volume 27, Issue 6, pp. 871–893.

Kaufman, Alexander; Nikom Petpha. 2016. Moral rice network, Dharma Garden temple, Yasothon province, Northeast Thailand, in Loconto, in A., Poisot, A.S. & Santacoloma, P. (eds.), *Innovative Markets for Sustainable Agriculture—How Innovations in Market Institutions Encourage Sustainable Agriculture in Developing Countries*. Rome: FAO/INRA. pp. 181–199 (Chapter 10).

Kumar, Praveen. In press. *Paticcasamuppada (The Theory of Dependent Origination): A Scientific Means of Changing Person*. The present volume, Chapter 8.

Macy, Joanna. 1991. *Mutual Causality in Buddhism and General Systems Theory: The Dharma of Natural Systems*. Albany: State University of New York Press.

Nayak, Amar K.J.R. 2019. *Transition Challenges and Pathways to Sustainable Community Systems: Design and Systems Perspectives*. The present volume, Chapter 1.

Polanyi, Karl. 1957. *The Great Transformation*. Boston: Beacon Press.

Preface

There has been increasing realisation that the mainstream logic of economic growth and development is in jeopardy and not sustainable. The key pillars of the economy, environment and society have gradually drifted away from their natural principles and values and have weakened, making communities and societies unstable, vulnerable and unsustainable.

While the intention of many of our initiatives towards economic growth and sustainability—whether in production systems, producer organisations, community governance or institutions—may have been noble, inconsistencies between the intention and approaches, methods, tools, techniques and actions often make these initiatives unsustainable over time. The inherent technical inconsistencies and tensions in the designs, and lack of synergy across different sub-systems, perpetuate lock-in effects and greater external control than freedom.

In the light of the above impending/socio-economic–environmental crisis, this book is an effort to discuss the possible transition strategies and pathways towards facilitating Sustainable Community Systems. It presents theoretical, practical and policy perspectives and empirical evidences of the key dimensions of any community systems: relationships, institutions, production, organisation and governance.

The journey towards this exploration began around 2005, when I had finished my doctoral studies on foreign direct investment strategies of some of the most successful multinational corporations and had begun an extensive study on the explosive growth strategies of large Indian multinational enterprises. Subsequently, three international editions of my books on *Foreign Multinationals in India*, *Indian Multinationals* and *India in the Emerging Global Order* were published. My courses on International Business and Strategic Management of Indian Multinationals were doing very well with the students of a major business management school in India.

However, my past twenty years of engineering and management studies, experience of working with large corporations and scholarship on globalisation and foreign direct investment in India and Japan, prior to 2005, had raised several queries and doubts in my mind as to whether the mainstream theory, policy and practice would stand the test of time from a sustainability point of view.

I wondered whether the very design of large corporations that helped these firms to grow was at the root cause of unsustainability of communities and society at large. This inner voice of dissent regarding what I was taught and was teaching successfully led me to formulate an action research project on developing organisational design for sustainable community enterprise systems.

This small action research project was initiated in a remote and neglected tribal community in Odisha, India, in 2007, using my own limited personal savings. It subsequently received the attention of development agencies like NABARD and Rabobank Foundation that supported the action research. This study was gradually complemented with other related studies by several other organisations. The interesting and surprising outputs of these studies led to a series of seminars, workshops and round-table discussions from 2012 onwards, and a very large symposium in January 2017. The scholarly contributions of the 2017 symposium and "Transition Strategies for Sustainable Community Systems" form the contents of this volume.

Based on the ongoing research and deliberation on the five key dimensions of sustainability, this book has been presented in five parts, focusing on relationships, institutions, production, organisation and governance. Chapter 1 provides an overall design and systems perspectives of the above five dimensions and on respective factors associated with each of the dimensions. The last chapter is a synthesis of the valuable contribution of all the chapters under different dimensions and the possible way forward.

Bhubaneswar, India   Amar K. J. R. Nayak

Acknowledgements

This book has been an outcome of an engaging symposium of senior academics, practitioners, policy-makers and young researchers in sustainability. Over 500 members participated in different forums over a period of about two years, and nearly 150 members participated in the three-day symposium in January 2017. I wish to record the contributions of all these members in making this book. Its chapters emerged from papers presented during the three-day symposium in 2017. I would like to express my deep gratitude to all the authors for their scholarly contributions.

I would like to express my sincere gratitude to His Excellency, Dr. S. C. Jamir, Governor of Odisha, Prof. V. S. Vyas, Prof. Y. K. Alagh, Prof. A. Vaidyanathan, the late Dr. Pushpa M. Bhargava, Dr. Peter Kenmore, Prof. K. V. Raju, Prof. G. Krishnamurthi, Mr. Subhash Mehta and Prof. Anup Dash for their scholarly guidance and support.

Several institutions have supported this cause, and I would like to thank and record the contributions of the Xavier Institute of Management, Xavier University Bhubaneswar, the National Bank for Agriculture and Rural Development, the TATA Trusts, the Development Management Institution, Patna, the National Institute of Agricultural Marketing, Jaipur, the Indian Council of Agricultural Research, the Government of India, IMAGE, the Government of Odisha, NISWASS Bhubaneswar, and the Sustainability Trust, Bhubaneswar.

I record deep appreciation for Dr. H. K. Bhanwala, Dr. U. S. Saha, Mr. Shobhana K. Pattanayak, Mr. R. Venkataramanan, Dr. (Mrs.) Irina Garg, Dr. Trilochan Mohapatra, Shri M. V. Ashok, Dr. K. C. Panigrahi, Mr. Manoj Ahuja, Dr. Bimal Kumar Das, Mr. Jitendra Nayak and Mr. Pradyut Kumar Bag for their financial support and co-operation. I greatly appreciate (Dr.) Paul Fernandes, SJ, E. A. Augustine, SJ, (Dr.) Donald D'Silva, SJ, (Dr.) Lourduraj Ignacimuthu, SJ and (Dr.) V. Arockia Das, SJ for all their administrative support and encouragement.

I would like to thank my research support team of Mr. Bibhuti Sahoo, Mrs. Radha Rani Ray and Mrs. Sunita Mishra, and my doctoral colleagues, Mr. Asish Panda, Mr. Amar Patnaik and Mrs. Usha Padhee, for their valuable and timely contributions. I also thank Dr. Johanna Schwarz of Springer Nature and Dr. Hans Günter Brauch of AFES-PRESS, the editor of this book series, for their support and co-operation in getting this book published. Without their meticulous and anonymous review process, this book could not have reached this standard before publication.

I would like to thank Dr. Hans Günter Brauch, who was the first to show interest in this conference compilation. He showed great patience and commitment to get it peer reviewed before publication. Without the consent from Dr. Johanna Schwarz of Springer, this book would not have gone through the process. I express my deep gratitude to English Language Editor Dr. Vanessa Greatorex for helping to make this book reader-friendly. I also thank Mr. Arulmurugan Venkatapalam, the Project Coordinator of Springer, and his devoted and superb technical team in Chennai, Tamil Nadu, for his efficient coordination and support.

Bhubaneswar, India   Amar K. J. R. Nayak
July 2018

Contents

Abbreviations

AKRSP	Aga Khan Rural Support Programme
APEDA	Agriculture Products Export Development Authority
ATMA	Agriculture Technology Management Agencies
BCI	Better Cotton Initiative
BPL	Below Poverty Line
BR	Bio-physical Resilience Index
BRLPS	Bihar Rural Livelihoods Project Society
BT	Bacillus Thuringiensis
CAB	Cotton Advisory Board
CACP	Commission for Agricultural Costs and Prices
CCS	Credit Co-operative System
CEDEC	Centre for Development Education and Communication
CICR	Central Institute for Cotton Research
CLCuV	Cotton Leaf Curl Virus
CSO	Central Statistical Organisation
DAS	Days After Showing
DCF	Discounted Cash Flow
DEA	Data Envelopment Approach
DES	Directorate of Economics and Statistics
DFID	Department for International Development
DIME	Design, Implementation, Monitoring and Evaluation
DPAP	Drought Prone Area Programme
EF	EcoFarming
ERI	Economic Resilience Index
FAO	Food and Agriculture Organisation
FGD	Focus Group Discussions
FICCI	Federation of Indian Chambers of Commerce and Industry
FPC	Farmer Producer Companies
FPO	Farmer Producer Organisations
FSR	Farming System Research

FYM	Farm Yard Manure
GM	Genetically Modified
GOI	Government Of India
GOTS	Global Organic Textile Standard
GP	Gram Panchayat
GSDP	Gross State Domestic Product
GWF	Grey Water Footprint
HDP	High Density Planting
HH	Households
HRI	Health Resilience Index
IAD	Institutional Analysis and Design
IARC	International Agency for Research on Cancer
IARI	Indian Agricultural Research Institute
ICAC	International Cotton Advisory Committee
ICAR	Indian Council of Agriculture Research
IIT	Indian Institute of Technology
IRDP	Integrated Rural Development Programme
IRMA	Institute of Rural Management
ISOT	Indian Standard for Organic Textiles
ITDA	Integrated Tribal Development Authority
IWC	In situ Water Conservation
JAS	Japanese Agricultural Standard
KVIC	Khadi and Village Industries Commission
KVK	Krishi Vigyan Kendras
KVS	Kisan Vikas Samiti
LDL	Low-Density Lipoprotein
LP	Linear Programming
LRI	Livelihoods Resilience Index
MANAGE	National Institute of Agriculture Extension Management
MCDM	Multiple Criteria Decision-Making
MDG	Millennium Development Goals
MGNREGS	Mahatma Gandhi National Rural Employment Guarantee Scheme
MPI	Multidimensional Poverty Index
MPWPCL	Madhya Pradesh Women Poultry Producers Company Ltd
MSP	Minimum Support Price
NABARD	National Apex Bank for Agricultural and Rural Development
NAIP	National Agricultural Innovation Project
NAIS	National Agricultural Innovation System
NARS	National Agricultural Research System
NCDC	National Co-operative Development Corporation
NCF	National Commission for Farmers
NDC	National Development Council
NDDB	National Dairy Development Board
NFDB	National Fisheries Development Board
NHB	National Horticulture Board

NISWASS	National Institute of Social Work and Social Sciences
NOP	National Organic Program
NRA	Natural Resource Accounting
NSSO	National Sample Survey of Organisation
OBC	Other Backward Classes
OCS	Organic Content Standard
ODF	Open Defecation Free
OLS	Ordinary Least Squares
OUAT	Orissa University of Agriculture and Technology
PAN	Pesticide Action Network
PIN	Professional Institutional Network
PO	Producer Organisations
PRA	Participatory Rural Appraisal
PRI	Panchayati Raj Institutions
RCP	Real Cost of Production
RRA	Rapid Rural Appraisal
RWH	Rain Water Harvesting
SADA	State Agricultural Development Agency
SAU	State Agriculture Universities
SC	Scheduled Caste
SDG	Sustainable Development Goal
SEEA	System of Environmental Economic Accounting
SFA	Stochastic Frontier Approach
SFAC	Small Farmers Agribusiness Consortium
SFDA	Small Farmer Development Authority
SGSY	Swarnajayanti Gram Swarozgar Yojana
SHG	Self-help Groups
SI	Simpson Index
SLA	Sustainable Livelihoods Approach
SLF	Sustainable Livelihoods Framework
SMART	Specific, Measurable, Attainable, Realistic and Time-bound
SNA	Systems of National Accounts
SRI	System Root Intensification
SRI	Social Resilience Index
SSE	Social and Solidarity Economy
TFP	Total Factor Productivity
TLC	Total Literacy Campaign
TSC	Total Sanitation Campaign
TUA	Tank Users Associations
USDA	United States Department of Agriculture
VOC	Volatile Organic Carbons
WHO	World Health Organisation
WHS	Water Harvesting Structures

List of Figures

List of Tables

Chapter 1
Introduction: Transition Challenges and Pathways to Sustainable Community Systems: Design and Systems Perspectives

Amar K. J. R. Nayak

Abstract This chapter provides a background to the various dimensions of sustainability that we understand as the critical pillars of sustainability for any community system. Based on these five key dimensions, the contributions in this book have been categorised into five broad sections. This chapter discusses these five critical dimensions – relationships, institutions, production, organisation and governance – from design and systems perspectives for the sustainability of any community system. Under each dimension, critical factors are presented as a spectrum; where characteristics of each factor at one end of the spectrum tend to create locking-in effects, control by a few, inequity and unsustainability in the long run, while at the other end characteristics of the same factors facilitate equity, freedom and sustainability. With reference to characteristic of each factor, the chapter inherently argues that design for sustainability needs to be anchored on the natural principles of interconnections and interdependence.

Discussions in the chapter uncover that although sustainability appears to be fuzzy, blurred and impossible; making our communities and society sustainable is within our collective choices in the way we choose the direction of each of these factors under the five critical dimensions. Figures 1.1 and 1.2 of this chapter summarise the critical dimensions and the spectrum of each of the twenty-five factors. In the final analysis, this chapter implies that if people in small communities and various external stakeholders, including government officials, development and corporate executives, recognise the current design flaws and simultaneously make efforts to unlock themselves from the various lock-in effects, our transition to sustainability will be easily achievable.

Keywords Transition strategies · Sustainable communities · Relationships Institutions · Production · Organisation · Governance · Design Systems science

Amar K. J. R. Nayak, Professor of Strategic Management, Xavier Institute of Management, Xavier University Bhubaneswar, Odisha, India; Email: amar@ximb.ac.in

© Springer Nature Switzerland AG 2019
A. K. J. R. Nayak (ed.), *Transition Strategies for Sustainable Community Systems*,
The Anthropocene: Politik—Economics—Society—Science 26,
https://doi.org/10.1007/978-3-030-00356-2_1

1.1 Background

While the intentions in many of our initiatives towards sustainability have been noble; inconsistencies between the intent and approaches, methods, tools, techniques and actions often make these initiatives unsustainable over time. The inherent inconsistencies and tensions in the designs, and lack of synergy across different sub-systems – even in small communities – perpetuate lock-in effects, path dependencies and greater external control than freedom, leading to inequity and unsustainability.

The chapter has five parts: *First*, it discusses the basic postulate underlying the design and systems perspective in terms of the nature and state of relationships in society. *Second*, it presents the rationale behind the choice of the five major dimensions of sustainability around which this chapter has been structured. *Third*, it discusses the factors within each dimension, the spectrum of each factor, inherent tension within the spectrum of each factor, dynamics of relationships and lock-in processes within a dimension and across the dimensions. *Fourth*, it proposes a potential narrow but simple path to unlock sustainability. *Fifth*, it indicates the potential areas of research to understand sustainability better from design and systems perspectives.

First, with regard to the basic premises underlying the design and systems perspectives in terms of the nature and state of relationships in society, there are indeed deep interconnections and a high degree of interdependence between nature and our society. One can look at the workings of a clock to understand the interconnections and interdependence. The connection between the second, minute and hour hands can indicate the nature of interconnections, but the reality could be much more complex. Interconnected gears in a machine can illustrate it a little better. However, the interconnections that actually occur in society across time and space appear to be much deeper and more complex. Deep interconnections and interdependence in natural systems and the web of human life need to be accordingly considered in our new designs from a systems perspective.

There have been a number of studies that suggest there are growing economic inequalities between people across the world, weakening their socio-economic-political relationships. Piketty (2014) shows that global income inequality has increased in the last hundred and ninety years or so, and that the *Gini coefficient* has been increasing during this period. Income inequality in the emerging countries has been growing much faster in recent decades. For instance, in India, income inequality was lower during the 1950s to 1980s, when it was a social economy. Subsequent to the 1980s, when it started liberalising, inequality in terms of the *Gini coefficient* has been rising steadily. Further, wealth inequalities across countries have been much greater than income inequalities. We often categorise these phenomena as development paradoxes. Should we not discover why this happens and what causes these increases in inequalities; factors of unsustainability?

These inequalities do not seem to be only a phenomenon of the last two hundred years. Rousseau (1762) wrote about the inequalities over three hundred years ago. In his book, *The Social Contract*, Rousseau wrote: "Man is born free and everywhere he is in chains. One man thinks himself the master of others but remains more of a slave than they are." Income and material inequality has been a constant source of fear and loss of control. As far back as four thousand years, *Chakra Vyuha* and *Labyrinths* from across the world show the symbols of fear and loss of control by the common man. Today's control mechanisms may not just be physical or material controls, but could also entail psychological and mental control of the individual by market forces.

Globally, the seventeen noble goals specified in the Sustainable Development Goals 2030 of the United Nations and ratified by the nation states were formulated to transform the world. However, in the current state of increasingly unequal relationships in our communities, nations and society at large, can these goals be achieved?

The significance of relationships for the sustainability of our communities is expressed in the working definition of sustainability:

> Sustainability is a dynamic state of deep relationships among the people and all the constituents both living and non-living within a micro ecological unit that strongly values the acts of sacrifice, reciprocity and love for each other; where the priority is to strengthen the weakest, and the spirit of high internal competition with high external co-operation not only drives its own ecological unit to peace, joy and happiness but also inspires other micro ecological units for such deeper inter relationships (Nayak 2011: p. 117).

In our analysis, relationships are studied from the systems science perspective and spirituality (including philosophical and theological) perspective. Further, relationships are core to all other dimensions of study: institutions (both formal and informal) at different levels of our society, production systems (with specific reference to sustainable agricultural systems), organisations (including producer organisations), and governance (with special reference to grass-root-level community governance).

Second is the rationale behind the dimensions and factors, the boundaries in terms of design thinking and systems perspectives. These have evolved during the last two decades through the various deductive and inductive studies, engaging action research with rural agricultural communities and involvement in policy formulation. They strongly present a compelling need to understand the lock-in mechanisms of our designs and systems within and across these five dimensions.

Although our research and action research have largely focused on production in the primary sector – namely the agricultural sector and the organisational design of *farmer producer organisations* (FPOs) as community enterprise systems – the issues of institutions and community governance have been critical to the sustainability of the first two dimensions. Further, the factors of social capital, trust and co-operation have been fundamental to the sustainability at all these levels. Accordingly, five dimensions – addressing Relationships (I), Institutions (II),

Production (III), Organisation (IV) and Governance (V) – have been chosen for analysis and discussion.

A community system not only consists of a certain number of families in a cluster of villages but also a certain extent of geographical spread with its ecosystem. Accordingly, a micro-watershed, the basic unit of ecology, forms the technical base of our community system. However for optimisation of size for economic viability (especially under a diverse production system), and easier identification of physical boundaries which affect the people, a *Gram Panchayat* (GP), or Ward that may consist of two to four micro-watersheds depending on whether it is in a hilly, plain or coastal region, has been used as a community system for the analysis.

Interestingly, countries across the world are structured according to similar primary boundaries of their respective local communities; examples include the Gewog in Bhutan, the Parishad Council in Bangladesh, the Grama Niladari in Sri Lanka, and the Union Council in Pakistan. Countries of Africa have a similar political/administrative governance structure. For instance, Tanzania has Ward at the lowest level and District/Municipality at the next upper level. The United Kingdom does not have such structures but has the traditional, informal institutional structure of Parish at the lowest level of community organisations. Advanced industrial economies usually have Ward at the lowest level and District at the higher level. In other words, as small community systems exist in countries around the world, it would be prudent to facilitate holistic sustainability initiatives in such small communities in the respective nations to attain the gradual but overall sustainability of a nation.

Third relates to the factors within each dimension, the spectrum of factors in each dimension, tensions within the spectrum, dynamics of relationships among factors within a dimension and lock-in processes. Figure 1.1 shows a list of the factors within different dimensions. Interestingly, it also shows the order of layering, from core to periphery and high interdependence among these five different dimensions.

Each of the five critical dimensions of a sustainable community system is being studied from a few specific factors. The key factors in each of the five dimensions are discussed briefly in this section.

1.1.1   Relationships

The nature and state of relationships often shape the sustainability of our endeavours in institutions, production systems, organisations and community governance. Indeed, relationships seem to have been at the core of human engagements and our endeavours towards sustainability. We explore such relationships from the systems science and spirituality perspectives.

From a systems science perspective, we deliberate on the notions of *interdependence* and identifying self with others around us, and on the notions of *capital*

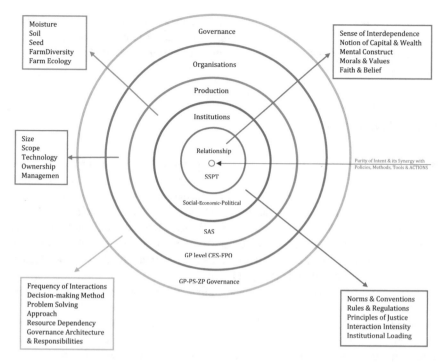

Fig. 1.1 Transition strategies for sustainable community systems. *Source* The author, 28 October 2016; www.xub.edu.in/NABARD-Chair/activites.html

and wealth. The spirituality perspective has been discussed through philosophical and theological lenses. The broad factors have been interpreted in terms of the *morals and values* and deep-rooted *faith and beliefs* that probably shape our behaviour, and actions that are either in sync with or contrary to the principles of sustainability. The factor *mental construct* has been studied from both a systems science perspective and a spirituality perspective.

1.1.2 Institutions

The nature and type of institutions at the district, state, national and global levels can either facilitate or destroy sustainable principles adopted in the other four inner layers, namely, community governance, production organisation, production systems, and culture of relationships at the core. Even if there are inconsistencies in the institutions at the higher levels from district to global level, coherence of institutions within a district that is at district, block or GP-Ward level can greatly facilitate sustainability at the other four inner layers – Gram Panchayat (GP), FPOs, Production Systems and Relationships. Accordingly, analysis in this dimension includes five key factors of institutions – *norms and conventions, rules and regulations, principles of justice, interaction intensity,* and *institutional loadings* – such that external institutions facilitate communitarian principles at lower levels of governance.

1.1.3 Production

While production would often include primary, secondary and tertiary sectors, we have just used agricultural production in the primary sector for our analysis. Agriculture is greatly being impacted by climate changes and is increasingly becoming unsustainable across the world. It appears that in the course of taking agriculture forward for greater productivity through intensive external inputs, we have made agriculture unsustainable by making small farmers unviable, increased risks to food safety, and accelerated environmental degradation through our current industrial production methods of agriculture.

Agricultural production science could holistically adopt the principles of *seed, soil, moisture, diversity and ecology* in line with the principles of agroecology that can enable a small farmer's agricultural field to become sustainable and the farmer viable in the short term and sustainable in the long term. Against this backdrop, transition strategies related to the above principles need to be explored with regard to agricultural research, practices, ecosystem services, and policies. Similarly, it is imperative for secondary and tertiary sectors to align themselves more closely with natural principles, as in the primary sector of agricultural production.

1.1.4 Organisations

Organisations have been the key engines of economic growth in the history of human enterprise systems. However, today's organisational designs seem to greatly facilitate private financial capital creation rather than social wealth creation for the whole of society. In the primary sector, although *Farmer Producer Organisations* (FPOs) are initiated on the principles of social capital formation, they gradually seem to adopt the design of organisations for private wealth creation. Therefore there is a need to redesign FPOs so that they can evolve into community enterprise systems rather than private enterprises. FPO for our analysis includes different forms of collectives: primary co-operatives, *Self Help Groups* (SHGs), farmers' clubs, producer organisations, and producer companies.

Accordingly, organisational design factors that can facilitate a higher frequency of interactions among the members/owners, a greater number of transactions throughout the year, and help members find greater value through these interactions and transactions need further exploration. The design is to facilitate not only financial capital in the short term but also greater social capital formation in the long run that can ensure a sustainable wealth creation process. The key design factors include *size, scope, technology, ownership* and *management.* Size refers to the number of members and geographical extent. Scope refers to the number and type of activities that an FPO can engage in. Technology refers to the processes and product technology suitable for an FPO. Ownership refers to the shareholding structure of the FPO, and Management refers to the management structure and type of managerial skills appropriate for an FPO.

1.1.5 Governance

The community in our analysis consists of a Gram Panchayat or a Ward. This has been chosen for analysis based on the technicality of the watershed, economic viability (keeping diversity as the basis of production), and the existence of a politically and socially recognisable boundary.

The analysis here includes key factors, namely, *frequency of interactions, decision-making method, problem-solving approach, resource dependency,* and *governance architecture and responsibilities.* While the desirable direction of each of these factors may be readily understood, the studies need to figure out ways to overcome the challenges of community governance that can facilitate sustainability in community enterprise systems, sustainable agricultural systems, and deepen relationships among members within the community. Figure 1.2 provides the broad spectrum of each of the factors within the different dimensions.

1.2 Discussions

The issues that are most interesting to explore are the spectrums of each factor within the different dimensions and the inherent tensions within the spectrum of each factor. Figure 1.2 provides the list of factors and presents the extreme positions possible in the spectrum of each factor. The inherent tension in each spectrum of a factor comes out clearly in the process. From a transition strategy point of view, the possible intermediate positions that lie between the two extremes of the spectrum can be discerned.

Further, the characteristics of factors within each dimension seem to show a dynamic relationship with each other. As an illustration, let us see how the five factors of organisational design – size, scope, technology, ownership and management – are dynamically interconnected with each other in the specific case of, say, a farmer producer organisation (a co-operative or producer company).

As we increase the organisational size in terms of membership and geographic spread, the scope of activities in the organisation tends to become limited. In other words, as the size increases, the organisation tends to become specialised. As it becomes more specialised, it can't help but rely more on technology, which leads to

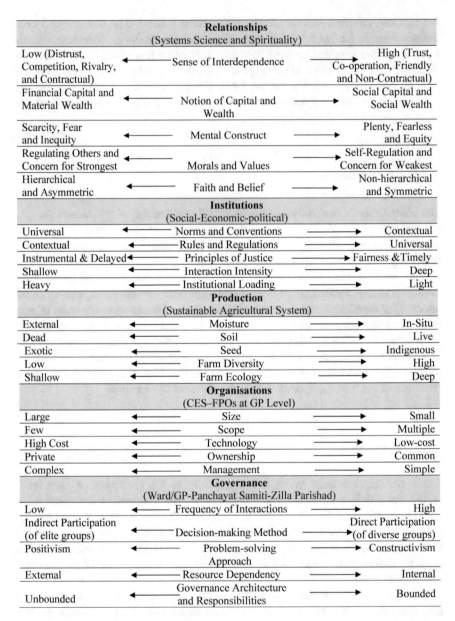

Relationships		
(Systems Science and Spirituality)		
Low (Distrust, Competition, Rivalry, and Contractual)	← Sense of Interdependence →	High (Trust, Co-operation, Friendly and Non-Contractual)
Financial Capital and Material Wealth	← Notion of Capital and Wealth →	Social Capital and Social Wealth
Scarcity, Fear and Inequity	← Mental Construct →	Plenty, Fearless and Equity
Regulating Others and Concern for Strongest	← Morals and Values →	Self-Regulation and Concern for Weakest
Hierarchical and Asymmetric	← Faith and Belief →	Non-hierarchical and Symmetric
Institutions		
(Social-Economic-political)		
Universal	← Norms and Conventions →	Contextual
Contextual	← Rules and Regulations →	Universal
Instrumental & Delayed	← Principles of Justice →	Fairness &Timely
Shallow	← Interaction Intensity →	Deep
Heavy	← Institutional Loading →	Light
Production		
(Sustainable Agricultural System)		
External	← Moisture →	In-Situ
Dead	← Soil →	Live
Exotic	← Seed →	Indigenous
Low	← Farm Diversity →	High
Shallow	← Farm Ecology →	Deep
Organisations		
(CES–FPOs at GP Level)		
Large	← Size →	Small
Few	← Scope →	Multiple
High Cost	← Technology →	Low-cost
Private	← Ownership →	Common
Complex	← Management →	Simple
Governance		
(Ward/GP-Panchayat Samiti-Zilla Parishad)		
Low	← Frequency of Interactions →	High
Indirect Participation (of elite groups)	← Decision-making Method →	Direct Participation (of diverse groups)
Positivism	← Problem-solving Approach →	Constructivism
External	← Resource Dependency →	Internal
Unbounded	← Governance Architecture and Responsibilities →	Bounded

Fig. 1.2 Spectrum of dimensions and factors for optimising internal designs and systems consistency. *Source* Nayak (2017)

subsequent technology intensification in the organisation. This process and product technology intensification doesn't come free to the organisation, but requires huge investment. These investments are brought in by some investors who would like to take an ownership position in the organisation, either directly or indirectly through greater management control. With more capital inflow to the firm, the ownership and management structure gradually gets modified, often in favour of major investors or owners. As all the four factors shift towards one side of the spectrum (to the far left, as in the table) – that is, size increases, specialisation increases, technology intensifies, and ownership-management becomes concentrated – the producer organisation is obliged to adopt a very complex management structure to reduce transaction costs. Interestingly, these interconnected changes could be initiated by a change in any one of these five factors; a process or product technology intensification, for instance, can lead to a subsequent shift towards specialisation, and so on.

When the size grows, specialisation increases, technology intensifies, ownership-management gets concentrated and management becomes complex, an organisation would often evolve into a large multinational corporation. Unfortunately, large corporations seem to show signs of unsustainability in the future. Several historical research studies (Schumpeter 1943; Vernon 1971, 1977, 2009; Nayak 2009) on large multinational corporations around the world tend to make this point. Is the recent bankruptcy of Fagor, the flagship unit of Mondragon Co-operative Group, the result of such dynamics? Are the internal tensions in India's best co-operative, AMUL, due to design flaws? Would this lens of analysis be useful to understand and resolve the challenges of large corporations in the global economy?

Conversely, when all these factor positions are at the other end of the spectrum (to the far right, as in Fig. 1.2), we get small informal organisations, such as self-help groups (SHGs) or small primary co-operatives in India. While SHGs groups have been good social units among economically poor women, they do not seem to be technically viable to undertake more than a few limited functions. So it appears that one of the challenges in the drive to achieve sustainability of producer organisations has been to find the optimal positions of the design factors. Small is beautiful (Schumacher 1973) indeed; however, in the current challenging context, what would be the optimal organisational size to facilitate transition towards a sustainable community system?

It may also be interesting to see the inter-industry dynamics across organisations in primary-secondary-tertiary sectors. Greater volume of production of a crop in a given geography or collection by a producer organisation, which may be initiated by itself or triggered from outside, often requires processing or manufacturing activity that gradually leads to more technology dependency, first on process technologies, then on product technologies. The increased transaction cost due to larger operation is then dealt with by further scaling-up processes. With scaled-up production capacity, a higher price-signalling mechanism is often used to source inputs by the processing or manufacturing units in the secondary sector, which initially comes as a boon in terms of higher prices to farmers, but gradually becomes a bane to the small producers in the primary sector, as observed in commercial farming across the world.

With greater demand to deal with the complexities in the secondary sector, the tertiary sector flourishes in terms of global trade investments and all supporting services, including education and training in various fields, such as technical, management, economics, and diplomacy, that are often path-dependent on the existing organisations. In the given context, the criteria for efficiency are different for individuals/organisations in different sectors. For instance, diversity is efficient for farmers, specialisation is efficient for processing/manufacturing units, and scale and scope are required for retail organisations. With such technical contradictions across sectors, while pursuing their individual goals, they together perpetuate greater asymmetries across the primary-secondary-tertiary sectors in an economy (Nayak 2018), the long-term impacts of which seem to play out more clearly in highly industrialised economies.

Organisation has been perceived as the engine of growth in the present market economy and, accordingly, governance and institutions – the other critical layers of a society – increasingly appear to get shaped by the demands of the leading organisations. Inherent design deficiencies in such organisations can fuel further asymmetries, more unequal relationships, deeper division among people and per-petuate the vicious cycle of unsustainability in society.

The above brief analysis of very complex dynamics indicates that the factors within a dimension are interdependent and hence influence each other in a manner which drives not only its own dimension (say organisation) but also the other four dimensions in a particular direction. Analysis of the factors of each dimension individually reveals a similar dynamic relationship within and outside.

For instance, the factors of agricultural production – moisture, soil, seed, diversity, and ecology – are deeply interconnected. On the one hand, sustainable agricultural systems thrive and grow under deep ecology and high diversity of crops, horticulture and farm animals. With abundant foliage and the soil not being exposed to direct sunlight due to deeper ecology on the farm, the water requirement of the farm is reduced. Rainwater run-off is slowed down and soil erosion is reduced. Under these conditions, the soil is richer with organic life that enhances the soil health and there is no need for inorganic fertilizers for the soil. Harmful pests on such diversified farms have been observed to be low. Genetically stable indigenous seeds, carefully selected by the farmer when planted on such a farm, provide very high yields.

On the other hand, single crop production in a season with little farm diversi-fication requires high-yielding seeds, such as genetically-modified seeds, large amounts of inorganic fertilizers, and a large amount of water. Farm ecology is not much of a concern on such a farm. The system of operation for such a farm becomes complex over time and, more importantly, the net income of the farmer remains more uncertain.

What is, however, important for our analysis here is that if a farmer were to adopt one of the production factors of any of the above two paradigms on a farm, she/he has to gradually adopt the other associated factors over time. The specialised industrial production and centralised governance-market system often push farmers towards a high external input intensive mono-cropping agricultural system.

The factors of community governance are also deeply interconnected. Although it is desirable for local communities to self-govern themselves, the positivist problem-solving approach at the higher level of governance in a State drives the system towards centralised governance. The positivist orientation typically relies on professionals over volunteers for achieving efficiency in public service delivery. While this orientation and these principles and methods fit better with centralised decision-making, they gradually reduce the direct participation of people in the community. With lesser people's participation and ownership by the people in local governance, issues of transparency and accountability arise. The idea of a community, its culture, heritage, social wealth and common pool resources erodes, gradually leading to increasing costs and the failure of the governance system.

The factors of institutions are equally deeply interconnected. The large size of a State and its commensurately large budget would logically lead it to divide its responsibilities between separate departments and functions. Accordingly, the policy signals of the state institutions for regulatory and administrative mechanisms tend to become compartmental in nature.

The overall development approach of such a state is dispersed rather than saturation. With resource constraints which make it difficult to meet all the requirements of the various constituencies of the population, the government is forced to design specialised programmes and schemes. Further, to be perceived positively by the general public across constituencies, the government often implements these schemes in a dispersed manner rather than in the holistic manner that is technically required for resolving the various issues of people in a given small community.

The compatible institutional architecture to execute the policies to achieve the targets of each ministry or state department accordingly requires a top-down hierarchical institutional architecture that is largely market-determined rather than bottom-up community-driven. The institutional architecture arising out of these compulsions serves more as a delivery agency of public services; and the institutional cost of providing services can only increase over time, placing a greater fiscal burden on any government in power.

With the above policy signal, developmental approach and institutional architecture of the State, the institutional support will tend to become piecemeal rather than being holistic. The input-output market network in a given community will also adopt a borderless loose market network that is in line with the overall competitive market economy framework.

In other words, the institutions evolve and consolidate with policy signals that are compartmental rather than convergent in nature, the development approach being dispersed rather saturation, the institutional architecture being top-down and market-determined rather than bottom-up community-driven, institutional support being piecemeal rather than holistic, and input-output networks being borderless loose market networks rather than deep local market networks.

However, taking a bottom-up view of institutions may be more universal in nature, in line with sustainability principles. Accordingly, the key factors of institutions include norms and conventions, rules and regulations, principles of justice, interaction intensity, and institutional loadings (Nayak 2018).

The dimension of relationships appears to have been at the core of the other four dimensions – institutions, production, organisations, and governance – mentioned earlier. The factors of relationships are deeply interconnected and often the most challenging to change. We explore the dimension of relationships from a systems science perspective and a spirituality perspective. The spirituality perspective can be further analysed from the perspectives of theology (religious studies) and philosophy. The five broad factors of the relationship dimension address relationships at different levels. Faith and belief are at the core of an individual. Morals and values are an outcome of relationships at family level. Mental constructs are an outcome of our education, training and experience. Sense of interdependence is an outcome of the relationship with neighbours and one's small community. The notion of capital and wealth is the overall societal orientation towards what is perceived as capital and wealth.

Depending on the faith and belief system, a person may be rooted in hierarchical and asymmetric beliefs at the core or otherwise. Accordingly, morals and values may be directed at regulating others and inherently favour the strongest, or may be otherwise. Mental constructs that are not only developed through upbringing but more through education, training and work experience could shape thinking in terms of scarcity, fear and inequity versus feelings of plenty, fearlessness and equity. If the sense of interdependence is low for a person, he/she is likely to have a feeling of distrust, competition and hate, and if the sense of interdependence is high, he/she is likely to exhibit trust, co-operation and love in the local community. Depending on a person's notion of capital and wealth, an individual may see them in terms of financial capital and material wealth versus social capital and social wealth.

If all twenty-five factors of the five dimensions were to be simultaneously considered; the positions of these factors may be spread widely across their respective spectrums. Such a spread has been the cause of the complex control mechanisms and lock-in systems and processes of our present time; a context where apparently no single individual, organisation or institution can possibly reverse the vicious cycle of unsustainability. If most factor positions are to the left of their respective spectrums, societal collapse would be imminent. However, if most of the factor positions are towards the right, it is possible to move a few factors from far left to right, making small communities transit towards sustainability. As more small communities transit towards sustainability, this could gradually lead the larger society towards sustainability.

Fourth, under the above highly interconnected complex dynamics, where our own designs and systems at different levels appear to perpetuate lock-in effects and external control, the chances of our becoming unsustainable seem to be far greater than becoming sustainable. Under these circumstances, what could be the way forward to transit to sustainability?

One of the steps could be to review our designs at each level of institutions, production, organisation and governance, and reflect on the core of all these dimensions – our culture of relationships. While it will be hard to remove all inconsistencies at one go; being aware of these inconsistencies and sharing this knowledge with others could be a starting point. *Two*, ensure that the design of each sub-system is based on the principle of interconnections and interrelationships. *Three*, identify factor positions in each dimension that are consistent with sustainability principles. It may be noted here that following sustainability principles at the core – that is to say, relationships – can have a healing effect at higher levels of engagement, and following sustainability principles at higher levels – i.e. governance – can facilitate sustainability at the lower levels of engagement. *Four*, facilitate synergy of factor positions across all five dimensions in relatively simpler community systems where the present lock-in effects are relatively low (Nayak 2014).

While these steps may be considered only as initial starting points, developing details that are context-specific needs greater deliberation, and engaged action research into different dimensions with the stakeholders in a given community and context will help the cause of sustainability.

Fifth, on research possibilities in the domain of sustainable community systems, the field appears to have been largely unexplored and neglected. Therefore, it presents a huge opportunity – or rather presents a crying need – to study and facilitate the transition process in both practice and policy. Indeed, there is no dearth of studies relating to individual dimensions – i.e. relationships, agricultural production systems, organisational design, governance and institutions. However, most of these studies have broadly followed the reductionist approach of science and hence do not seem to have taken a systems view, either within the dimensions or across these dimensions.

Further, past studies relating to factors in each of the above dimensions has often been limited to studying these factor positions at one end of the spectrum, leaving out the multitude of alternatives. In other words, the language, logic and values of analysis have largely been limited to the perspective of competition or a hybrid of competition and co-operation rather than exploring sustainability from the perspective of co-operation (Nayak 2014).

This chapter, based on nearly two decades of research, proposes only twenty-five core factors for study and deliberation. However, there seem to be a much larger number of sub-factors and possible combinations of these factors and sub-factors under different contexts that need to be studied to facilitate the process of transition to sustainability across communities.

The methodologies of research to explore alternatives for sustainability are likely to be different from the current methodologies of research and inquiry. Empiricism may not provide clues to sustainability, as in the present reality most practices and policies do not seem to be internally consistent with sustainability principles. Interdisciplinary research, systems thinking, action research and holistic implementation processes could be some potential ways to explore new alternatives towards successful transition processes.

The present volume is an effort to stitch together the various efforts and initiatives towards sustainability from the design and systems framework towards facilitating sustainable communities in different contexts, regions and countries across the world.

References

Nayak, A.K.J.R., 2009: "Optimizing Asymmetries for sustainability: design issues of producer organizations", *XIMB Sustainability Seminar Series, Working Paper* 1.0, January, Bhubaneswar.

Nayak, A.K.J.R., 2011: "Efficiency, effectiveness and sustainability: the basis of competition and cooperation", *XIMB Sustainability Seminar Series, Working Paper 3.0*, January, Bhubaneswar.

Nayak, A.K.J.R., 2014: "Logic, language and values of co-operation versus competition in the context of recreating sustainable community systems", in: *International Review of Sociology*, 24(1): 13–26.

Nayak, A.K.J.R., 2017: *National Conference on Transistion Strategies for Sustainable Community Systems* (Bhubaneswar: XIM Bhubaneswar).

Nayak, A.K.J.R., 2018: "Economies of scope: context of agriculture, smallholder farmers, and sustainability", in: *Asian Journal of German and European Studies*, 3:2, https://doi.org/10.1186/s40856-018-0024-y.

Piketty, Thomas, 2014. *Capital: In the Twenty First Century* (Cambridge, Mass: Harvard University Press).

Rousseau, J.J., 1762: *The Social Contract*.

Schumacher, E.F., 1973: *A Study of Economics as if People Mattered* (London: Bonde and Briggs).

Schumpeter, J.A., 1943: *Capitalism, Socialism and Democracy* (London: Routledge).

Shah, T., 1996: *Catalysing Co-operation: Design of Self-governing Organizations* (New Delhi: Sage).

Vernon, R., 1971: *Sovereignty at Bay: The Multinationals Spread of US Enterprises* (New York: Basic Books).

Vernon, R., 1977: *Storm Over the Multinationals: The Real Issue* (London: Macmillan).

Vernon, R., 2009: *In Hurricane's Eye: The Troubled Prospects of Multinational Enterprises* (Cambridge, MA: Harvard University Press).

Part I
Relationships

The nature and state of relationships often shape the sustainability of our endeavours in institutions, production systems, organisations and community governance. Relationships seem to have been at the core of human engagements and their sustainability. In this section, we intend to explore and understand "relationships" from three perspectives: systems science, philosophical and theological.

From philosophical and theological perspectives, this section will deliberately focus on the deep-rooted *faith and beliefs, and morals and values* that probably lead us to behaviours and actions that are contrary to the principles of sustainability. From a systems science perspective, it delves into the notions of *interdependence* and *identifying self with others* and the *notions of capital and wealth*. Faith and mental constructs could probably be analysed from all three perspectives.

Chapter 2
'Good Anthropocene': The *Zeitgeist* of the 21st Century

Anup Dash

Abstract As we enter into the Anthropocene, the debate around sustainable community systems assume great significance as a response to the current economic, political, cultural and ecological crises. The Anthropocene isn't just a new geological epoch; it is a crisis of the *EARTH SYSTEM* with 'human agency' at the centre. Hence, it is a very provocative concept. It challenges us to build a 'New Global Ethos' for the 'Good Anthropocene' – to bend curves in directions that are good for people and the planet, navigating the threats through a new moral practice and changing the course for the future using our unlimited creativity. But this, as the present chapter argues, involves a much deeper philosophical exercise to question, destabilise, and de-centre the 'human' in an anthroposised planet, and to work on theoretical innovations based on alternative ontological foundations (against the hegemonic order of 'the universal rationalist paradigm') for an era of post-human evolution. As a system of thought, Anthropocene erodes the very philosophical foundation of 'Modern' and 'modernity', and inspires the social sciences to build upon new 'forms of cognition' for a paradigm shift. This chapter argues that such a paradigm shift is already taking place in the *Social and Solidarity Economy* (SSE) landscape, with a compulsive pressure on us to *rethink* the 'anthropos', rejecting the notion of the *homo economicus* as the core ontological assumption and creating opportunities for sustainable futures through a new communitarian revolution sprouting up from epistemic ruptures in anthropocentrism, atomism, positivism and the Cartesian dualisms.

Keywords Good Anthropocene · Capitalocene · Ontological shift
Homo economicus · Social and solidarity economy (SSE) · Sustainability

The modern 'Satanic' civilisation, M.K. Gandhi claimed more than a century ago, referring to the modern western civilisation, 'is such that one has only to be patient and it will be self-destroyed' (Gandhi 1909: 3). The 'Gandhian Moment' has come. Already the paradigm that has governed our lives over the last two and half

Anup Dash, formerly Professor of Sociology, Utkal University, Bhubaneswar, Odisha, India;
Email: dashanup@hotmail.com

© Springer Nature Switzerland AG 2019
A. K. J. R. Nayak (ed.), *Transition Strategies for Sustainable Community Systems*,
The Anthropocene: Politik—Economics—Society—Science 26,
https://doi.org/10.1007/978-3-030-00356-2_2

centuries is threatening to crash, with signs of social, economic and environmental collapse. In the global struggle for more (development?), we are exploiting and running down the Earth's natural capital to an extent that the earth system itself is severely impaired, and the ecosystem services – the very foundation of our life and well-being – are irreversibly damaged. The global economy is in significant eco-logical overshoot, and we need to discover ways to reduce humanity's overall ecological footprint. Man, the *homo sapiens faber*, has acquired an enormous capacity to leave a significant and durable impact on the earth's ecosystem on a planetary scale – on its geological, biotic and climatic processes – marking the onset of the *Anthropocene* (Crutzen/Stroermer 2000; Crutzen 2002; Steffen et al. 2011). The term 'Anthropocene', coined by Nobel prizewinner Crutzen/Stroermer (2000), is a new conceptualisation of geological time, where human agency has become the main geological force shaping the face of the earth. 'Human pressures' are pushing the conditions of biospheric stability – climate and biodiversity above all – to breaking point. The 'planetary life support systems' are eroding, and we are fast moving out of the 'safe boundaries' of these life support systems. We have already crossed three of the critical seven 'planetary boundaries' – for climate change, rate of biodiversity loss, and changes to the global nitrogen cycle (Rockström et al. 2009). In simple Green Arithmetic formulation, it can be put as: *Human action + Nature = Planetary Crisis*.

Of course, the Anthropocene was not made in a day, nor was it created uni-formly, and it has deep roots (Ellis et al. 2016: 192). However, Crutzen suggested that the start date of the Anthropocene be placed roughly about the time that the industrial revolution began. But the 'planetary dashboard', now updated to 2010 by Steffan and his team, shows the 'Great Acceleration' (echoing Karl Polanyi's 'Great Transformation') in human activity since 1950 (Steffen et al. 2015). The post-1950 acceleration in the human activity (human consumption and production) driving the global economic system, without doubt, has seen the most rapid transformation of the human relationship with the natural world. The dominant feature of the socio-economic trends is that the human economic activity continues to grow ever more rapidly.

'Anthropocene' is a very provocative concept. The Anthropocene isn't just a new geological epoch; it is a crisis of the Earth System with the 'human agency' at the centre. This immediately raises the question of 'responsibility'. The concept lumps together all 'anthropos' as an *undifferentiated mass* and ascribes the same responsibility to them in this newly defined 'geological force'. The 'anthropos' responsible for shaping the planet, whom the geologists push to the centre of the geo-story of the Anthropocene, is not the same as the Indian *adivasis*, or the indigenous communities around the world who have lived in harmony with nature for centuries. It is a 'being' with a different history and culture, philosophy and *weltanschauung*. 'Man' does not mean humans, but a particular kind of 'being' – 'the anthropological monster', to use the language of Bourdieu (1997: 61) – invented by European Enlightenment thought (in the name of 'reason' and 'pro-gress') and brought into operation by the forces of modernisation and economic growth. It was this 'Man', firmly rooted in materialism and anthropocentrism, who set out to arrogantly conquer nature, colonise territories to plunder, exploit, dominate and control, not only the nature; but also the people who lived close to and in harmony with nature for centuries. The European invader, with all his 'superior' knowledge and science, culture and capacities, skills and ideas, epistemology and *weltanschauung*, did not consider the 'inferior' indigenous native people as *Homo sapiens*. Their lands could be usurped as *terra nullius* – land empty of people, 'vacant' and 'unused'. Their 'modes of thought', their epistemology, morality and sensibility were downgraded as *cognitively inferior* and discarded as 'non-rational', 'primitive', 'magic', and 'unscientific'.

The hegemonic order of 'the universal rationalist paradigm', 'scientific knowledge' and 'modernism' oppressed and destroyed all the other 'rationalities', the multiple modes of thought as well as the traditional knowledge forms of the peripheral communities. Therefore the treatment of humanity as an aggregated whole masks the deep structures of global inequality, which badly distorts the distribution of the benefits of the Great Acceleration (e.g. Malm/Hornborg 2014). As Biermann rightly says, "the Anthropocene is an epoch that sees the human species with extreme variations in wealth, health, living standards, education and most other indicators that define well-being" (2014: 58). Therefore, in the updated planetary dashboard of 2010, Steffen (*op.cit.*) addressed the equity issue and deconstructed the socio-economic trends. He shows that "[t]he bulk of economic activity, and with it, the lion's share of consumption, remain largely within the OECD countries... Insofar as the imprint on the Earth System scales with consumption, most of the human imprint on the Earth System is coming from the OECD world" (p. 11). Should all the 7.5 billion people in the world today live western lifestyles, we'd need several more planets.

If the European enlightenment shaped the 'anthropos' for the Anthropocene, the 'New enlightenment' – free market as the road to prosperity – caused the 'great acceleration'. The 'New Enlightenment', driving the current spur of globalisation – based on the logic of Capitalism and deeply embedded in the psycho-social dynamics of neoclassical economics and its ontological construct of the *homo economicus* – is changing, in very fundamental ways, our economies and polities, our societies and cultures, institutions and values, lifestyles and *weltanschauung*. There is no doubt that greed-driven capitalism imposes a relentless pattern of violence on nature, humans included, to plunder and loot *as if there is no tomorrow*. Questions of anthropocentrism, dualist framings of 'nature' and 'society,' and the role of capitalism, with its extractive logic fuelled by greed in the process of wealth creation, are all frequently bracketed by the dominant Anthropocene perspective. Therefore social scientists have advanced the concept of the *Capitalocene* to emphasise the capitalist world-ecology – capitalism as a way of organising nature (Moore 2014a, b, 2015; Haraway 2014). It refers to the historical era shaped by relations privileging the endless accumulation of capital. The alternative to the 'Age of Man' (the Anthropocene) is the 'Age of Capital' (the Capitalocene). *Capitalocene* is '*neoliberal* Anthropocene'. The shock of the Anthropocene reminds us of the prediction Karl Polanyi made, as far back as 1944 in his influential book, *The Great Transformation*, of the imminent 'breakdown of our civilization' (pp. 3–5). The *ethos*, *logos* and *pathos* of the rhetoric of capitalist growth are fast losing their steam. This basic illusion of the age of capital has come to an end. As Arthur Miller wrote, 'An era can be said to end when its basic illusions are exhausted' (1974/75: 30).

The good news is that humanity has become aware of the dawn of the Anthropocene and its dark prognosis for the next century. We are aware of the challenges facing the anthroposised planet, and we can see our responsibility as its stewards. We can identify the opportunities they offer for shaping the future. We have the capacity to master this huge epochal shift, and to bend the curves. As Rees (2016) stated, "[t]he Anthropocene epoch could inaugurate even more marvellous eras of evolution". Human societies could navigate the threats, achieve a sustainable future, and inaugurate an era of post-human evolution through a new moral practice, and a culture that grows with Earth's biological wealth instead of depleting it. That is the "New Global Ethos" (Crutzen and Schwägerl 2011), to grow in different ways than with our current hyper-consumption, to live in ways to protect and sustain the resources and share resources across species as we are all interconnected as *one giant living organism*. The guiding principle of human life has to be the meeting of needs, not greed. Gandhi's famous quote, "the earth provides enough to satisfy every man's needs, but not every man's greed" is such a perfect summation of the principles of ethics and justice to carve out the 'Good Anthropocene'. That is the *zeitgeist* of the 21st century.

'Good Anthropocene' thus gives us the best chance *for bending curves in directions that are good for the people and the planet.* It is a historic opportunity to change the course for the future using our unlimited creativity and our sense of moral purpose. The Anthropocene, Seielstad (2012) argues, is humanity's defining moment that reminds us as a community that we can be agents of positive change. It becomes increasingly clear that shallow environmentalism (e.g., as evident from the UN climate change talks) cannot create the 'Good Anthropocene', leaving the deep conflict of values and interests at the core of the dominant paradigm untouched and unchallenged. Especially, we know that the dominant interests embedded in the institutional structures (a market economy, an industrial system, modern science and technology, the system of state bureaucracy) will hardly ever change, given their rigidities and the anthropocentric vision around which they have grown. Being products of 'the promethean project to which the Enlightenment gave birth', they would not challenge Baconian science, Cartesian dualism, or the Newtonian mechanistic construction (Dobson 1990: 9).

Therefore the tragedy of the Anthropocene administers a massive jolt to the sociological imagination, especially because sociology itself is a child of the European enlightenment. – founded on the idea of 'reason as the engine of progress'. '*Sapere aude!*', Kant declared as his great motto of the Enlightenment. Reason was celebrated as the repository of the 'moral instinct' which formulates an absolute value for its own sake, as an end in itself. Rational principles rest on a rule that demands moral perfection as this derives from 'the rational concept of perfection' (Kant 1990: 59). Kant predicts that reason will progress to more reason; produce ever greater quantities of reason, culminating in a 'rational kingdom of ends'. However, Max Weber was disenchanted with Kant's reason, because it failed to produce Kant's utopia, and instead, realised 'the iron cage' of mechanisation and regimentation; and led to the *irrationality* of reason itself. Instead of Reason producing more reason, as Kant predicted, Weber saw modernity as a process driven by (increasing) 'rationalisation' – one in which ultimate values rationalise and devalue themselves, and are replaced increasingly by the pursuit of materialistic, mundane ends. Reason was crushed by the forces of its own creation, the materialistic determinism of capitalism. The *Wertrationalität* gave way to the *Zweckrationalität* – an increasingly systematic and calculating approach to thought and behaviour, marking the eclipse of the *Wertrationalität*. The conflict of these types of rationality (formal-calculative and value-substantive) has played a particularly fateful role in the unfolding of rationalisation processes in the West, leading to the 'death of reason' of the modern man.

As a system of thought, Anthropocene erodes the very philosophical foundation of 'modern' and 'modernity', and inspires the social sciences to build upon new 'forms of cognition' for a paradigm shift... In the words of Bruno Latour, 'Anthropocene' is 'a gift' from the natural sciences (albeit 'a poisonous gift') to the social sciences. The potential gift of the Anthropocene is its compulsive pressure on us to *rethink* the 'anthropos' so as to be relevant and useful for the new reality (Latour 2014). It holds the promise of new philosophical insight for scientific renewal. It challenges the basic assumptions of the orthodox social science disciplines and demands an 'ontological turn' for new conceptual tools to combine the ecological and the sociological, ecology and economy, nature and culture – away from anthropocentrism, atomism, and positivism, and to interrogate established binaries of human/non-human, subject/object, and nature/culture, positive/normative, fact/value, matter/spirit – and to question the hegemonic Newtonian-Baconian ontological-epistemological edifice. At the heart of the idea of Anthropocene is the crisis of the ontological construct of 'the human' as an autonomous, selfish, atomistic, cold and rational being.

Hence, 'Good Anthropocene' creates a very significant epistemic rupture. It advances the concept of agency and self to include the 'non-human others', and builds on the 'post-humanism' and 'transhumanism' as an alternative theoretical tool to 'de-centre' the human, to reject individualism and 'human exceptionalism', privileging the human to the detriment of other agential entities and other ontologies, to subvert all kinds of dualism and binaries, and "to slow down reasoning" (de Lima Costa et al. 2017). As Seielstad (*op.cit.*) argues, the first thing we must do is re-align our thinking, our perspectives and priorities to develop adaptive responses and deepen post-growth dialogues.

One such new perspective is an emerging paradigm in the social sciences which seeks to decolonise economics – the *primus inter pares* in the social sciences – and to de-centre man. Economics has powerfully led the advancing social science paradigms based on atomistic ontology, Cartesian dichotomy, logical positivistic epistemology and methodological individualism, but also guided policy-making (through mathematical modelling) based on a *narrow economism*, distorted world-views and commodification of social lives at large. This is an emerging perspective to reinvent economics for alternative visions of another – better – economy from the wreckage of orthodox economics. This captures the spirit and motto of the world social forum: *Um outro mundo é possivel* and seeks to create 'the future we want'. A new wind is already blowing for a 'post-capitalist' and 'post-materialist' world that imagines a different life beyond capitalism as an alternative to the oxymoron of Growth. If the neoliberal capitalist economy caused 'the Great Acceleration' in the Anthropocene during the later part of the twentieth century, the drive towards 'post-capitalist' society takes place as the neoliberal paradigm itself decomposes from inside and its 'side-effects' (to quote Beck et al. 1994: 175) give rise to growing global turbulences with the 'great sustainability challenge'. It is not the paradigm any more, but its 'side-effects' – endogenous to the paradigm itself – that have now become the motor of social history.

Already the seeds of the 'Good Anthropocene' are sprouting up in their rich diversity on the margins of the capitalist hegemony. We can see the contours of another economy, a qualitatively different economy, in the shape of new communitarian movements through which local communities resist and respond to the multiple crises of global capitalism, and innovate alternative ways to meet economic needs within their *local solidarity-based associational space*. These diverse forms of economic expression, often lumped together under the rubric of *Social and Solidarity Economy* (SSE), are important social innovations in varied forms of democratic social designs in 'associational economics', expressed through the blooming institutions, practices and modes of meeting human needs through 'social provisioning' based on reciprocity, co-operation, solidarity and non-economic incentives as the alternative to 'market provisioning' based on profit and competition (Dash 2016). As Julie Matthaei, a co-founder of the US Solidarity Economy Network emphasises, these types of transformative economic organising insert 'solidaritous values' into our relationships with people and the environment through solidarity production practices, solidarity transfer and exchange, as well as solidarity use and consumption practices. What is new in this emerging framework is the way in which they are envisioned and articulated as part of a different system, and part of a growing 'movement' for a post-capitalist value change. The SSE landscape is replete today with such innovations designed to 'correct the flaws' in the capitalist system, and seeking to bring *social content, moral purpose, environmental focus* and a *democratic character* to capital.

Organisationally, the SSE *blends* values (economic with social, moral and environmental). With a democratic governance structure, these organisations are rooted in the local social fabric, based on solidarity, trust, co-operation and community spirit, that drives bottom-up practices of sustainable development, essentially looking for a multidimensional rationality beyond the greed-driven market. SSE seeks to subordinate profit to people and the planet, and, as Volkmann (2012: 102) explains, 'includes aspects of solidarity and fairness in opposition to pure profit-maximising'. As an alternative system to capitalism, Social and Solidarity Economy includes forms of economy built on relationships and ethics of care, co-operation and solidarity, instead of competition and individualism (Miller 2009). Economic analysis loses most of its relevance as a method of inquiry to explain the working of the economy outside the system of price-making markets that are based on non-utilitarian motives, non-market relations and non-monetary transactions. Therefore, the rich mosaic of the SSE landscape does not fit within the theoretical-conceptual frameworks and the analytical tools of conventional orthodox economics. As a result, they have been pushed aside by the orthodox policy regime as inefficient (on a scale of wrongful comparison with the single bottom-line profit-maximising enterprises), ignoring their impact in terms of social/ environmental returns. Thus, a 'poor social imaginary' about the SSE has been constructed through the distorted lens of 'imperial' orthodox economics.

Thus 'Good Anthropocene' advances an alternative to the hegemonic capitalist rationality and seeks to overcome the *rationality deficit* that has caused our civilisational crisis. It articulates a distinctly different rationality, principle and focus of the economy for sustaining people, communities, and nature as core goals. It is a paradigm built on relationships and ethics of care, co-operation and solidarity, instead of competition, individualism and pure profit-maximising, which rotates on a different orbit of the *intent* and the *content* of economic life. If the era of capitalism produced its own economics based on four value-laden concepts (instrumental rationality, efficiency, profit and competition), as a post-capitalist movement the 'Good Anthropocene' articulates an economy whose DNA is constituted by what Razeto (1998) characterises as 'the factor C' – co-operation, community spirit and collective action. If the capitalist economy moved towards greed, growth and globalisation, the 'Good Anthropocene' changes course and gravitates towards localism, reciprocity and sustainability. It seeks to reduce our ecological footprint, emancipates rather than subjugates people, and moves beyond the *narrow economism* of GDP by emphasising the multidimensionality of well-being (Dash 2013). It represents a 'new ontological turn' in shaping the 'post-capitalist' system, defined by reciprocity, co-operation and solidarity (away from atomism, greed and competition) based on a *multidimensional* rationality. It is rooted in the *local social fabric* (based on trust and co-operation) and a *new conception of wealth* (in contrast to the toxic happiness of carbon materialism) in which equity, cohesion and sustainability in the reproduction of life is central.

The building block of the huge monolithic economic edifice and the micro foundation of orthodox economics is the *homo economicus* – the ontologically cold, calculative, instrumentally rational, atomistic man with a 'separative self' (England 1993). With its ontological assumption of the *homo economicus*, orthodox economics grossly neglects both the logical possibility and empirical reality of economic practices based on different 'other rationalities', 'relational capital', as well as 'co-operative logic' for the creation of 'psychic income', 'social profit' and 'ecological well-being'. The model of *homo economicus* has changed not just how individuals think of themselves and their preferences, but how they relate to each other in creating an economy of thin ties. With its 'performative' power, orthodox economics is so strongly institutionalised and has so deeply ingrained economic rationality into our way of being and our subjectivity that it has become the genetic essence of the tribe of *homines economici*. As a result, the oxymoron of growth is today dangerously out of sync with our social and environmental well-being; it erodes our solidarity with nature and the future and impairs the moral framework governing our cohesive community life, increasingly creating conditions for us to question the ability of the growth-driven paradigm, championed by the neoclassical hegemony, to create 'the future we want'.

Orthodox economics severely constrains our cognitive abilities to imagine economic alternatives which *put a moral brake on capitalism.* Therefore, advancing the theory and practice of the 'Good Anthropocene' essentially involves *deconstruction* of the neoclassical paradigm – a highly complex philosophical, political, social and moral exercise questioning the thought, the science and the institutions that create this 'iron cage' of greed and instrumental rationality in which the paradigm of toxic growth is locked up. A 'poor social imaginary' of man, with emphasis on self-interest and maximisation as prime movers of human action governed by the principle of competition, strip the *homo economicus* of any morality and substantive rationality, and create a 'thin theory of human action' (Taylor 1988). Hirschman (1986: 109) has called this position the self-destruction thesis. Capitalism would undermine its own basis if it did not reproduce the social and moral preconditions on which it rests. Today we are victims of a deep systemic crisis of capitalism, fuelled by unregulated greed, and this crisis is symptomatic of a 'systemic failure of the economics profession' (Colander et al. 2009). Many would now call orthodox economics a 'dismal' science and even a 'failed' science, questioning not only the moral quality of the capitalist economy (Crouch 2012), but also the tyranny of the orthodoxy in economics (Colander *op.cit.*; Freeman 2009).

Orthodox economics is passing through a crisis and a period of 'unrest' (Fullbrook 2003), because of its failure to cope up with its own ontology and epistemology. The worst excess of neoclassical economics is the loss of 'the moral minimum' from our social life. The Nobel Economist Amartya Sen laments that the nature of economics 'has been substantially impoverished by the distance that has grown between economics and ethics' (1997: 7). Coyle sums up the critics: economics is crude and 'too narrow in its focus, caring only about money; too dry and robotic in its view of the human nature; too reductionist in its methodology' (2007: 2). In the garb of a 'hard' science, economics is not only empirically empty and intellectually bankrupt, but is also a dangerous cultural failure. There is something fundamentally wrong in economic orthodoxy; many of its assumptions are fatally flawed, many of its 'dogmas' do not hold in reality. Economics has increasingly become 'an arcane branch of mathematics rather than dealing with real economic problems' (Friedman 1999: 137) and, as Coase summarises, the theoretical system in economics 'floats in the air' and 'bears little relation to what actually happens in the real world' (1999: 4). Economics, in fact, is a colossus with feet of clay – with a dubious methodological status and a preference for *doxa* over *episteme* – and has become a highly contested discipline, as many 'are worried about the increasing adoption of its suspiciously narrow and distorting world views as part of the questionable cultural trend of… commodification of our social lives at large' (Mäki 2005: 212).

The *homo economicus* is an abstract construction, and does not exist in real-life situations. Social life is hardly ever fully utilitarian, and people do not actually optimise utility through consistent and precise cost-benefit calculations (Homans 1990: 77). Even the father of modern economics, Adam Smith, famously said: "How selfish so ever man may be supposed, there are evidently some principles in his nature, which interest him in the fortune of others, and render their happiness necessary to him, though he derives nothing from it except the pleasure of seeing" (1759/1790, ch.I.I.1). John Stuart Mill, the originator of the idea of the *homo economicus*, himself admitted that it is "an arbitrary definition of man" and par-tially, a "very thin slice of human nature" separated out for analysis by political economy, which, as an abstract and mental science (akin to geometry), "reasons, and... must necessarily reason... from assumptions, not from facts" (Mill 2004: 110). The 'Rational Economic Man' model has been decisively disconfirmed by experimental economics. For example, based on laboratory experiments, Gintis (2000) claims that in many circumstances economic actors "are *strong reciproca-tors* who come to strategic interactions with a propensity to cooperate, respond to co-operative behaviour by maintaining or increasing co-operation, and respond to non-co-operative free-riders by retaliating against the 'offenders', even at a personal cost" (Gintis 2000: 313). Therefore, as Bourdieu argues, *the homo economicus,* as ontologically constructed, is a "kind of anthropological monster" (Bourdieu *op.cit.*).

A better economy requires a better economics. What is needed is "a change of skin" (Leff 2009: 105), an ontological-epistemological revolution in economic science, to understand and explain 'the other economy' – economic behaviour based on different logic, values and motivation as a means to social-relational, psychic-emotional, moral-ecological well-being with a focus on sustainability. The articulation of a better economics for the 'Good Anthropocene' helps us think against the grain of our impoverished social vision and impaired lives, and the *Weltanschauung* which is especially locked up in the belief that 'there is no alternative' to global capitalism. It questions the reductionism of the neoclassical economics, its narrow economism and the rational choice theory. It rejects the triad of the colossal neoclassical default – positivism, methodological individualism and the ontological construction of the *homo economicus* – conceptualises the human agency with considerable ontological sophistication, and offers a theory of con-trastive explanation of the *personae* of the 'human agency' by focusing on a more refined theory of the basic constitution of social life and human nature. By 'so-cialising' the *homo economicus*, it moves beyond orthodox economics to the model of multidimensionality and relatedness, in which both material and non-material motivations drive human behaviour. It focuses on the interpersonal ties, social capital, trust and co-operation, collective action and so on which not only lubricate and sustain, but also give meaning, substance and purpose to economic actions. In stark contrast to the neoclassical paradigm, it is conceptually anchored in the 'sustainable livelihood' approach, which seeks to promote ways for people to make a living from their social and natural environment in a socially just, environmentally responsible, and morally correct way. It is ontologically based on the value of a relationship of solidarity and non-instrumental motivations.

Such a theory obviously draws on a rich diversity of sociological and philo-sophical traditions and deontological ethics, which offers us a more refined understanding of the complex reality of the multidimensionality of human action. Max Weber, for example, famously made the typology of rational, affective, and traditional action. More importantly, he made a distinction between two types of rational actions, namely action based on economic and instrumental rationality (*zweckrationalität*), and action based on value or substantive rationality (*wertrationalität*). Weber's *wertrationalität* is non-economically rational yet economically non-rational. It is not reducible to *zweckrationalität*. Thus, rational behaviour can include not just the purely instrumental ends such as utility, profit or wealth, but also social and moral ends (Granovetter 1985), nor are non-rational actions necessarily ontologically less legitimate or empirically less sound than rational action. Schumpeter (1991: 337) admits that our social life, including our economic life, is often ontologically irrational. Economic action, far from being utility maximising behaviour, is "constrained and conditioned by social relations" (Aspers 2011: 175). The issue is, as Sen puts it, "whether there is a plurality of motivations or whether self-interest alone drives human beings" (1987: 19). It rejects the self-interest thesis of economic orthodoxy and advocates a theory of ethically driven, deontological, other-directed and multidimensional motivation.

Human beings are 'less than perfectly rational' and have different behavioural dispositions for their long-term, as against short-term, interests and rationality. The long-term rationality, with strongly ingrained norms about fairness, reciprocity, and co-operation, often overrides the short-term cold and a calculated 'rationality'. Humans have culturally evolved an elaborate system of ethics and morality, and a code of individual and collective conducts, which enable them to take decisions, not simply for short-term gains, but for long-range benefits as well. Horton rightly argues that "the evolution of ethics, morals, fairness, and justice in human relationships, including economic relationships, has buttressed our long-term survival and evolutionary success. ...To monitor reciprocity and fairness, humans have developed acute abilities to detect cheating, free-riding, and unfairness" (Horton 2011: 474). Francis Fukuyama very convincingly argues that the substantive conclusions of new evolutionary biology are more supportive of *homo sociologus* than *homo economicus* (quoted in Horton *ibid.*).

In making a forecast, Thaler says, "rationally, I realise that the forecast most likely to be right is to predict that economics will hardly change at all" (Thaler 2000: 134), but clearly predicts that *homo economicus* will evolve into *homo sapiens*. Building on the critique of neoclassical economics, Horton (*op.cit.*: 475) predicts that "homo economicus will become extinct". *Homo economicus* is a sociopath – designed to cheat, lie, and exploit. *Homo reciprocans* presents a more realistic and biologically correct behavioural model than *homo economicus*. Moral reasoning is not a cultural artifact invented for convenience and opportunity. Morality and ethics provide the glue that binds our species, while the social skill of co-operation creates and furthers the common good, and, over the long run, enables us to live in peace by cooperating with unrelated others, and protects mankind from destroying itself. In creating 'the future we want', we need to develop the ontology of sociality, of *homo reciprocans* and *homo sociologicus*.

The promise of the 'Good Anthropocene' calls for developing a robust paradigm, with an alternative epistemological foundation built around a superior social ontology of inter-relationality, as well as philosophical principles different from logical positivism (monism) and individualism (atomism), for us to gain the confidence and capacity to think of the 'Good Anthropocene' more boldly. This is the challenging scientific project to be more developed to shape economic practices and policies more coherently in the twenty-first century.

References

Aspers, P., 2011: *Markets* (Cambridge: Polity).

Beck, U. et al., 1994: *Reflexive Modernization: Politics, Tradition and Aesthetics in the Modern Social Order* (Cambridge: Polity).

Biermann, F., 2014: "The Anthropocene: A governance perspective", in: *The Anthropocene Review*, 1: 57–61.

Bourdieu, P., 1997: "Le champ economique", in: *Actes de la recherche' en sciences socials*, 119: 48–66.

Coase, R., 1999: "Interview with Ronald Coase", in: *Newsletter of the International Society for New Institutional Economics*, 2: 3–10.

Colander, D. et al., 2009: "Financial crisis and the systemic failure of academic economics" Working Paper 1489, Kiel, *Kiel Institute of the World Economy*.

Coyle, D., 2007: *The Soulful Science* (Princeton: Princeton University Press).

Crouch, C., 2012: "Sustainability, neoliberalism, and the moral quality of capitalism", in: *Business and Professional Ethics Journal*, 31: 63–374.

Crutzen, P.J., 2002: "Geology of mankind: the Anthropocene", in: *Nature,* 415: 23.

Crutzen, P.J.; Schwägerl, C., 2011: "Living in the Anthropocene: Toward a New Global Ethos", in: *Yale Environment* 360, *Yale School of Forestry and Environmental Studies*, 24 January; at: https://e360.yale.edu/features/living_in_the_anthropocene_toward_a_new_global_ethos (3 January 2019).

Crutzen, P.J.; Stroermer, E.F., 2000: "The Anthropocene", in: *IGBP Newsletter*, 41(12): 17–18.

Dash, A., 2013: "Building smart cooperatives for the 21st century", in: Roelants, Bruno (Ed.), *Cooperative Growth for the 21st Century* (Brussels: CICOPA/International Cooperative Alliance).

Dash, A., 2016: "An epistemological reflection on social and solidarity economy", in: *Forum for Social Economics,* 45(1): 61–87.

de Lima Costa, C. et al., 2017: "On the Posthuman", in: *Ilha do Desterro*, 70(2): 9–14.

Dobson, A., 1990: *Green Political Thought: An Introduction* (London: Unwin Hyman).

England, P., 1993: "The separative self: Androcentric bias in neoclassical assumptions", in: Ferber, Marianne A.; Nelson, Julie A. (Eds.), *Beyond Economic Man: Feminist Theory and Economics* (Chicago: Chicago University Press).

Freeman, A., 2009: "The economists of tomorrow", MPRA Paper No. 15691 (Munich Personal RePEc Archive); at: http://mpra.ub.uni-muenchen.de/1569/ (16 November 2016).

Friedman, M., 1999: "Conversation with Milton Friedman", in: Snowden, B.; Vane, H. (Eds.), *Conversations with Leading Economists: Interpreting Modern Macroeconomics* (Cheltenham: Edward Elgar).

Fullbrook, E., (Ed.), 2003: *The Crisis in Economics: The Post-autistic Economics Movement–The First 600 Days* (London/New York: Routledge).

Gandhi, M.K., 1909: *Hind Swaraj or Indian Home Rule* (Ahmedabad: Navajivan).

Gintis, H., 2000: "Beyond homo economicus: Evidence from experimental economics", in: *Ecological Economics*, 5: 311–322.

Granovetter, M., 1985: "Economic action and social structure: The problem of embeddedness", in: *American Journal of Sociology*, 91: 481–510.

Haraway, D., 2014: "Anthropocene, Capitalocene, Chthulucene: Staying with the Trouble", Keynote given at the conference on Arts of Living on a Damaged Planet, University of California, Santa Cruz.

Haraway, D. et al., 2016: "Anthropologists Are Talking – About The Anthropocene", in: *Ethnos: Journal of Anthropology*, 81, 3: 535–564.

Hirschman, A.O., 1986: *Rival Views of Market Society and Other Recent Essays* (Cambridge, Mass.: Harvard University Press).

Homans, G., 1990: "The Rational Choice Theory and Behavioural Psychology", in: Calhoun, C.; Meyer, M.W.; Scott, W.R. (Eds.), *Structures of Power and Constraint. Papers in Honour of Peter M. Blau* (Cambridge: Cambridge University Press).

Horton, T.J., 2011: "The coming extinction of homo economicus and the eclipse of the Chicago school of antitrust: Applying evolutionary biology to structural and behavioural antitrust analysis", in: *Loyola University Chicago Law Journal*, 42: 469–522.

Kant, I., 1963: *On History* (New York: Bobbs-Merrill).

Kant, I., 1990: *Foundations of the Metaphysics of Morals* (London: Macmillan).

Latour, B., 2014: "Anthropology at the Time of the Anthropocene. A Personal View of What Is to Be Studied". Distinguished lecture at the American Anthropologists Association meeting in Washington, December 2014; at: http://www.bruno-latour.fr (18 October 2016).

Leff, E., 2009: "Degrowth, or deconstruction of the economy: Towards a sustainable world", Occasional Paper Series No. 6, in: *Contours of Climate Justice, Critical Currents* (Uppsala: Dag Hammarskjöld Foundation).

Mäki, U., 2005: "Economic epistemology: Hopes and horrors", in: *Episteme*, 1: 211–222.

Malm, A.; Hornborg, A., 2014: "The geology of mankind? A critique of the Anthropocene narrative", in: *The Anthropocene Review*, 1: 62–69.

Matthaei, J., "Solidarity Economy" page of the Transformation Center web page ; at: http://avery. wellesley.edu/Economics/jmatthaei/transformationcentral/index.shtml (9 November 2016).

Mill, J.S., 2004: "Essay V: On the Definition of Political Economy and the Method of Investigation Proper to It", in: *Essays on Some Unsettled Questions of Political Economy* (Pennsylvania State University, Electronic Classic Series): 93–125.

Miller, E., 2009: "Solidarity economy: Key concepts and issues", in: Kawano, E. et al. (Eds.), *Solidarity Economy 1: Building alternatives for people and planet* (Amherst, MA: Center for Popular Economics).

Moore, J.W., 2014a: "The Capitalocene, Part I: On the Nature and Origins of Our Ecological Crisis" (unpublished), Binghamton University, Fernand Braudel Center.

Moore, J.W., 2014b: "The Capitalocene, Part II: Abstract Social Nature and the Limits to Capital" (unpublished), Binghamton University, Fernand Braudel Center.

Moore, J.W., 2015: *Capitalism in the Web of Life: ECOLOGY AND THE ACCUMULATION OF CAPITAL* (New York: Verso).

Razeto, L., 1998: "El 'Factor C': La Fuerza de la Solidaridad en la Economia. ['Factor C': The force of solidarity in the economy']; at: http://www.luisrazeto.net/content/el-factor-c-la-fuerza-de-la-solidaridad-en-la-economia-envista (December 2016).

Rees, M., 2016: "The Anthropocene epoch could inaugurate even more marvellous eras of evolution", in: *The Guardian*, 29 August.

Rockström, J. et al., 2009: "Planetary boundaries: exploring the safe operating space for humanity", in: *Ecology and Society* [online], 42(2): 32; at: http://www.ecologyandsociety.org/vol14/iss2/art32/ (14 January 2016).

Schumpeter, J.A., 1991: *The Economics and Sociology of Capitalism,* Swedberg, Richard (Ed.), (Princeton, NJ: Princeton University Press).

Sen, A., 1987: *On Ethics and Economics* (Oxford: Basil Blackwell).

Seielstad, G.A., 2012: *Dawn of the Anthropocene: Humanity's defining moment* (Alexandria, VA: The American Geosciences Institute).

Smith, A., 1759/1790: *Theory of Moral Sentiments* (London: A. Miller).

Steffen, W. et al., 2004: *Global Change and the Earth System: A Planet Under Pressure* (The IGBP Book Series, Berlin, Heidelberg, New York: Springer Verlag).

Steffen, W. et al., 2011: "The Anthropocene: Conceptual and historical perspectives", in: *Philosophical Transactions of the Royal Society* A, 369: 842–867.

Steffen, W. et al., 2015: "The trajectory of the Anthropocene: The Great Acceleration", in: *The Anthropocene Review,* 2: 81–98.

Steffen, W.; Crutzen, P.J.; McNeill Jr., 2007: "The Anthropocene: Are humans now overwhelming the great forces of Nature?" in: *Ambio*, 36: 614–621.

Taylor, M., 1988: "Rationality and revolutionary collective action", in: Taylor, Michael (Ed.), *Rationality and Revolution* (Cambridge: Cambridge University Press).

Thaler, R.H., 2000: "From homo economicus to homo sapiens", in: *Journal of Economic Perspectives*, 14: 133–141.

Volkmann, K., 2012: "Solidarity economy between a focus on the local and a global view", in: *International Journal of Community Currency Research*, 16: D-97-105. *Theory and Practice* (New Jersey: Prentice Hall).

Chapter 3
Sustainable Communities and Moral Values

Bijayananda Kar

Abstract This chapter discusses sustainable community systems from a design and systems perspective. The socio-political and political backgrounds of communities do, of course, vary from country to country. Although a sense of general commonality is believed to exist despite the diversity of and individual differences between different political set-ups, any peculiar distinctive features should not be ignored. However, instead of focusing on the differences, it is crucial to emphasise the importance of maintaining balance and harmony between sustainability and the way systems are designed and built. For clarity, the relationship between sustainable communities and moral values is also analysed here.

Keywords Sustainable Community · Moral value · Surmise · Barren Religious

3.1 Introduction

Although, at a social level, there is an overall intention that different sectors, such as agriculture, production, community governance, and mutual cultural relationships, should develop in a sustainable way, in practice the rise in sustainability is not that impressive. This is due to several factors, such as inconsistencies between the aim and approach, and the adoption of methods and tools which are not practical or sustainable in the real-life situation of the field concerned.

In this regard, there are several impediments, such as communal bureaucracy, with its peculiar negative measures which foil the goal by the use of delaying tactics, and the age-old social taboos that are poured into the minds of the masses, who are usually lethargic when urged to adopt any constructive move towards prospective change. It is good that an attempt is being made in certain circles to re-examine the various factors involved in order to eliminate the negative elements,

Bijayananda Kar, former Professor and Head, Post Graduate Department of Philosophy, Utkal University, Bhubaneswar, Odisha, India; Email: bkar.nkar@gmail.com

© Springer Nature Switzerland AG 2019   31
A. K. J. R. Nayak (ed.), *Transition Strategies for Sustainable Community Systems*,
The Anthropocene: Politik—Economics—Society—Science 26,
https://doi.org/10.1007/978-3-030-00356-2_3

so that unsustainability in designs and systems can be avoided. If such an attempt is sincerely carried out, the lock-in measure mostly being identified in certain unhealthy devices can be checked and eliminated. There are, in fact, some sound plans already being adopted by concerned sections and organisations which have made notable progress.

This article therefore humbly proposes to make the vital point that the development of community progress needs to be pursued sustainably and can be accomplished mainly by placing the emphasis on change which is pragmatic and which also corresponds to socio-individually acceptable moral norms which are reasonably identified as sustainable in the particular context and situation. In other words, despite differences and diversities among groups and communities within politically established countries, a common general form of modality can be brought out for fructifying sustainability in connection with a systems perspective from the angle of a socio-individual reasonable sense of moral norms suitable for the occasion.

3.2 Tradition and Modernity

It goes without saying that in today's world, the prime rationale for formulating systems in any socio-individual concern is to secure the sustenance of human life. The rise and development of science and technology has had a considerable impact on the information, knowledge and skills operational in various fields and sectors. As a result, there have been changes and modifications to man's likes and dislikes and also in conceptual thinking and expectation.

That is why there is a remarkable gap between the respective outlooks of ancient and modern man. Either ancient is totally given up or it is considerably modified in tune with present-day needs and requirements which are barely necessary on either political or moral grounds. Of course, there are some drawbacks. A few instances may profitably be cited in this regard. As stated before, communal and official bureaucracy in certain governmental and also non-governmental circles is often a delaying tactic which prevents fruitful objectives being attained. Accordingly, sustainable rise and development is rather adversely affected. The bureaucratic mechanism is found to be a boosting affair for red-tapism, thereby causing havoc for sustainable development at the socio-human level.

Further, in the Indian context, it is a fact that, in general, citizens in that country are not socially of the same or similar standard. There is not merely a huge economic or financial disparity between rich and poor or socially upper and lower classes, tribes and races, but also a lack of generally expected and desirable social awareness of socio-individual public safety devices in streets and other public places, such as businesses and enclosed meeting areas. Even private residential areas are not regularly cleaned due to the apathy of their inhabitants. Not only are such areas sadly neglected or abused with minimal regard for a socio-individual sense of morality, but this level of neglect is against legal rules and regulations.

Such deviations are simply frustrating for any effective democratic governance, which is in accordance with independent India's well approved and established constitutional set-up. While a minority of educated, skilled, technically advanced people – who mostly live in small towns, cities and metros – have tolerable scientific awareness, a large number of citizens residing in rural or semi-rural slum areas are neither satisfactorily literate nor have minimum soundness for healthy living and neat sanitation. Those people are found to be not only careless and negligent about themselves but also to have the least concern for public safety and hygienic conditions.

Rather, on the contrary, they deliberately, on many occasions, are found to disrupt public sanitary systems by using drains, or open fields that are surrounded by bushes and plants, as toilets. Such acts are not only illegal but highly immoral and unhygienic. Such instances are said to indicate that it is not only impossible, but almost insurmountable, to develop among Indians as a whole the sense of mutual affordable relationships that are not only legal and moral, but also move a long and satisfactory way towards furthering a sustainable community as the backdrop of a systems perspective. Human survival is facilitated by a mutual sense of one's own dignity and that of others, as well as other reciprocative actions, and it is a practical necessity and requirement to avoid the type of rigid passions and emotions that can lead to negligence.

Moral aspiration is not simply one ideal utopia, but also a living necessity in the practical sphere. It is rooted in pragmatic requirements. In order to have positive expectations, it is counterproductive to worry that satisfactorily attaining morality is almost impossible and may not be workable on a factual plane.

Here two issues may be looked into. The moral ideal which is instantiated or instantiable as a matter of clear possibility in the form of actuality needs to be distinguished from the idea of morality that can only be viewed as a visionary ideal. When one moves into the analysis of moral concepts, the socio-moral dimension cannot be overlooked. The sense of ideality is meaningful and plausible only at the level of practical actuality. Conversely, someone in the grip of an overpowering emotion, being imbalanced, may prefer to be plunged into visionary utopianism. As hinted before, morality is a socio-human concern that is very much needed on the practical plane. Any sort of theological pushing is neither prudent nor practically compelling. The idea of sustainable community is quite plausible in this reading.

3.3 Socio-individual Awareness and Responsibility

A sense of awareness need not be circumscribed by either individual gain at the cost of a social community/group or vice versa. A mutual sense of co-operation and a positive mark of solidarity are necessary. Industrial and mechanical tools and implements are innovated and advanced when they are found to be progressive and beneficial to the people who are actually involved in using them; otherwise there is untoward distrust between the people at one end of the scale who advocate industrialisation/mechanisation and the people at the other end who solely depend on working out something as per age-old customs and traditions that are outmoded, unprogressive and unproductive vis-à-vis the demands of the present time.

To overcome such mismanagement, as well as unhealthy situations, it is the clear responsibility of governance on all fronts, and also of the different social organisations which deal with such issues, to make a sincere effort to induce people who lack a fitting sense of self-awareness and self-responsibility to adopt the new plans and devices that are necessary today.

It is not enough that, as a citizen of India, one is politically independent and now relatively more developed and progressive. There are still many fronts where much care and responsibility for all the people of India are needed. For that, intensive educative measures, both formal and informal, are essential among all sections of people. All this is of human concern on the factual plane. It is also important to foster human awareness of spiritual values and socio-individual co-existence in the earthly empiric setting. If spiritual values and ideals are injected in youngsters (in particular) by their elders, the utmost care needs to be taken as to how far and to what extent such values and ideals are of pragmatic significance as per the present requirement, so that the general sense of progress on all socio-individual fronts is not adversely affected, inflicting country-wide indiscipline and disaster.

Blind belief, dogma, superstition and refusal to move forward in any discussion or debate do not simply exist among one section of society, but are found everywhere. Hence, immediate corrective measures are essential.

Sustainable development of community progress can be fruitfully pursued and accomplished primarily by focusing on pragmatic changes and socio-individually acceptable moral norms which can reasonably be viewed as suitable on the empiric plane. Any hankering or pining for trans-empirical values or the idiosyncratic attitudes and feelings of an individual are in no way to be encouraged or entertained at communal level to disturb and hamper the social need for development and progress. This is also more or less applicable in the context of a design and systems perspective.

Mankind has traditionally had two attitudes towards nature: holistic, or somewhat dual, in the sense that man feels entitled to control and dominate nature and its diverse manifestations for his own purpose and use. Some of the classical religious scriptures of both East and West have more or less indicated that man has dominion over animals. Nature is thus considered in certain circles to be subservient to man.

Despite the rise of scientific knowledge and progress, exploitative attitudes towards nature have continued. This could perhaps be regarded as an unforeseen aspect of the religious legacy of the past. In modern times, the rapid growth of industrialisation has led to greater use and exploitation of nature and the extinction of animal species, with disastrous consequences for the environment. Careless disposal of waste products and the overuse of chemicals have heavily polluted the entire environment – air, land and water – in both populated and remote places, such as hills and jungles.

When dealing with a sustainable community as the setting for a design and systems perspective, it is obvious that one cannot ignore the typical technological innovations which enable it to function. But, at the same time, one also cannot ignore or avoid their grave consequential effect on the environment. That is why there is now a powerful urge in both science and social science to keep a firm balance between industrial and technological use, with the sincere move to have a clean and unpolluted natural environment. In that context the traditional outlook of morality as the background of a religious setting needs a thorough reappraisal for the required changes to take place.

There is now an increasing demand for investigation into environmental and ethical issues. There is likewise a legitimate move to acknowledge the moral obligation to preserve and protect nature, plants and trees, which have sustainability rights alongside the human need to use animals and plants for their survival. In other words, there is now a pressing call for a reciprocal balance and harmony between the two issues.

The traditional sense of religious morality is now seriously re-examined and re-assessed. The extent to which religious attitudes, theological beliefs and doctrines are viewed as the necessary backbone of moral reasoning is the main point of challenge.

3.4 Morality and Theology

It is obvious that religion continues to be one of the living institutions in India and abroad. Morality is firmly embedded in traditions, customs and religious practices of the relevant cultural communities of different countries. In that context, in certain dominant quarters, supernatural concepts like God are thought to be the final custodian of the sense of morality and it is therefore taken for granted that He is the supreme authority to whom the individual must surrender completely in all actions, deeds and thoughts. However, believing in no sense of god/gods, some noted religions like Buddhism are acknowledged as spiritual beliefs without a theistic God or theological attitude. This shows that there is no necessary relation between morality and religion/theology. Though most religions maintain that they contain moral ideas to be followed and practised, non-theists are not automatically immoral and unethical. In other words, non-theological beings or non-believers of any religious type can, and often do, exhibit a bona fide sense of morality.

Whether most of us agree on this point or not, it is a fact that in some of the major developed countries in the world today a considerable number of ordinary people, as well as scientists and social scientists, are critical of traditional religions and, as such, are definitely more convinced by reason and evidential verification. Many internationally reputed writers and thinkers from various fields, such as Freud (1927) with *The Future of an Illusion*, Dawkins (2006) with *The God Delusion*, Christopher Hitchens (2008) with *God is not Great*, and others, have produced books and lectures which forcefully and convincingly argue against theological/ spiritual beliefs, taboos, superstition and prejudices.

That is why, before trying to achieve a balance, with sustainable community as the background of a design and systems perspective, it is essential to continuously and critically note how far the existing so-called moral and theological sayings with a spiritual basis are conducive towards clarity and knowledge or are merely the outcome of unaccounted surmise and baseless quixotic speculations. Quite a notable number of common people, mediocre intellectuals of any discipline, are still found to be victim to the stupidity of claiming their unfounded belief to be pure knowledge and wisdom.

In order to acquire transcendental, spiritual knowledge (*Divya-Jñāna*) and a retainable, sustainable community, it is necessary to identify a practical and efficacious tool, suited to the needs and requirements of the existing context/situation. Thus it is necessary to rethink socio-individual plans at an empirical level. This requires scientific investigation, observation and experiments made in a spirit of rational curiosity about any new discoveries, findings or inventions which materialise. In order to extend knowledge, openness – not sterile rigidity – is essential. If the experimental skill and enquiry are found to be inoperative, dysfunctional or unworkable, then that type of investigation needs to be withdrawn or modified.

So flexibility is the criterion for acquiring the knowledge and practical skills which are essential to undertake a sustainable community plan or programme. Any such endeavour is based on rational enquiry, investigation and probing. In an earthly setting there is no scope for the pre-acceptance of mere surmise without any reasonable basis. Therefore, so-called theological-spiritual backing is not necessary; rather it is redundant.

But, at least in India today, the closed mindset of people in general as well as mediocre intellectuals demonstrates that they are disposed towards some sort of dogmatic reluctance to accept any reasonable practical or pragmatic device for rational advancement, and instead still stoop to a theological set-up, and thus are not at all rationally prone.

In such a situation, there is scant likelihood of other-worldly people attempting to formulate effective designs and systems in the real world. Instead there is a tendency to make conjectures about some peculiar extra-sensory, transcendental vision that is never found to have any positive bearing on the concrete pragmatic plane. It is nevertheless notable that Sigmund Freud, referring to the state of rational probing in the ancient period, commented, "The voice of reason was small but very persistent." In the modern period, the stoppage and arrest of dogmatic theologism is probably mostly due to the progress of skilled scientific research. If, in the course of further studies and research, a specific scientific research finding is found to be inadequate and requires some change or modification, the scientific research is not totally abandoned; a different research theorisation is identified which is found to offer a more convincing explanation with regard to the situation involved. This shows that scientific research is open to revision and review. That is how, on the earthly plane, knowledge is primed for innovation.

3.5 Socio-human Requirement and Theological Necessity

Before concluding, let it be pointed out that, both in lexicographical and commonly used sources, there is more than one use of the terms 'spirit/spiritual/spirited'. These include the theological sense of spiritual (though this is usually about some specific type of religious belief). Apart from that, there is a different use of 'spiritual' in the common sense which is not used in that specific sense of theology (often in ordinary use referring to something factual not meta-factual). For instance, an interviewer might say to an interviewee, "Oh, you are quite spirited and you react to our questions promptly." Here 'spirited' implies something active or energetic and not something holy or sacred which evokes submissive reverence. Such a use of 'spirit' or 'spirited' is quite an ordinary mode of communication and cannot be abandoned since this usage is perfectly proper and quite expressive. All that is argued here is that when used in its theological sense, the term 'spiritual' possesses bona fide connotations of morality which are not implicit in its everyday use. However, morality is not necessarily linked with religion. It is a socio-human requirement at the empirical level and its legitimacy is never reasonably questioned. In that sense, it is quite proximate both in ordinary and technical use. And, in that way, morality is linked with sustainable community. Any design for system-building cannot be gainsaid in that situation. Thus there is no point in linking sustainable community with religion-cum-spiritual usage, which is not a competing force in social structure, either in a general sense or in science/social science.

The present governmental system, which is run by diverse political groups and parties in the Indian democratic set-up, does not seem to be that effective at eradicating the social malady of bungling religious, theological dogmatic rigidity within the socio-individual level of human welfare and solidarity, and thus arrests, or rather jeopardises, human development and progress on the concrete empirical plane.

References

Dawkins, Richard, 2006: *The God Delusion* (London: Bantam Books).
Freud, Sigmund, 1927: *The Future of an Illusion* (London: Hogarth Press, second edition, W.W. Norton & Company).
Hitchens, Christopher, 2008: *God is not Great* (London: Emblem Editions).
Marx, Karl, 1844: *Critique of Hegel's Philosophy of Right* (Paris: Deutsch-Franzosische, Jahrbucher).
Thiroux, Jacques, 1990: *Ethics – The Theory and Practice, Fifth Edition* (New Jersey: Prentice Hall).

Chapter 4
The Buddhist Perspective on Sustainable Development

K. T. S. Sarao

Abstract The present-day profit-orientated global economic system in which moral sentiments are viewed as irrelevant is overwhelmingly controlled and run by consumerism and salespersons. It is fuelled by greed, profiteering, competition and selfishness, leading to environmental degradation, wastage and inequality. In this paper, an attempt has been made to show that, from a Buddhist perspective, the direction in which the world is moving would make it impossible to attain sustainable development. There is an urgent need not only to examine our attitudes and lifestyles but also our policies. From a Buddhist perspective, the primary criterion governing policy formulation must be the well-being of the members of society as a whole. Buddhism views the vulgar chase of luxury and abundance as the root-cause of suffering and encourages restraint, voluntary simplicity and contentment with the minimum. A new relationship must be established between people and nature – one of co-operation, not exploitation. Production must serve the real needs of the people, not the demands of the economic system.

Keywords Buddhism · Sustainable · Globalisation · Consumerism
Dependent origination

4.1 The Context of Sustainable Development

Sustainable development, as defined in *The Brundtland Report* (1987), is "development that meets the needs of the present without compromising the ability of future generations to meet their own needs" (World Commission on the Environment and Development 1987: http://www.un-documents.net/our-common-future.pdf). This report, for the first time, also referred to the need for the integration of economic development, natural resources management and protection, social equity and inclusion with the purpose of meeting human needs without undermining the "integrity, stability and beauty" of natural biotic systems. Before *The*

K. T. S. Sarao, Professor of Buddhist Studies, Delhi University, New Delhi, India;
Email: ktssarao@hotmail.com

© Springer Nature Switzerland AG 2019   39
A. K. J. R. Nayak (ed.), *Transition Strategies for Sustainable Community Systems*,
The Anthropocene: Politik—Economics—Society—Science 26,
https://doi.org/10.1007/978-3-030-00356-2_4

Brundtland Report such an apprehension was well expressed in the influential book *Limits to Growth* (1972), which examined five variables (world population, industrialisation, pollution, food production, and resource depletion) on the computer modelling of exponential economic and population growth with finite resource supplies. The findings were that even if new resources are discovered over a period of time and the current reserves therefore change, still resources are finite and will eventually be exhausted. The book predicts that changes in industrial production, food production, and pollution are all in line with the economic and societal collapse that will take place within the twenty-first century itself (Meadows et al. 2004; Hecht 2008). To put it simply, "the laws of thermodynamics are absolute and inviolate. Unless phytomass stores stabilize, human civilization is unsustainable.... There is simply no reserve tank of biomass for planet Earth. The laws of thermodynamics have no mercy. Equilibrium is inhospitable, sterile, and final" (Schramskia et al. 2015: 9511). How do we come to grips with the problem spelt out above and attain sustainable growth? From a Buddhist perspective, humankind has chosen a wrong path (*agatigamana*) to development and there is the urgent need for two corrective measures. Firstly, there is the need to put a system in place which can not only design and develop non-pollutive alternative technologies needing minimal specialist skills and which use only renewable resources, such as wind and solar power, but can also minimise the social misuse of such technologies. Secondly and more importantly, there is the need to sensitise humanity to the practical understanding of the issue whereby human *Weltanschauung* can be changed and the revival of spirituality that treats nature with respect can take place.

The present-day profit-orientated global economic system in which moral sentiments are viewed as irrelevant is overwhelmingly controlled and run by consumerism and salespersons. In a system such as this, the corporate sector plunders and pollutes on the back of rampant consumerism with the acknowledged goal of *profit maximisation*, which in turn almost always degenerates into *expropriation of wealth*. Organisations of enormous size monopolise the production and distribution of goods. Through the use of clever means, these organisations create an insatiable craving among the masses to possess more and more. A high-consumption lifestyle is aggressively promoted through advertisements, and psychological pressure in various forms is employed to intensify the craving for maximum consumption. One is lured into buying as much as possible irrespective of whether one needs it or has saved enough to pay for it. Thus, goods are bought not because people need them but because they want them. In fact, a consumer society is characterised by the belief that owning things is the primary means to happiness, thus consumption is accepted "as a way to self-development, self-realisation, and self-fulfilment" (Benton 1997: 51–52). As a matter of fact, consumerism has become so ingrained in modern life that it is viewed by some as a new world religion whose power rests in its extremely effective conversion techniques (Loy 1997: 283). This religion, it has been pointed out, works on the principle that not only will growth and enhanced world trade be beneficial to all, but growth will also not be constrained by the inherent limits of a finite planet. Its basic flaw is that it depletes rather than builds "moral capital" (Loy 1997: 283). Fritjof Capra has pointed out that "the health hazards created by the economic system are caused not only by the production process but by the consumption of many of the goods that are produced and heavily advertised to sustain economic expansion" (Capra 1983: 248). Similarly, Erich Schumacher, the author of *Small is Beautiful*, has warned that a materialistic attitude which lacks ethical inhibitions carries within itself the seeds of destruction (Schumacher 1973: 17–18, 56, 119). As pointed out by Erich Fromm, the profit-orientated economic system is no longer determined by the question "What is good for Man?" but by the question "What is good for the growth of the system?" Moreover, consuming has ambiguous qualities: It relieves anxiety, because what one has cannot be taken away; but it also requires one to consume ever more, because previous consumption soon loses its satisfactory character. Actually, this globalising profit-orientated system works on the principle that egotism, selfishness, and greed are fundamental prerequisites for the functioning of the system and that they will ultimately lead to harmony and peace. However, egotism, selfishness, and greed are neither innate in human nature nor are they fostered by it. They are rather the products of social circumstances. Moreover, greed and peace preclude each other (Fromm 2008: 5–8, 23). From a Buddhist perspective, more production of material goods, their increased consumption, and craving (*taṇhā*) for them does not necessarily lead to an increase in happiness. Buddhism teaches that in order to arrive at the highest stage of human development, one must not crave possessions.

One major flaw of the current globalising consumer system is that it promotes competition rather than co-operation. A competitive and adversarial attitude, or the continuous feeling that one has to work against something, not only generates conflict and resentment, but also invariably results in unhealthy side-effects. At international level, mutual antagonisms between nations have resulted not only in billions of dollars being wasted each year on the production of armaments, but also in a major chunk of scientific manpower and technology being directed at the war industry. For instance, military activities in the world engage approximately twenty-five per cent of all scientific talent and use forty per cent of all public and private expenditure for research and development (see Pavitt/Worboys 1977). Sadly, not only do economists look with some apprehension to the time when we stop producing armaments, but also "the idea that the state should produce houses and other useful and needed things instead of weapons, easily provokes accusations of endangering freedom and individual initiative" (Fromm 1955: 5). However, as Bertrand Russell once pointed out, "The only thing that will redeem mankind is co-operation, and the first step towards co-operation lies in the hearts of individuals" (Russell 1954: 204). It has been seen that individuals with co-operative skills are more creative and psychologically better adjusted. With its emphasis on co-operation and interdependence, Buddhist practice can inspire the building of partnership societies with need-based and sustainable economies.

Political leaders and business executives often take self-serving decisions. Moreover, "the general public is also so selfishly concerned with their private affairs that they pay little attention to all that transcends the personal realm.... Necessarily, those who are stronger, cleverer, or more favoured by other circumstances... try to take advantage of those who are less powerful, either by force and violence or by suggestion… [Conflict in society] cannot disappear as long as greed dominates the human heart" (Fromm 2008: 10–11, 114). A society driven by greed loses the power of seeing things in their wholesomeness. Consequently, we do not know when enough is enough. "The hope…that by the single-minded pursuit of wealth, without bothering our heads about spiritual and moral questions, we could establish peace on earth, is an unrealistic, unscientific, and irrational hope…the foundations of peace cannot be laid by…making inordinately large demands on limited world resources and… [putting rich people] on an unavoidable collision course – not primarily with the poor (who are weak and defenceless) but with other rich people" (Schumacher 1973: 18–19). In the present economic system, points out Schumacher, anything that is not *economic* is sought to be obliterated out of existence. "Call a thing immoral or ugly, soul-destroying or a degradation of man, a peril to the peace of the world or to the well-being of future generations; as long as you have not shown it to be 'uneconomic' you have not really questioned its right to exist, grow, and prosper" (Schumacher 1973: 27).

4.2   Sustainable Development from a Buddhist Perspective

In this regard, Buddhism regards greed (*lobha*: Morris/Hardy 1995–1900: iv.96) and egotism (*avaññattikāma*: Morris/Hardy 1995–1900: ii.240; iv.1; *asmimāna*: Oldenberg 1879–1883: i.3; Rhys Davids/Carpenter 1890–1911: iii.273; Trenckner/ Chalmers 1888–1896: i.139, 425; Morris/Hardy 1995–1900: iii.85) as leading to suffering. The real problem lies in the human tendency to have – to possess – which Buddha called craving (*taṇhā*). It may be pointed out that Buddhism does not mind wealth and prosperity as long as they are acquired and used in accordance with the ethical norms. Moreover, from a Buddhist perspective, apart from taking into account the profitability of a given activity, its effect upon people and the environment, including the resource base, is equally important.

Another flaw of the current globalising consumer system is that it is widening the division between the rich and the poor. According to the Credit Suisse *Global Wealth Report*, the richest one per cent of people in the world now own half of the planet's wealth, and, at the other extreme, the poorest fifty per cent of the world's population owns just 2.7% of global wealth (Kentish 2017). This type of stark poverty and inequality, leading to the marginalisation and exclusion of the majority of the world's population, has implications for social and political stability among and within states. It will be unrealistic to expect spiritual, psychological, and social harmony in the world while it remains materially divided. As a member of a common human family, each individual must have access to a reasonable share of the resources of the world so that s/he is able to fulfil his/her basic needs to realise his/her potential as a productive and respected member of the global family. This means that there is an urgent need for equitable access to resources not only between nations, but also between humans, irrespective of gender and nationality. As the desperate poverty of the poor has been responsible to some extent for the overuse of limited resources, economic justice and social equity are important. However, affluent societies are the real problem children of today's world. For instance, it has been estimated that the birth of an American baby represents more than fifty times as great a threat to the environment as the birth of an Indian baby (Jones 1993: 14). Well-documented research has shown that world hunger caused by scarcity of food is a myth because the amount of food produced in the world at present is sufficient to provide about eight billion people with an adequate diet. The main culprit is agribusiness in a world marred by inequalities (see Capra 1983: 257–258). "Without a revolution in fairness, the world will find itself in chronic conflict over dwindling resources, and this in turn will make it impossible to achieve the level of co-operation necessary to solve problems such as pollution and overpopulation" (Elgin 1993: 42).

In this regard, it may be said that Buddhism promotes a wide distribution of basic necessities so that no one has to suffer deprivation, as deprivation is the root cause of social conflict. Thus, talking about the cause of social conflict, Buddha pointed out that "goods not being bestowed on the destitute, poverty grew rife; from poverty growing rife, stealing increased, from the spread of stealing violence grew apace, from the growth of violence, the destruction of life became common" (Rhys Davids/Carpenter 1890–1911: iii.67). From a Buddhist perspective, an ideal society would follow the motto of happiness and welfare for the maximum number of people (*bahujanahitāyabahujanasukhāya*: Oldenberg 1879–1883: i.21). In such a society one would not look for one's own satisfaction in ways that may become a source of pain/suffering (*aghabhūta*) for others (Feer 1884–1898: iii.189). Hoarding wealth in any form is looked down upon in Buddhism (Morris/Hardy 1995–1900: iii.222) and if a wealthy person were to enjoy his wealth all by himself, it would be a source of failure for him (Fausböll 1985: 102). In fact, someone working for the sake of wealth (*dhanahetu*: Fausböll 1985: 122), craving wealth (*dhanatthiko*: Fausböll 1985: 987; *bhogataṇhā*: Sarao 2009: 355), or taking pride in wealth i.e., displaying economic snobbery (*dhanatthaddho*, Fausböll 1985: 104) is considered to be a fallen human being and an ignoramus who hurts himself as well as others. Thus, in a Buddhist approach to social and economic development, the primary criterion governing policy formulation must be the well-being of members of the society as a whole i.e., production should be orientated towards serving the real needs of the people instead of the people serving the demands of the economic system. As emphasised by Fromm, Buddhism supports every human being's right to be fed without qualification in the way that a nourishing mother feeds her children, who do not have to achieve anything in order to establish their right to be fed, and it opposes the mentality of hoarding, greed, and possessiveness. In such a perception, people's "income is not differentiated to a point that creates different experiences of life for different groups" (Fromm 2008: 69).

The present system believes that fulfilment of the material needs of humankind will lead to peace and happiness. But this is a mistaken view. As Erich Fromm points out, an animal is content if its physiological needs – hunger, thirst and sexual urges – are satisfied because, being rooted in the inner chemistry of the body, they can become overwhelming if not satisfied. Inasmuch as man is also an animal, these needs must be satisfied. But inasmuch as one is human, the satisfaction of these instinctual needs is not sufficient to make one happy because human happiness depends on the satisfaction of those needs and passions which are specifically human. These essential needs which modern civilisation fails to satisfy are "the need for relatedness, transcendence, rootedness, the need for a sense of identity and the need for a frame of orientation and devotion" (Fromm 1955: 25, 28, 65, 67, 134). From the Buddhist perspective, economic and moral issues cannot be separated from each other because the mere satisfaction of economic needs without spiritual development can never lead to contentedness among people. By pointing out that the vulgar chase of luxury and abundance is the root-cause of suffering, Buddhism encourages restraint, simplicity and contentment. This way of life embraces frugality of consumption, a strong sense of environmental urgency, a desire to return to human-sized living and working environments, and an intention to realise our higher human potential – both psychological and spiritual (Elgin/ Mitchell 1977: 5). This type of enlightened simplicity would integrate "both inner and outer aspects of life into an organic and purposeful whole.... outwardly more simple and inwardly more rich...and living with balance in order to find a life of greater purpose, fulfillment, and satisfaction" (Elgin 1993: 25). Enlightened simplicity is essential to attain sustainable development and to solve global problems of environmental pollution, resource scarcity, socio-economic inequities, and existential/spiritual problems of alienation, anxiety, and lack of meaningful lifestyles. Thus, the need of the hour for developed nations is to follow what Arnold Toynbee called the *Law of Progressive Simplification* i.e., progressively simplifying the material side of their lives and enriching the non-material side (Toynbee 1947: 198). Taking a position akin to Buddhism, Elgin (1993: 32–35) has suggested that one choosing to live a life of simplicity would not only lower the overall level but also modify the patterns of one's consumption by buying products that are long-lasting, easy to repair, serviceable, energy-efficient, and non-polluting in their use as well as manufacture. Besides believing in deep ecology, one would show an ardent concern for the poor and the needy. One would prefer a smaller-scaled and human-sized living and working environment that fosters a sense of community and mutual caring. One would shift one's diet in favour of one that is more natural, healthy, simple, and suitable for sustaining the inhabitants of Planet Earth. One would not only recycle but also downsize by owning only those possessions that are absolutely required. One would develop personal skills to handle life's ordinary demands for enhancing self-reliance, minimising dependence upon others, and developing the full range of one's potentials. One would also spare time on a regular basis to volunteer to help in improving the quality of life of the community. Enlightened simplicity requires being content (*saṃtuṭṭhi*: Rhys Davids/Carpenter 1890–1911: i.71; Trenckner/Chalmers 1888–1896: i.13; Fausböll 1985: 265; Sarao 2009: 204; Morris/Hardy 1995–1900: ii.27, 31, ii.219) with little, avoiding wastefulness i.e., few desires (*appicchatā*: Oldenberg 1879–1883: iii.21; Rhys Davids/Carpenter 1890–1911: iii.115; Trenckner/Chalmers 1888–1896: i.13; Feer 1884–1898: ii.202). Contentment, which is viewed in Buddhism as the best wealth (*saṃtuṭṭhiparamaṃdhanaṃ*: Sarao 2009: 204), is the mental condition of a person who is satisfied with what he has or the position in which he finds himself (*saṃtussamānoitarītarena*: Fausböll 1985: 42).

"Private property", as once pointed out by Karl Marx, "has made us so stupid and partial that an object is only ours when we have it, when it exists for us as capital...Thus all the physical and intellectual senses have been replaced by...the sense of having" (Bottomore 1963: 159). Thus, as pointed out by Erich Fromm, people acquire things, including useless possessions, because they "confer status on the owner" (Fromm 1955: 133). In the *Having Mode of Existence*, a person's relationship with the world is one of possessing and owning, treating everybody and everything as property. The fundamental elements in the relationship between individuals in this mode of existence are competition, antagonism and fear. In such a mode, one's happiness lies in one's superiority over others, in one's power and capacity to conquer, rob, and kill. The peril of the having mode is that even if a state of absolute abundance could be reached; those who have less in terms of physical health and attractiveness, gifts and talents bitterly envy those who have *more* (Fromm 2008: 66–67, 91–92). In the *Being Mode of Existence* one's happiness lies in aliveness and authentic relatedness to the world, loving, sharing, sacrificing, and giving. The difference between these two modes of existence is that whereas the *having* mode is centred around persons, the *being* mode is centred around things (Fromm 2008: 15, 21, 66).

There is an urgent need to sensitise people to the fact of the interconnectedness and interdependence of all living beings, including humans, and resources. The earth is not only teeming with life, but seems to be a living being in its own right. A wide-ranging, objective, well-documented, and value-free piece of scientific research shows that each living creature has its place in the biosphere, whereby it plays its unique role as part of the collective balance. As pointed out by Capra, all the living matter on Earth, together with the atmosphere, oceans, and soil, forms a complex system that has all the characteristic patterns of self-organisation. Thus, "the earth is a living system and it functions not just like an organism but actually seems to be an organism – Gaia, a living planetary being" (Capra 1983: 284–285). From a Buddhist perspective, not only that life is inherently valuable but human and other forms of life are also interdependent and reciprocal. Thus, nature and humanity on the one hand and humans amongst themselves on the other are seen as mutually obligated to each other. A living entity can neither isolate itself from this causal nexus nor have an essence of its own. In other words, as part of the Dependent Origination (*paṭiccasamuppāda*), humans are seen as affecting their environment not only through the purely physical aspects of their actions, but also through the moral and immoral qualities of such actions. It is thus said that if a king and his people act unrighteously, this has a bad effect on the environment and its gods, leading to little rain, poor crops and weak, short-lived people (Morris/Hardy 1995–1900: ii.74–76). This message is also strongly implied by the *AggaññaSuttanta* of the *DīghaNikāya* (Rhys Davids/Carpenter 1890–1911: iii.80–98) which shows how, in the beginning, nature was bountiful but became less so when humans began to take greedily from it. When they began to harvest more rice than they needed, it was not naturally able to grow quickly enough. This necessitated cultivation, which in turn caused division of land into private fields, so that property was invented. The origin of private property became the root cause of different social and economic ills. Thus, one is not surprised that from a Buddhist point of view, consumer-orientated modernity "is rejected because it is seen as a form of life that has in a short period of time despoiled the landscape and done irreparable damage to the environment" (Lancaster 2002: 1–2).

Just as poverty is the cause of much crime, wealth too is responsible for various human ills. Buddhism views material wealth as being required only for meeting the bare necessities and must only be earned through righteous and moral means. Generosity (*dāna*) and liberality (*cāga*) are always linked in Buddhism with virtue (Sarao 2009: 177). Moreover, by giving, one gets rid of greediness/selfishness (*macchariya*) and becomes more acceptable to others because "one who gives makes many friends" (Fausböll 1985: 187; Morris/Hardy 1995–1900: iii.273. v.40, 209; Rhys Davids/Carpenter 1890–1911: iii.234). Above all, it is not necessary to have much to practise generosity because giving even from one's meagre resources (*dajjāappampi*) is considered very valuable (Feer 1884–1898: i.18; Sarao 2009: 224). Generosity is one of the important qualities that make one a gentleman (Morris/Hardy 1995–1900: iv.218). Buddha is the spiritual friend (*kalyāṇamitta*) (Feer 1884–1898: v.3) par excellence and the saṃgha members who are his spiritual heir (*dhammadāyādo*) are also expected to act as such (Rhys Davids/Carpenter 1890–1911: iii.84; Feer 1884–1898: ii.221). Buddha compares the person who earns wealth righteously and shares it with the needy to a person who has both eyes, whereas the one who only earns wealth but does no merit is like a one-eyed person (Morris/Hardy 1995–1900: i.129–130). In other words, if a healthy society is to be built, liberality and generosity must be fostered as its foundation pillars.

Avoidance of wastage, which is one of the most serious stumbling blocks in the path to sustainable development, is an important aspect of Buddhist enlightened simplicity. The fig-tree glutton (*udumbarakhādika*) method decried by Buddha (Feer 1884–1898: iv.283) – the method of shaking down an indiscriminate amount of fruit from a fig-tree in order to eat a few – is exactly the same as the one employed in drift-net fishing, where much more aquatic life is destroyed than utilised.

4.3   Need for a Fundamental Change

Humanity cannot continue to consume the planet's limited resources at the rate to which it has become accustomed. Through unbridled expansion, the economy is not only absorbing into itself more and more of the resource base of the extremely fragile and finite ecosystem, but is also burdening the ecosystem with its waste. As human population grows further, the stress on the environment is bound to rise to even more perilous levels. Exploding population levels wipe out what little is accomplished by raising living standards. As pointed out by Paul and Anne Ehrlich, considering present technology and patterns of behaviour, our planet is grossly overpopulated now and the limits of human capability to produce food by conventional means have also very nearly been reached. Attempts to increase food production further will tend to accelerate the deterioration of our environment, which in turn will eventually reduce the capacity of the earth to produce food. The Green Revolution "is proving ecologically unsustainable, dependent as it is on an economically and socially vulnerable, high cost, petrochemical agriculture" (Jones 1993: 13). Its dark side is reflected in crops' vulnerability to pest problems, loss of genetic diversity through mono-cropping and neglect of local varieties, fertilizer-induced increase of weeds, the threat of fertilizer pollution in fragile soils, toxicity through pesticides leading to cancer and adverse effects on the body's natural immune system, erosion accelerated by multiple cropping, and the mindless squandering of water resources. "Yet these alarming results have barely affected the sale and use of fertilizers and pesticides" (Capra 1983: 257). Through the degradation of the environment, the future is clearly being undermined by the rich in emulation of the developed world and by the poor to stay alive by salvaging the present by savaging the future. In fact, we are faced with "the prospect that before we run out of resources on any absolute basis we may poison ourselves to death with environmental contaminants" (Elgin/Mitchell 1970: 5). Global warming is now irreversible, and nothing can prevent large parts of the planet becoming too hot to inhabit, or sinking underwater, resulting in mass migration, famine and epidemics. "Signs of potential collapse, environmental and political, seem to be growing…while politicians and elites fail to recognise the basic situation and focus on expanding their own wealth and power" (Ehrlich/Ehrlich 2009: 68). As suggested by Stephanie Kaza, the environmental impact is accelerated by the rapidly rising population numbers, increasingly efficient technologies, and consumption rates beyond the planet's capacity. These three have been linked by the equation $I = PAT$, or environmental Impact = Population size multiplied by Affluence (or degree of consumption) multiplied by Technology. Reduce any one of these, and the impact drops; increase one or all three, and the impact rises, in some cases dramatically (Kaza 2000: 23). Since human beings are social creatures who naturally come together for common ends, this means that a social order guided by Buddhist principles would consist primarily of small-scale communities with localised economies in which each member can make an effective contribution. From the perspective of Buddhist-economics, "production from local resources for local needs is the most rational way of economic life" (Schumacher 1973: 42). To attain sustainable development, what we need most of all is streamlining and downsizing. Only small-scale and simple technology would not drain natural resources, as in it production would be aimed principally at local consumption, so that there is direct face-to-face contact between producers and consumers. Large-scale technologies are dehumanising and morally wrong, as they become impersonal and unresponsive, making individuals functionally futile, dispossessed, voiceless, powerless, excluded and alienated. "Wisdom demands a new orientation of science and technology towards the organic, the gentle, the non-violent, the elegant and beautiful" (Schumacher 1973: 20).

4.4   The Buddhist Economic Perspective

Buddhist values mean that environment should not be over-exploited and "non-renewable goods must be used only if they are indispensable, and then only with the greatest care and the most meticulous concern for conservation...The Buddhist economist would insist that a population basing its economic life on non-renewable fuels is living parasitically" (Schumacher 1973: 43–44).

Thus, from a Buddhist perspective, a new relationship must be established between people and nature, one of co-operation not of exploitation or domination. The driving force of such an economy would be to make a distinction between a state of utmost misery (*daḷiddatā*: Feer 1884–1898: v.100, 384, 404), being sufficient (*yāpanīya*: Oldenberg 1879–1883: i.59, 212, 253), and glut (*accogāḷha*: Morris/Hardy 1995–1900: iv.282). There would be a balance between material excess and deprivation i.e., avoidance of both mindless materialism and needless poverty, leading to a balanced approach to living that harmonises both inner and outer development. It would be unbuddhistic to consider goods as more important than people and consumption as more important than creative activity. To build a sustainable future, affluent members of society will need to make dramatic changes in their overall levels and patterns of consumption. We must choose levels and patterns of consumption that are globally sustainable, i.e., use the world's resources wisely and not overstress the world's ecology, i.e., consume in ways that respect the rest of life on this planet. Such an aim was made explicit in the Green Buddhist Declaration, prepared by members of the international Buddhist community for discussion at the World Fellowship of Buddhism in Colombo (1980): "We believe that since world resources and the ecosystem cannot support all peoples at the level of the consumption of the advantaged nations, efforts towards global equity must be coupled with efforts towards voluntary simplicity, in one's individual lifestyle and through democratically-determined policies. The economic structures which encourage consumerist greed and alienation must be transformed."

From a Buddhist perspective, it is also important for policies to be grounded in moral and ethical values that seek the welfare of humankind as a whole. As suggested by Durning (1992), the linked fates of humanity and the natural realm depend on us, the consumers. We can curtail our use of ecologically destructive things and cultivate the deeper, non-material sources of fulfilment that bring happiness: family and social relationships, meaningful work, and leisure. Implementation and realisation of the spirit underlying the Buddhist Eight-fold Path (*aṭṭhaṅgika-magga*) encompassing wisdom (*paññā*), morality (*sīla*), and meditation (*samādhi*) in eight parts can truly offer a path leading to sustainable development. Right View (*sammā-diṭṭhi*) and Right Thought (*sammā-saṃkappa*) constitute wisdom; Right Speech (*sammā-vācā*), Right Conduct (*sammā-kammanta*) and Right Livelihood (*sammā-ājīva*) constitute morality; and Right Effort (*sammā-vāyāma*), Right Mindfulness (*sammā-sati*) and Right Concentration (*sammā-samādhi*) form the practice of meditation (Rhys Davids/Carpenter 1890–1911: ii.311–315). By following this path of wisdom, morality, and meditation, one can grow inwardly and follow a life of enlightened simplicity. By following this path, humans can aim at harmonious living (*dhammacariya, samacariya*: Trenckner/ Chalmers 1888–1896: i.289; Feer 1884–1898: i.101) and compassion (*karuṇā*) with "the desire to remove what is detrimental to others and their unhappiness" (Fausböll 1985: 73). This would form the basis of the *Weltanschauung* of the well-adjusted and balanced person, who would seek inner peace (*ajjhattasanti*, Fausböll 1985: 837) and inward joy (*ajjhattarata*, Sarao 2009: 362; Rhys Davids/Carpenter 1890– 1911: ii.107; Feer 1884–1898: v.263) by exercising a degree of restraint, limiting his/her needs, and avoiding being greedy (*ussuka*: Sarao 2009: 199) because one can never become worthy of respect if one is envious, selfish, and fraudulent (*issukīmaccharīsaṭho*: Sarao 2009: 262).

It is time for each of us to choose a way of life that is materially simple, inner-directed, and ecology-friendly. The fundamental issue is of the Earth's finite capacity to sustain human civilisation. "Lifeboat ethic" must be replaced by "spaceship earth ethic." Mindful living opens our perception to the interdependence and fragility of all life, and our indebtedness to countless beings, living and dead, from the past and the present. Let me conclude with Elgin:

> To live sustainably, we must live efficiently – not misdirecting or squandering the earth's precious resources. To live efficiently, we must live peacefully, for military expenditure represents an enormous diversion of resources from meeting basic human needs. To live peacefully, we must live with a reasonable degree of equity, or fairness, for it is unrealistic to think that, in a communications-rich world, a billion or more persons will accept living in absolute poverty while another billion live in conspicuous excess. Only with greater fairness in the consumption of the world's resources can we live peacefully; and thereby live sustainably, as a human family (Elgin 1993: 41–42).

References

Anonymous, 1983: *Discussion Points for United Nations University Project on Perceptions of Desirable Societies in Different Religions and Ethical Systems* (Tokyo: United Nations University).

Benton, Raymond, 1997: "Work, Consumption, and the Joyless Consumer", in: Goodwin, N.R.; Ackerman, F.; Kiron, D. (Eds.), *The Consumer Society* (Washington, DC: Island Press).

Bottomore, T.B. (ed. and trans), 1963: *KarlMarx: Early Writings* (New York: McGraw-Hill).

Capra, Fritjof, 1983: *The Turning Point: Science, Society, and the Rising Culture* (Toronto: Bantam Books).

Durning, Alan, 1992: *How Much is Enough? The Consumer Society And The Future Of The Earth* (New York: W.W. Norton).

Ehrlich, P.R.; Ehrlich, A.H. 2009: "Population Bomb Revisited", in: *The Electronic Journal of Sustainable Development*, 3(1).

Elgin, D., 1993: *Voluntary Simplicity: Toward a Way of Life that is Outwardly Simple. Inwardly Rich*, rev. ed. (New York: Quill).

Elgin, D.; Mitchell, A. 1977: "Voluntary Simplicity," *CoEvolution Quarterly*, Summer, no. 14, 21 June 1977: 4–19.

Fausböll, V. (Ed.), 1985: *The Sutta-Nipāta* (London: Pali Text Society).

Feer, M.L. (Ed.), 1884–1898: *The Saṃyutta Nikāya*, 5 vols. (London: Pali Text Society).

Fromm, Erich, 1955: *The Sane Society* (New York: Henry Holt and Company).

Fromm, Erich, 1976, 2008: *To Have or To Be?* (New York: The Continuum Publishing Company).

Hecht, Jeff, 2008: "Prophesy of economic collapse 'coming true'," *New Scientist*, 17 November; at: www.newscientist.com/article/dn16058-prophesy-of-economic-collapse-coming-true. html#.UtYbRL7rZjo (16 March 2015).

Hinüber, von O.; Norman, K.R. (Eds.), 1994: *The Dhammapada* (Oxford: Pali Text Society).

Jones, Ken, 1993: *Beyond Optimism: A Political Buddhist Ecology* (Oxford: Jon Carpenter).

Kaza, Stephanie, 2000: "Overcoming the Grip of Consumerism", in: *Buddhist-Christian Studies*, 20: 23–42.

Kentish, Ben, 2017: "Richest 1% own half the world's wealth, study finds", in: *The Independent*, 15 November, London.

Lancaster, L., 2002: "The Buddhist Traditions in the Contemporary World: History and Critique", A keynote speech made at the opening ceremony of the Fourth Chung-Hwa International Conference on Buddhism, 18–20 January 2002.

Loy, David, 1997: "The Religion of the Market", in: *Journal of the American Academy of Religion*, 65(2): 275–290.

Marx, K.; Engels, F., 1845: *Selected Works (1845–1849)*, vol. I., in: Marxist Internet Archive; at: www.marxists.org/archive/marx/works/sw/ (15 January 2014).

Meadows, D.H.; Meadows, D.L., et al., 2004: *The Limits to Growth*, updated version (White River Junction, VT: Chelsea Green Publishing Company and Earthscan).

Merchant, Carolyn, 1980: *The Death of Nature*. New York: Harper & Row.

Morris, R.; Hardy, E. (Eds.), 1995–1900: *The Aṅguttara Nikāya*, 5 vols. (London: Pali Text Society).

Oldenberg, V. (Ed.), 1879–1883: *The VinayaPiṭaka*. 5 vols. (London: Pali Text Society).

Pavitt, K.; Worboys, M., 1977: *Science, Technology and the Modern Industrial State* (London: Butterworth, SISCON).

Rhys Davids, T.W.; Carpenter, J.E. (Eds.), 1890–1911: *The Dīgha Nikāya*, 3 vols. (London: Pali Text Society).

Russell, Bertrand, 1954: "Human Society", in: *Ethics and Politics* (London: George Allen and Unwin).

Sarao, K.T.S. (trans.), 2009: *The Dhammapada: A Translator's Guide*. (New Delhi: Munshiram Manoharlal).

Schramskia, John R.; Gattiea, David K.; Brown, James H., 2015: "Human domination of the biosphere: Rapid discharge of the earth-space battery foretells the future of humankind", in: *Proceedings of the National Academy of Sciences of the United States of America*, vol 112, no. 31: 9511–9517. www.pnas.org/content/early/2015/07/14/1508353112.full.pdf (27 Nov. 2016).

Schumacher, E.F., 1973: *Small is Beautiful: Economics as if People Mattered*. 1999 print with Commentaries (Point Roberts, WA: Hartley & Marks).

Toynbee, Arnold, 1947: *A Study of History*, vol. 1 (New York: Oxford University Press).

Trenckner, V.; Chalmers, R. (Eds.), 1888–1896: *The Majjhima Nikāya*. 3 vols (London: Pali Text Society).

World Commission on the Environment and Development, 1987: *Our Common Future [The Brundtland Report]* (Oxford: Oxford University Press); at: www.UN-documents.net (16 July 2013).

World Fellowship of Buddhists, 1984: *Green Buddhist Declaration* (Bangkok: World Fellowship of Buddhists).

Chapter 5
Paticcasamuppada: The Theory of Dependent Origination: A Scientific Means of Changing Outlook and Behaviour

Praveen Kumar

Abstract *Paticcasamuppada* is one of the central concepts of Buddhist philosophy, propounding that each *dhamma* (mental phenomenon) is intrinsically connected to other *dhammas* (*mental phenomena*) *philosophically*. This argument is used to explain *dukkha* (suffering) and *samasara* (world), which continue the cycle of birth incessantly. In the present scenario this argument becomes pertinent in the context of sustainable development. Development of anything is centred around man and the value system held by him. The greed in human beings is detrimental to them and nature, in the context of sustainable development. *Paticcasamuppada* explains the origin of greed in man and how it can be reduced or eliminated. Thus *paticcasamuppada* becomes relevant in the context of sustainable development.

Keywords *Avijja* (ignorance) · *Sanikhara* (karma formations) · *Vinnana* (consciousness) · *Nama-rupam* (mind and matter) · *Salayatanam* (six faculties) *Phasso* (contact) · *Vedana* (sensation) · *Tanha* (desire) · *Upadana*m (clinging) *Bhavo* (process of becoming) · *Jati* (birth) · *Jara-marana* · *Dukkha domanassa* (decay, death, lamentation, pain)

5.1 Introduction

In Buddhist philosophy *paticcasamuppada* is the law of dependent origination. It explains *dukkha* and *samsara*, the cycle of birth and death i.e. repeated existence. It is a causal chain with twelve links, which are: 1. *avijja* (ignorance); 2. *sankhara* (Karma formation); 3. *vinnana* (consciousness); 4. *nama-ripam* (mind and matter); 5. *salayatanam* (six faculties); 6. *phasso* (contact); 7. *vedana* (sensation); 8. *tanha* (desire); 9. *upadanam* (clinging); 10. *bhavo* (process of becoming); 11. *jati* (birth); and 12. *jara-marana, dukkha donanasa* (decay, death, lamentation, pain).[1]

Associate Professor, Niswass College, Bhubaneswar, Odisha, India; Email: Praveenkumar. kumar1976@gmail.com

© Springer Nature Switzerland AG 2019
A. K. J. R. Nayak (ed.), *Transition Strategies for Sustainable Community Systems*,
The Anthropocene: Politik—Economics—Society—Science 26,
https://doi.org/10.1007/978-3-030-00356-2_5

These twelve links extend over three lives and span them. The first two belong to the past life, the last two represent the future (i.e. rebirth), and the rest of the links – i.e. the eight links from *vinnanam* to *bhava* – represent the present life. Each link is an effect of the preceding link, which acts as a cause.[2]

It means every link of this chain originates depending on the preceding link and gives rise to the succeeding link. Here nothing arises independently. Everything has a cause to arise. *Imasmim sati idam hoti* (this being that becomes). *Imass uppada idam uppajjati* (from this arises the next). This law also speaks about the opposite. Nothing happens without a cause. Nothing happens fortuitously. *Imasmim asatim idam na hoti* (if an event does not occur then the subsequent event does not takes place). *Imassa nirodha idam nirujjhati* (from the cessation of this, that ceases). When the cause ceases, the effect also ceases to exist. In the complete absence of cause there is no possibility of any effect.[3]

Buddha propounded this theory in the context of the cycle of birth. This theory also explains the origination of greed in man and how it can be reduced or eliminated. Thus, *paticcasamuppada* has becomes relevant in the context of sustainable development.

The first link of the law of dependent origination explains this *avijjapaccayasankhara* (ignorance gives rise to formations) and *vedanapaccayatanha* (sensations produce craving). When we do not know the real nature of things we crave pleasant sensations and want them to continue and unpleasant ones to stop. This is how we react in ignorance. We like pleasant sensations and want them to continue and, on the other hand, we do not like the unpleasant ones and want them to stop. We reacts in ignorance and multiply our misery because it is not possible for all cravings to be fulfilled.

According to Buddhist philosophy, craving or greed is the primary source of all evils. Greed is a defilement which is very deeply rooted in the human mind and it is perhaps the basic defilement from which all other defilements proceed.[4]

It has been very graphically shown in the *Aggana sutta* (palitext) of the *Digha Nikaya* how man lost his innocence because of his greed. This text shows how greed gives rise to ugliness, arrogance and lust and how it leads to hoarding, stealing, lying, censoring, perishing and many other evil things. So long as this defilement is not removed, a man will steal, crave more and more, and commit all sorts of crimes. In short, Buddhist philosophy calls for the removal of greed, which is seen as a basic defilement.[5]

How greed can affect ecology has been very graphically shown by the Buddha in this *sutta*.

---

[1] Amgraj Chaudhary, 2009. *Aspects of Buddha Dhamma*. Delhi.

[2] *Ibid.*

[3] *Ibid.*

[4] *U kolay, Pathiaka Vagga pali, Yangoa Myanmar* (see *Agganna Sutta*).

[5] Chaudhary, *Aspects of Buddha Dhamma*.

In the beginning man was without any greed. The earth with its savour was spread out in the waters even as a scum forms on the surface of boiled milky rice that is cooking. Eventually it became endowed with colour, and taste.

When greed appeared in some people they tasted the savoury earth, liked it and developed a craving for it. Thus, they started taking lumps of it with their hands. They began to hoard it for future use. Then it so happened that they lost their radiance and many other bad things appeared in them. As a result, passion developed in them. They also became jealous of those people who were beautiful and lust developed in them. Out of greed they began to hoard rice. This made them not only greedy but also lazy. Their greed, jealously, lust and laziness drastically affected ecology and at a later stage when the rice that was clean and fragrant developed powder and husk because of his greed.

When people were not greedy they cut stems which grew again and again but when greed developed in them the rice stubbles stood in clumps.

With the passage of time when rice stubbles ceased to grow they thought of dividing the rice fields and set boundaries. This is how they made private property. What was common to all became private to some. This was because of their greed, which greatly affected not only ecology but also the minds of the people.[6]

The story narrated in the *Agganna sutta* may be taken as a fable which realistically describes what happens to man when he is too greedy and full of cravings. He meets a tragic fate, as did the man in the fable who killed the goose that laid a golden egg every day or the *Midas* of Greek mythology, who had plenty of gold but no food to eat.

How greed affects ecology has been shown here. In modern times man has become so greedy that he establishes big industries to manufacture different kinds of goods to make money, but its results are horrible. The entire atmosphere is polluted by industrial areas. The water of the nearby river is polluted. He has mental stress all the time, and he is condemned to breathe in carbon dioxide and carbon monoxide gases, which are dangerous to lungs.

How to reduce greed? We can get rid of defilements like greed only when we realise on an experiential level that nothing is permanent. Knowing this on an intellectual level will not do. Hearing from others that everything is impermanent also will not do. It will be effective only when we practise *vipassana*. When we practise *vipassana* we see very clearly that nothing lasts forever. The pleasant sensations we feel when we like somebody or something do not last forever. We also realise that unpleasant sensations do not last forever. With this realisation, again and again and time and again, we develop non-attachment or *nirveda*. Why pursue things which do not last forever?[7]

Buddhist philosophy teaches us how to develop *bhavanamaya panna*, which enables one to see at the experiential level that things that we crave are impermanent. And what is impermanent is suffering "*yadaniccam tam dukkham*" (What is impermanent is an object of suffering). Why does Buddha want us to develop *bhavana maya panna*? Because it is very effective for transforming the mind and

[6]*Agganna Sutta.*

[7]Chaudhary, *Aspects of Buddha Dhamma.*

filling it with good qualities. We understand at the experiential level, not just at the intellectual level, the impermanence of things of the world, which helps us to develop non-attachment (*nirveda*). This realisation will enable us not to crave for things any more. And it is craving which causes suffering in life and makes us move in the cycle of birth and death.[8]

But this is not easy to develop. It is necessary to experience it again and again, time and again. Time and over again, to be able to develop a kind of *samvega*, a religious awe and fear not to fall prey to cravings. In the *tipitaka* we find words such as *asevati* (to do), *vaddheti* (to grow), *pavaddheti* (to make effort), *bruheti* (to develop), *bhaveti* and *bahulikaroti* (to move ahead). All these words, which are synonymous, serve a special purpose. What is that purpose? The purpose is that if we want to realise the true nature of things we have to do it again and again.

Only then can we realise it. And when we thus see how we burn with desire, how could we go on asking for more fire? In this way we develop what is called *bhavanamaya panna*, which is true understanding or insight or wisdom.

Anyone who practises *Vipassana* sees at the experiential level that all objects of the world are impermanent. Because they are impermanent they cause suffering, and knowing this leads to disillusionment and disenchantment. The individual no longer thinks of developing cravings for or aversions to the objects of the world.[9]

As individuals become free from craving and hatred, they develop many good qualities. They will no longer be the creators of problems in society but will be able to solve them. They will burn like a candle, giving light and removing darkness. People around them will also be lit by them. Thus many candles will burn to give light in the society. Such people will practise four *Brahmaviharas* such as *metta* – loving kindness for all; *karuna* – compassion for the poor and the downtrodden; *mudita* – sympathetic joy in the good fortune of others; and *upekkha* – equanimity.

Vipassana is one of the means through which the law of *paticcasamuppada* and the twelve links of *paticcasamuppada* can be understood. One has to drive out defilements from one's mind in order to be fit to experience the truth of these laws.

The eightfold path prescribed by the Buddha is an important means through which one can understand the importance and significance of *vipassana*. Implementation and realisation of the spirit underlies the Buddhist eightfold path (*attharigika-magga*). Encompassing wisdom (*panna*), morality (*sila*) and meditation (*samadhi*) in eight parts can truly offer a path leading to sustainable development. Right-view (*samma-ditthi*) and right-thought (samma-samkappa) constitute wisdom. Right-speech (samma-vaca), right-conduct (*samma-kammanta*) and right-livelihood (*samma-ajiva*) constitute morality. Right-effort (*samma-vayama*), right-mindfulness (*samma-sati*) and right-concentration (*Samma-samadhi*) form the practice of meditation (Rhys Davids/Carpenter 1890–1911: 11, 311–315). By following this path of wisdom, morality, and meditation one can grow inwardly and follow a life of enlightened simplicity. By following this path humans can also aim

---

[8]*Ibid.*
[9]*Ibid.*

at a harmonious life (*dhamma cariya, sama cariya*) (Trenckner/Chalmers 1888–1896 1.289 Feer 1884–1898 i.101) and compassion (*karuna*) with "the desire to remove what is detrimental to others and their unhappiness" (Fausboll 1985: 73). This would form the basis of the *Weltanschauung* of the well-adjusted and balanced person, who would seek inner peace (*ajjhattasanti*: Fausboll 1985: 837) and inward joy (*ajjhattarata*: Sarao 2009: 362; Rhys Davids/Carpenter 1890–1911: 11.107; Feer 1884–1898: V.263) by exercising a degree of restraint, limiting his/her needs, and avoiding being greedy (*ussuka*: Sarao 2009: 199) because one can never become worthy of respect if one is envious, selfish, or fraudulent (*issuki macchari satho*: Sarao 2009: 262).

This text has demonstrated that greed is the main cause of all evils in society, how it originates, and how it can be eliminated or reduced. The moment we destroy our greed, we will be able to achieve peace, prosperity and security in society. Thus we can achieve sustainable development.

Vipassana is the only means through which *paticcasamuppada* can be understood practically. It also helps in understanding the twelve links of *paticcasamuppada* and how different links are related with one another. *Vipassana* meditation is very much required to remove our defilements, such as *lobha* (craving), *dosa* (aversion) and *moha* (ignorance). The moment we remove our defilements there will be peace, prosperity, security and sustainable development.

References

Anonymous, 1983: *Discussion Points for United Nations University Project on Perceptions of Desirable Societies in Different Religions and Ethical Systems.*

Bottomore, T.B. (ed. and trans), 1963: *Karl Marx: Early Writings* (New York: McGraw-Hill).

Capra, F., 1983: *The Turning Point: Science, Society, and the Rising Culture* (Toronto: Bantam Books).

Durning, A., 1992: *How Much is Enough? The Consumer Society and the Future of the Earth* (New York: W.W. Norton).

Ehrlich, P.R.; Ehrlich, A.H., 2009: "Population Bomb Revisited", in: *The Electronic Journal of Sustainable Development*, 3, 1; at: http://www.d0cst0c.c0m/d0cs/12166078/Population-Bomb-Revisited (23 December 2013).

Ehrlich, P.R.; Ehrlich, A.H.; Holdren, J.P., 1977: *Ecoscience: Population, Resources, and Environment* (San Francisco: W.H. Freeman).

Elgin, D., 1993: *Voluntary Simplicity: Toward a Way of Life that is Outwardly Simple. Inwardly Rich*, rev. ed. (New York: Quill).

Elgin, D.; Mitchell, A., 1977: "Voluntary Simplicity", in: *Co Evolution Quarterly*. Summer, 14, 21 June 1977: 4–19.

Fausboll, V. (Ed.), 1985: *The Sutta-Nipata* (London: Pali Text Society).

Feer, M.L. (Ed.), 1884–1898: *The SamyuttaNikaya*. 5 vols (London: Pali Text Society).

Fromm, E., 1955: *The Sane Society* (New York: Henry Holt and Company).

Fromm, E., 1976, 2008: *To Have or To Be?* (New York: The Continuum Publishing Company).

Goodwin, N.R.; Ackerman, F.; Kiron, D. (Eds.), 1997: *The Consumer Society* (Washington D.C: Island Press).

Hecht, J., 2008: "Prophesy of economic collapse 'coming true'", in: *New Scientist,* 17 November; at: www.newscientist.com/article/dnl6058-prophesy-of-economic-collapse-coming-true.html#. UtYbRL7rZjo (18 April 2015).

Jones, K., 1993: *Beyond Optimism: A Political Buddhist Ecology* (Oxford: Jon Carpenter).

Kaza, S., 2000: "Overcoming the Grip of Consumerism", in: *Buddhist-Christian Studies,* 20: 23–42.

Kiyosaki, R.; Lechter, S., 1997: *Rich Dad Poor Dad* (New York: Warner Books).

Lancaster, L., 2002: "The Buddhist Traditions in the Contemporary World: History and Critique," a keynote speech made at the opening ceremony of the Fourth Chung-Hwa International Conference on Buddhism, 18–20 January 2002.

Loy, D., 1997: "The Religion of the Market", in: *Journal of the American Academy of Religion,* 65, 2: 275–290.

Marx, K.; Engels, F., (1845–1849): *Selected Works (1945–1849),* I vols. Marxist Internet Archive; at: www.marxists.org/archive/marx/works/sw/ (15 January 2014).

Meadows, D.H.; Meadows, D.L. et al., 2004: *The Limits to Growth,* revised edition, (White River Junction, VT: Chelsea Green Publishing Company—Earthscan).

Merchant, C., 1980: *The Death of Nature* (New York: Harper & Row).

Morris, R.; Hardy, E. (Eds.), 1995–1900: *The Anguttara Nikaya.* 5 vols (London: Pali Text Society).

Oldenberg, V. (Ed.), 1879–1883: *The Vinaya Pitaka.* 5 vols (London: Pali Text Society).

Payne, R.K. (Ed.), 2010: *How Much is Enough? Buddhism, Consumerism, and the Human Environment* (Somerville, MA: Wisdom Publications).

Rhys Davids, T.W.; Carpenter, J.E. (Eds.), 1890–1911: *The Digha Nikaya.* 3 vols (London: Pali Text Society).

Russell, B., 1954: *Human Society in Ethics and Politics* (London: George Allen and Unwin).

Sarao, K.T.S. (transl.), 2009: *The Dhammapada: A Translator's Guide* (New Delhi: Munshiram Manoharlal).

Schumacher, E.F., 1973: 1999 edition with commentaries: *Small is Beautiful: Economics as if People Mattered* (Point Roberts, WA: Hartley & Marks).

Sen, A., 1997: "Economics, Business Principles and Moral Sentiments", in: *Business Ethics Quarterly,* 7, 3, July 1997: 5–15.

Sivaraksa, S., 1992: *Seeds of Peace* (Berkeley: Parallax Press).

Toynbee, A., 1947: *A Study of History,* 1 vols (New York: Oxford University Press).

Trenckner, V.; Chalmers, R. (Eds.), 1888–1896: *The Majjhima Nikaya,* 3 vols (London: Pali Text Society).

World Commission on the Environment and Development, 1987: *Our Common Future [The Brundtland Report]* (Oxford: Oxford University Press); at: www.UN-documents.net (16 July 2013).

World Fellowship of Buddhists, 1984: *Green Buddhist Declaration* (World Fellowship of Buddhists).

Chapter 6
We Are One After All!

S. Antony Raj SJ

Abstract The sustainability necessary to make the world a better place rests on five Ps – People, Peace, Prosperity, Partnership and Planet. Sustainable community systems are organic spaces where there is peace born of justice. Sustainability as peace is seen as the result of partnership, a partnership that is for betterment – in other words, for prosperity. Without people, peace, partnership, prosperity, the planet will not be taken care of. Can there be sustained peace, dialogue within different religious communities, religion acting as an instrument in maintaining communities that live in and with peace?

Keywords Sustained peace · Faith traditions · Religion · Sustainable community Kandhamal · Reconciliation · Forgiveness

6.1 Introduction

In an ecosphere that is disruptive and in chaos, how does one imagine sustainable community systems? My search leads me to my anchoring in my faith. If sustained relationships bring peace and harmony, what could be the role of a faith tradition in sustaining that relationship? Rooted in my Judeo-Christian tradition with its concern for the weakest, I seek that dialogue of action with persons of goodwill and skill that facilitates peace born of justice, a justice that seeks reconciliation, a reconciliation that demands forgiveness.

Sometimes we all fall into despair, particularly when the news is filled with the depressing information like "the earth hit a record temperature for third year in a row"[1] or when we read that "one per cent of Indians own fifty-eight per cent of the

---

[1]See at: https://www.nytimes.com/2017/01/18/science/earth-highest-temperature-record.html (17 January 2017).

S. Antony Raj SJ, Lecturer, Xavier School of Human Resources Management, Xavier University, Bhubaneswar, Odisha, India; Email: stonysj@xshrm.edu.in

© Springer Nature Switzerland AG 2019
A. K. J. R. Nayak (ed.), *Transition Strategies for Sustainable Community Systems*, The Anthropocene: Politik—Economics—Society—Science 26, https://doi.org/10.1007/978-3-030-00356-2_6

country's wealth, i.e., more than half of the entire country's total wealth. Fifty-seven billionaires in India possess as much wealth as the poorest seventy per cent of the country in the Report on Global Inequality released on Monday 16 January 2017 by OXFAM,[2] an international confederation of eighteen non-governmental organisations. Sometimes we all fall into negativity when the milieu is polluted by harsh political rhetoric, majoritarian democracy, narcissistic megalomania, insolent appropriation of symbols, greedy behaviour, and the atrocities we humans are committing against ourselves through hate crimes against those who are already marginalised, the other animals, the water bodies, forests and the body of the planet herself is bringing us all closer and closer to an environ-mental, physiological and psychological collapse. From one perspective, things seem to be getting worse and worse. But is that the only perspective, and do we have the power to change it, if it is? And, in this negativity, how do we locate sustainability? In the challenging words of Martin Luther King Jr, whose memory we celebrate on the third Monday in January, "how do we choose community over chaos?" (King 1967).

As I search for answers, I turn to my roots in my faith tradition. If relationships are at the core of human engagement and sustainability, both – relationships and sustainability – can be located within the Judeo-Christian world view and spiritu-ality based on love, compassion and mercy. I view sustainability as peace, the outcome of compassion. My context is my spirituality, rooted in my Judeo-Christian faith tradition as lived by me in India among the poor. In this spirituality, God is relational, a being in relation, a community of mutual self-giving, who revels in community, and establishes community, whether it is the community of freed slaves in the Old Testament or the community of followers of Jesus's way in the New Testament. *No one has a hotline to God.* It is arrogant and absurd for any religion or sect to claim a monopoly on God's truth. God reveals in community for community. He chooses; he chose the Jews, he chooses the church. The choice is towards universality, for universal good. This imperative of univer-sality is encapsulated in *Srimad Bhagvad Gita*:

'*sanniyama-indriya-gramam / sarvatra sama-buddhayah*

te prapnuvanti mam eva / sarva-bhuta-hite ratah…'[3]

Those who are able to control their senses, have equanimity of mind and rejoice in con-tributing to the welfare of all creatures are dear to me[4] (The Bhagavadgita 1960).

Sarva Bhuta hite ratah,[5] the passionate concern for our common destiny, the welfare of all, especially the least, the *daridra narayan,* the rise of the last, the *antyodaya,* is because of our interconnectedness in spite of our seeming separations

---

[2]See at: https://www.oxfam.org/en/pressroom/pressreleases/2017-01-16/just-8-men-own-same-wealth-half-world (17 January 2017).

[3]See at: https://asitis.com/12/3-4.html (17 January 2017).

[4]*The Bhagavadgita*, 1960: Chap. 12, verses 3 and 4 (Gorakhpur: The Gita Press).

[5]*Ibid.*

due to various conditionings of time and space, creed, colour, class and gender. This theme had been already summed up in the *loka samgraha* of the *Bhagavad Gita* 3:20 (*Bhagavad Gita* 1960) as the welfare of all. This welfare is inclusive.

If my pilgrimage through this earth is with similar co-pilgrims and if salvation is seen as a journey from god to god, our common journey binds us together.

In the parable of the onion by Dostoyevsky (1992),[6] Grushenka tells Alyosha, after saying that she is not all that bad and giving an onion to Rakitin:

> Once upon a time there was a woman, and she was wicked as wicked could be, and she died. And not one good deed was left behind her. The devils took her and threw her into the lake of fire. And her guardian angel stood thinking: what good deed of hers can I remember to tell God? Then he remembered and said to God: once she pulled up an onion and gave it to a beggar woman. And God answered: take now that same onion, hold it out to her in the lake, let her take hold of it and pull, and if you pull her out of the lake, she can go to paradise. The angel ran to the woman and held out the onion to her: here, woman, he said, take hold of it and I'll pull. And he began pulling carefully and had almost pulled her all the way out when other sinners in the lake saw her being pulled out and all began holding on to her so as to be pulled out with her. But the woman was wicked as wicked could be, and she began to kick them with her feet: 'It's me who's getting pulled out, not you; it's my onion, not yours.' No sooner did she say it than the onion broke. And the woman fell back into the lake and is burning there to this day. And the angel wept and went away (Dostoyevsky 1992).

What could be the lessons from this parable? *Salvation is by work. If you love me, show me, for love is manifested more by deeds than words. My salvation is dependent on and intricately linked to your salvation, for salvation is inclusive.* One must be committed to the salvation not just of oneself, but of *all* the people. You must love: you should desire to save *everyone*.

All people, regardless of their age, gender, race, religion, or past behaviours deserve the opportunity to gain spiritual insight. Every person – no matter how heinous and harmful his or her past actions – should have access to spiritual redemption. You can argue that the very worst people – those who have caused emotional and physical harm to others – need God's mercy the most.

So, when the spiritual aspirant feels God's might flowing in her veins, her intention must be the salvation of all beings. Any less universal or selfless a desire becomes the pawn of ego and self-aggrandisement, sending you assuredly back to 'hell'. An isolated, fragmented self is the ambience of hell and endless suffering. Greed and selfishness lead to every kind of distortion in the spiritual process. Hatred and exclusivity are the most offensive of spiritual impediments. You cannot find heaven, let alone relative peace, holding on to such feelings within your heart. When you try to kick others into hell, that is when the onion breaks, and you find yourself back in the self-induced hell of loving no one but yourself.

Such interconnectedness and communion, such interdependence is all-pervasive, embracing the self, others, all of created universe and God. God is sought and

---

[6]Dostoyevsky 1992: Parable of the onion, in: *The Brothers Karamazov* Part III, Book 7, Chap. 3 pp. 343–358 (London: Random House).

encountered in all his creation. This creation, including all of us, is God's gift to me to be its steward, this earth is my home.

In such a world-view, God is perceived as personifying unconditional love, who in compassion creates humans in his very image, who in compassion hears the cry of the poor and frees them (as in the Old Testament), who in compassion is born as a human being, who in compassion is brutally murdered on the Cross, validating the weakness of power and the power of the weak. Compassion asks us to go where it hurts, to enter into the places of pain, to share in brokenness, fear, confusion, and anguish. Compassion challenges us to cry out with those in misery, to mourn with those who are lonely, to weep with those in tears. Compassion requires us to be weak with the weak, vulnerable with the vulnerable, and powerless with the powerless. Compassion means full immersion in the condition of being human. To me, God is compassion, God is mercy. My life with the communities of the poor has revealed me who this God is. Compassion is a piece of vocabulary that could change us if we truly let it sink into the standards by which we hold ourselves and others.

This world-view takes cognizance of the perennial human desire and pursuit for genuine peace and happiness, a return to the idyllic state of harmony at four levels – within self, with others including creation, and with God.

The theme of this fourfold reconciliation is dominant in every strand of such spirituality. Peace is seen as the outcome of restoring justice, justice understood as fidelity to demands of relationship. I am related to my 'self', to others, to creation, and to God. This relationship gets broken because of self-interest. My belief and lived out experience is that true peace is born of justice, a justice that seeks reconciliation. This reconciliation is fourfold, and my peace/reconciliation is interconnected/inseparable – reconciliation with God, humanity, creation. Reconciliation is a work of justice, and justice is always discerned/enacted in local contexts.

We, especially those of us believers, are familiar with *Reconciliation with God.* It roots us in gratitude and opens us to joy. God is the creator, sustainer and sanctifier, one from whom all good things come, one because of whom every created reality is holy. I am reconciled with God, who is my source and destiny, only if I am reconciled with humanity and creation.

Reconciliation within humanity: healing forms of suffering,

(a) The displacement of people: refugees, migrants, internally displaced peoples.
(b) The injustices and inequalities experienced by marginalised peoples (Dalits and tribals), women.
(c) Fundamentalism, intolerance, and ethnic-religious-political conflicts as a source of violence.

In the words of Pope Francis, "Where injustices abound, and growing numbers of people are deprived of basic human rights and considered expendable, the principle of the common good immediately becomes, logically and inevitably, a summon to solidarity and a preferential option for the poorest of our brothers and sisters."[7]

Reconciliation with Creation: There are clear links between the environmental and the social crisis today – the flawed way we organise our economies (production and consumption). To respond to multi-faceted challenge, we need to promote personal and communal lifestyles that are responsible/sustainable; we cannot forget to celebrate creation, give thanks to the Supreme Being 'for all the good received'.

Integral ecology is a key concept in chapter four of *Laudato Si*, Pope Francis' encyclical on the environment. It flows from his understanding that "everything is closely related" and that "today's problems call for a vision capable of considering every aspect of the global crisis."[8]

Relationships take place at the atomic and molecular level, between plants and animals, and among species in ecological networks and systems. For example, he points out, "We need only recall how ecosystems interact in dispersing carbon dioxide, purifying water, controlling illnesses and epidemics, forming soil, breaking down waste, and in many other ways which we overlook or simply do not know about."[9]

Nor can the 'environment' be considered in isolation. "Nature cannot be regarded as something separate from ourselves or as a mere setting in which we live,"[10] writes the Pope. "We are part of nature."

As a result, if we want to know "why a given area is polluted," we must study "the workings of society, its economy, its behaviour patterns, and the ways it grasps reality." And in considering solutions to the environmental crisis, we must "seek comprehensive solutions which consider the interactions within natural systems themselves and with social systems."[11]

These interrelationships enable Francis to see that "we are not faced with two separate crises, one environmental and the other social, but rather one complex crisis which is both social and environmental."[12] As a result, "Strategies for a solution demand an integrated approach to combating poverty, restoring dignity to the excluded, and at the same time protecting nature."[13] In such an "economic ecology", the protection of the environment is then seen as "an integral part of the development process and cannot be considered in isolation from it" (Francis/ McDonagh 2016).

---

[7]See at: http://w2.vatican.va/content/francesco/en/encyclicals/documents/papa-francesco_ 20150524_enciclica-laudato-si.html (17 January 2017).

[8]See at: http://catholicclimatemovement.global/laudatosi/ (17 January 2017).

[9]*Ibid.*

[10]See at: https://catholicclimatemovement.global/overview-laudato-si/ (17 January 2017).

[11]See at: https://www.ncronline.org/blogs/faith-and-justice/integral-ecology-everything-connected (17 January 2017).

[12]See at: https://www.cssr.org.au/justice_matters/dsp-default.cfm?loadref=656 (17 January 2017).

[13]See at: http://www.marianites.org/uploads/files/Laudato-Si-ch4-(English).pdf (17 January 2017).

He also calls for a "social ecology" that recognises, "the health of a society's institutions has consequences for the environment and the quality of human life."[14] This includes the primary social group, the family, as well as wider local, national, and international communities. When these institutions are weakened, the result is injustice, violence, a loss of freedom, and a lack of respect for law – all of which have consequences for the environment.

Pope Francis also argues that it is important to pay attention to "cultural ecology" in order to protect the cultural treasures of humanity. But "Culture is more than what we have inherited from the past; it is also, and above all, a living, dynamic and participatory present reality, which cannot be excluded as we rethink the relationship between human beings and the environment" (O'Brien/Shannon 2016).

He complains that a consumerist vision of human beings, encouraged by globalisation, "has a levelling effect on cultures, diminishing the immense variety which is the heritage of all humanity" (O'Brien/Shannon 2016). New processes must respect local cultures. "There is a need to respect the rights of peoples and cultures, and to appreciate that the development of a social group presupposes an historical process which takes place within a cultural context and demands the constant and active involvement of local people from within their proper culture" (Charlton/Armistead 2016). This interconnectedness means that "environmental exploitation and degradation not only exhaust the resources which provide local communities with their livelihood, but also undo the social structures which, for a long time, shaped cultural identity and their sense of the meaning of life and community."[15] In various parts of the world, he notes, indigenous communities are being pressured "to abandon their homelands to make room for agricultural or mining projects which are undertaken without regard for the degradation of nature and culture."[16]

This fourfold reconciliation demands genuine forgiveness, starting with forgiving oneself, and others, forgiveness in a world which demands eye for an eye. Experientially, the inner healing of the heart is seldom a sudden catharsis nor an instant liberation from bitterness, anger, resentment and hatred. It is here that the anchoring in a faith tradition which emphasises reconciliation and forgiveness provides clarity, helping us understand triggers, the compassion that makes the forgiveness possible. Hence, such peace, justice, reconciliation and forgiveness are not only lofty ideals but definite possibilities.

My belief and praxis are rooted in my world-view and spirituality of the Judeo-Christian faith tradition lived out, tested and verified in dialogue of life with humans of goodwill and skill, especially my engagement in conflict resolution and

---

[14]*Ibid.*

[15]*Ibid.*

[16]'On Care for Our Common Home', *Laudato Si*, Pope Francis's historic encyclical on the environment, with reflections by Sean McDonagh, Available at: https://books.google.co.in/books?id=FZuMCwAAQBAJ&pg, paragraphs 141–146 (17 January 2018).

peacebuilding in communities of survivor victims and alleged aggressors after the anti-Christian violence of 2008–2009 in Kandhamal District, Odisha. After the initial relief operations and extended rehabilitation, and after a tedious conflict analysis with the local communities, we discovered the common denominator of poverty and lack of sustainable livelihood options that unite both the victim survivors and the perpetrators of the hate crime. We Jesuits in 2010 chose one affected village community of Sindhiponkal, in Mundigodo Gram Panchayat of Tumudibandha Block, Kandhamal District, Odisha as the nucleus of ten surrounding villages (allegedly filled with perpetrators of violence), and initiated with the village communities, conflict resolution and peace-building through sustainable livelihood options and activities. Our engagement in one Gram Panchayat has become a movement, with our operations now extending to four Gram Panchayats – Guma and Mundigodo in Tumudibandha Block, Madaguda and Subarnagiri in Subarnagiri Block. Our endeavour with the people in empowerment by activities like convergence through linkages with the government programmes for the marginalised and with the line departments in agriculture, horticulture and animal husbandry, strengthening of Panchayati Raj Institution Members, the School Managing Committees and Self-Help Groups, training of the area's youth in social responsibility, has motivated the entire community towards the common good, forgiving the painful past.

The work has not been easy. All it takes is commitment and persistence. This paradigm – peace born of justice that seeks reconciliation that demands forgiveness – must be practised by one's self first! Though the causes of protracted violence in Kandhamal are complex, yet in analyses of the violence one can find that the symptom/outcome of the strained relationship between two faith traditions have played into the hands of decisive communal forces, making the assailant and the retaliator both victims.

So, my wish and my prayers are let there be peace! Let it begin with me! Our culture is obsessed with perfection and with hiding problems. But what a liberating thing to realise that our problems are, in fact, probably our richest sources for rising to this ultimate virtue of compassion and evoking compassion for the suffering and joys of others. For, in my faith tradition, perfection is compassion. God of the Judeo-Christian tradition said, "Be holy for I am holy."[17] Jesus, who said, "Be perfect as the heavenly father is perfect,"[18] also said "Be compassionate as your

[17]The Bible, *Book of Leviticus,* 3:20–26. The New Revised Standard Version (Washington, DC: Division of Christian Education of the National Council of the Churches of Christ in the USA, 1989).

[18]The Bible, *The Gospel According to Matthew,* 5:48. The New Revised Standard Version (Washington, DC: Division of Christian Education of the National Council of the Churches of Christ in the USA, 1989).

heavenly father is compassionate."[19] Thus, holiness is perfection is compassion. Just as we are hard-wired to learn a language, I do absolutely think we are born with this redemptive capacity to be compassionate. Then again, we must start to practise it around each other. We have to start to embody it in front of our children and in our common life. I do think that it will be infectious.

The compassionate life is neither sloppy goodwill towards the world nor the plague of chronic niceness. The way of tenderness avoids blind fanaticism. Instead it seeks to see with penetrating clarity that the other is myself! We are not 'two' but 'one'. My God pitched his tent among us humans, sanctifying the human condition. He did this out of his bountiful compassion. His becoming human, this union which resolves the non-duality, is not only the condition of my life with God but also with my brothers and sisters. The compassion of God in our hearts opens our eyes to see the unique worth of each person, especially the poor, for the Supreme Being resides in him/her. This is the unceasing struggle of a lifetime. The way you are with self and others every day is the true test of faith. Liken yourself to a beautiful part of creation. Awareness of this fact and healthy self-love go together. When we love what God the Supreme being has given us and share it with others naturally and without expectations of gratitude we are truly people who have spiritual self-confidence and compassion; and isn't that a great way to live? Our human compassion binds us the one to the other – not in pity or patronisingly, but as human beings who have learnt how to turn our common suffering into hope for the future, redeeming the resilience that was etched into our DNA.

This compassion arising out of such a faith perspective is the virtue of the strong, for it is not giving into weakness. For if one is compassionate, one has to rise up and address issues building a community that is not just sustainable but equal. Communities can be sustainable through interdependence. But interdependence is mostly exploitative in our society, with some people expected to perform some hierarchical roles.

Hence, because of the definite possibility of living such a perspective, I end my reflections with plenty of hope. Pierre Teilhard De Chardin (1931), the Jesuit Palaeontologist and philosopher, says in *The Spirit Of The Earth* (1931, VI, 32, 33, 34): "Love is the most universal, the most tremendous and the most mystical of cosmic forces. Love is the primal and universal psychic energy. Love is a sacred reserve of energy; it is like the blood of spiritual evolution".[20] Thus, love we must! To quote Pierre de Chardin again, this time from his 'The Evolution of Chastity,' in *Toward the Future* (1936, XI, 86–87), "The day will come when, after harnessing

---

[19]The Bible, *The Gospel According to Luke,* 6:36. *The Gospel According to Matthew*, 5:48. The New Revised Standard Version (Washington, DC: Division of Christian Education of the National Council of the Churches of Christ in the USA, 1989).

[20]See at: http://www.teilharddechardin.org/teilharddechardin.pdf (17 January 2017).

space, the winds, the tides, and gravitation, we shall harness for God the energies of love. And on that day, for the second time in the history of the world, we shall have discovered fire."[21]

> We are one after all
>
> You and I
>
> Together we suffer,
>
> together exist,
>
> and forever will recreate each other.
>
> The world is round so that friendship
>
> may encircle it.[22]

6.2 Conclusion

'An eye for an eye', will leave the planet in and with darkness, agents of our own suffering. Religion as a way of life is an instrument of hope, faith and inspiration. A constant dialogue within and between religions will create an ambience of respect which sensitises communities and individuals, that which establishes or sustains peace for sustainable communities. My exploration and experience in Kandhamal has shown me that, though in chaotic times, we treat one another as adversaries, we are the victims of our won triumphalist thinking. There can be no peace without justice, as justice is born out of peace. This justice seeks reconciliation that demands genuine forgiveness.

References

Charlton, Matthew W.; Armistead, M. Kathryn, 2016: *The Prophetic Voice and Making Peace* (New York: Bookbaby).

Dostoyevsky, F., 1992: *The Brothers Karamazov.* (London: Random House).

King, M.L., 1967: *Where do we go from here: Chaos or Community?* (Boston: Beacon Press).

O'Brien, David J.; Shannon, Thomas A., 2016: *Catholic Social Thought: Encyclicals and Documents from Pope Leo XIII to Pope Francis.* 3rd Revised Edition (New York: Orbis Books).

Pope Francis; McDonagh, Sean, 2016: *On Care for Our Common Home, Laudato Si: The Encyclical of Pope Francis on the Environment* (New York: Orbis Books).

The Bhagavad Gita, 1960: Chapter 12, verses 3 and 4 (Gorakhpur: The Gita Press).

The Bible, 1989: The New Revised Standard Version (Washington, DC: Division of Christian Education of the National Council of the Churches of Christ in the USA).

---

[21] *Ibid.*

[22] See at: http://www.goodreads.com/quotes/75202-we-are-one-after-all-you-and-i-together-we (17 January 2017).

Chapter 7
Re-envisioning Development for Sustainable Community Systems: Art, Spirituality and Social Transformations

Ananta Kumar Giri

Abstract Transition to sustainable community systems calls for re-envisioning human development. Human development does not only mean economic, political and ethical development; it also means artistic and spiritual development. All these dimensions of development are interlinked, but we have not paid sufficient attention to artistic and spiritual bases and horizons of human development and social transformations. In order to develop both individually and collectively, it is necessary to develop one's artistic and spiritual potential. The essay explores relationships between art, spirituality and human development and describes examples from histories of ideas, art experiments and educational movements for creatively cultivating cross-fertilisation among them further.

Keywords Artistic transcendence · Practical spirituality · *aestheses*
Multi-*topial* hermeneutics · Multi-valued logic · Ontological epistemology of participation

7.1 Introduction and Invitation

Transition to sustainable community systems calls for re-envisioning and reconstituting practices of development. The discourse and practice of human development has been at a crossroads for a long time. For quite some time, critics and reflective practitioners in the field of development have raised ethical and moral issues in the vision and practice of development, such as poverty, hunger, displacement and production of underdevelopment by the very interventionist process

Ananta Kumar Giri. Professor, Madras Institute of Development Studies, 79 Second Main Road, Gandhi Nagar, Chennai, Tamil Nadu, India; Email: aumkrishna@gmail.com

© Springer Nature Switzerland AG 2019
A. K. J. R. Nayak (ed.), *Transition Strategies for Sustainable Community Systems*,
The Anthropocene: Politik—Economics—Society—Science 26,
https://doi.org/10.1007/978-3-030-00356-2_7

and logic of development. This has led to the rise of a vibrant critical field of development ethics in which scholars and activists such as Denis Goulet, Amartya Sen, Ashish Nandy, Des Gasper, Claude Alvares, Vandana Shiva, Arturo Escobar and Martha Nussbaum, among others, have taken part. This has led to a critical and creative broadening of the vision and practice of development from mere economic and infrastructural development to broad visions and practices of human development. Sen (1999) spearheaded such a shift as part of a broad reconstitution of development as freedom at the core of which lies issues such as capability and freedom. But issues of art, spirituality, human development and social transformations are rarely addressed in this discursive field, though Martha Nussbaum's work is an exception. She draws our attention to the way art can help to transcend existing boundaries through the work of what she calls "artistic transcendence" (Nussbaum 1990: page).[1] Nussbaum also draws our attention to the significance of art in education, where 'the artist's fine-tuned attention and responsiveness to human life is paradigmatic of a kind of precision of feeling and thought that a human being can cultivate' (Nussbaum 1990: 379). Reflections like Nussbaum's urge us to understand the integral significance of art in life and society as a locus of a different way of seeing, envisioning and relationship. Art here constitutes a source of utopian imagination and possibility in self, culture and society – a dimension which has been cultivated by movements such as the cultural and artistic movements of avant-garde in modernity (Strydom 1984, 1994).

7.2   Art, Spirituality and Human Development: Widening and Deepening the Universe of Discourse and Practice

At present, the existing meaning, understanding and realisation of human development, art, spirituality and society are broadening. Human development means not only having the capability to function; it also, at the same time, involves the art and practice of imagination. Imagination is simultaneously individual and social, material and spiritual, and art plays an important role in fostering and growth of our imagination, especially collaborative imagination as it weaves selves and movements across fixed positions and boundaries (see Giri 2016a).[2] Human development

---

[1]Artistic transcendence in Nussbaum refers to these processes as she writes: "there is a great deal of room for transcendence of our ordinary humanity… transcendence, we might say, of an *internal* and human sort …. There is so much to do in this area of human transcending (which I also imagine as a transcending by *descent*, delving more deeply into oneself and one's humanity, and becoming deeper and more spacious as a result) that if one really pursued that aim well and fully I suspect that there would be little time left to look about for any other sort" (Nussbaum 1990: 379).

[2]Imagination is an indispensable foundation and ever-present companion of life and it is linked to aspiration and the dynamics of creativity in self, culture and society. But our dominant models and methods of research in modernity, as in the linked larger field of modernity itself, have lacked cultivation of imagination and creativity. In many ways the predominance of mechanical models in

means not only economic, political and ethical development; it also means artistic and spiritual development. All these dimensions of development are interlinked, but in our conventional and dominant discourses of development we have not paid sufficient attention to artistic and spiritual dimensions of development. To develop, both individually and collectively, one's artistic and spiritual potential must also develop. But to develop ourselves artistically does not just mean looking beautiful externally; art is linked to living life as a work and meditation of art in one's inner life as well. As Michel Foucault challenges us: "What strikes me is the fact that, in our society, art is now linked to objects, rather than to individuals or life itself [...] But couldn't we ourselves, each one of us, make of our lives a work of art? Why should a lamp or a house become the object of art – and not our own lives?"

Foucault links the process of self-making as a work of art to an ethical project as he urges us to realise that "the search for an ethics of existence" must involve an "elaboration of one's own life as a personal work of art" (Foucault 1988: 49). Thus ethics here become linked to aesthetics, generating the border-crossing movement of aesthetic ethics, which is helpful for understanding integral links between aesthetics, ethics and human development. Foucault's agenda of aesthetic ethics is developed in the context of his discussion of ethical life and ethical ideals in Antiquity. But this is not meant only to be an archaeology of the past but to suggest a possible mode and ideal of ethical engagement for the present and the future. For Foucault, in Antiquity, "the search for an ethics of existence" was "an attempt to affirm one's liberty and to give to one's life a certain form in which one could recognise oneself, be recognised by others, and which even the posterity might take as an example" (Foucault 1988: 49). For Foucault, life as a work of art involves care of the self, a conversion to self, an intense relation with oneself. While ethics are usually conceived as care for others, for Foucault, ethics must, at the same time, help one to "take oneself as an object of knowledge and a field of action, so as to transform, correct, and purify oneself, and find salvation" (Foucault 1988: 42). Furthermore, aesthetic ethics as care of the self involve cultivation of appropriate values in the conduct of life. The most important task here is not to be obsessed with exercising power over others and to be concerned with discovering and realising "what one is in relation to oneself" (Foucault 1988: 85). This can give rise to a new hermeneutics of self and society in which the model of self and social

---

science, scholarship, society and State has killed our wings of imagination and creativity. Imagination works at the interstices of body, mind, spirit, society, nature and the divine. Imagination is not confined to the individual in a narrow sense; it can begin with the vibration of silence and solitude of soul, but it also arises in our practices of co-beings, collaborations and points and pathways of contestations and confrontations. Creative imagination, as the linked pathway of moral and social imagination, has an indispensable collaborative dimension in which, as we walk, work, dream, sing, argue and fight together, our wings and roots of imagination get sharpened and deeper.

formation is not just predicated upon socially produced *a priori* models of self, state and society.[3]

Art plays a role in giving birth to the new imagination and hermeneutics of striving for self and society and for giving birth to the new hermeneutics of self and society and a new relational aesthetics which urges us to cross different borders. As argued later, this hermeneutics is related to our walking and meditating across multiple landscapes, pathways and imaginations of self and social realisations – multi-*topial* hermeneutics – rather than just being trapped within fixed positions and models of self and society.

Such paths of aesthetic ethics and hermeneutics of self call for a new art of relationship between self and power where one embodies a different way of being with power, as, for example, suggested by Hannah Arendt, working and meditating in concert,[4] to give birth to a desirable new path of being and relationship rather than exercising one's domination over others (see Giri 2009). It also calls for realising the limits of power and realising that power is not the only foundation of life. Life has other integrally linked foundations, such as love and respect – what is called *sraddha* in Indic traditions (see Giri 2012). *Sraddha* calls for what Foucault himself calls self-restraint vis-à-vis one's work of power, including in acts of resistance to power, such as in *Satyagraha*. This is particularly salutary in the field of development, where agents of development have sought to impose their own will and models on the targets of development interventions. Through development of self-control, the actors of development can resist the temptation to unnecessarily meddle in the lives of those with whom they are in interaction and thus facilitate their self-unfolding and self-flourishing. For Robert Chambers, "it implies that uppers have to give up something and make themselves vulnerable" (Chambers 1997: 234). An engagement in self-control also enables actors of development to be aware of the hegemonic implications of a project of ethics which is primarily prescriptive. It enables them to continuously seek to transcend the world of separation between the creators of development and the beneficiaries of such a creation. In this context, Majid Rahenema, who has applied Foucault's insights in going beyond the impasse of contemporary development interventions, calls for a "bottom up aesthetic order" in development, at the heart of which lies a desire on the part of

---

[3]Here the following reflections of S. N. Eisenstadt also deserve our careful consideration: While the term 'parrhesia' as used by Foucault goes beyond the simple emphasis on resistance as due mainly to the inconvenience of being confined within the coercive frameworks of an order and denotes the courageous act of disrupting dominant discourses, thereby opening a new space for another truth to emerge – not a discursive truth but rather a 'truth of the self', an authentication of the courageous speaker in this 'eruptive truth-speaking' – it does not systematically analyse the nature of the agency through which such other truth may emerge, or how the emergence of such 'truth of the self' may become interwoven with the process of social change and transformation (Eisenstadt 2002: 38).

[4]As Foucault writes, "The political, ethical, social, philosophical problem of our day is not to try to liberate the individual from the state and its institutions but to liberate us both from the state and the type of individuality linked to state. We have to promote new forms of subjectivity [...]" (Foucault 2005: 526).

the actors to be true to themselves and develop their "inner world" and challenge the distinction between the makers of the worlds of beauty, truth and goodness and those who enjoy their benefits. In such a bottom-up aesthetic reconstruction of development, "Right action involving others starts always as a personal work on oneself. It is the fruit of an almost divine kind of exercise, which usually takes place in the solitude of thought and creation" (Rahenema 1997: 401).

Aesthetic ethics as a path of realisation of links between art and human development can also draw inspiration from philosopher and historian Frank Ankersmit's cultivation of pathways of what he calls aesthetic politics (Ankersmit 1996). For Ankersmit, while "ethics makes sense on the assumption of a (Stoic) continuity between our intentions, our actions and their results in the socio-political world", aesthetics draws our attention to the gaps and discontinuities among them (1996: 44). Ankersmit makes a distinction between mimetic representation, which denies this gap between representations, and represented and aesthetic representation, which acknowledge this gap and builds on it. For Ankersmit, mimetic representation is against representation itself as "representation always happens, so to, speak, between the represented and its representation; it always needs the presence of their distance and the ensuring interaction [...]" (Ankersmit 1996: 44). The problem with modernist politics, for Ankersmit, is it has been a hostage to the politically correct ideology of mimetic representation, whereby political representatives are required to mirror the expectations of their constituency. This creates a compulsion for politically correct mimetic representation rather than a representation which is based on one's autonomous self-identity and negotiation between this identity and the aspirations of the represented. For Ankersmit, acknowledgment of this gap becomes an aesthetic work par excellence where actors learn to develop an appropriate political (and I would add spiritual) style) in the midst of fragmentation rather than a valorised united whole. Aesthetic political representation urges us to realise that "the representative has autonomy with regard to the people represented" but autonomy is not then an excuse to abandon one's responsibility. Aesthetic autonomy requires cultivation of 'disinterestedness' on the part of actors, which is not the same as indifference. To have disinterestedness i.e., to have "comportment towards the beautiful that is devoid of all ulterior references to use – requires a kind of *ascetic* commitment; it is the 'liberation of ourselves for the release of what has proper worth only in itself'" (Osborne 1997: 135).

In aesthetic politics, the development of appropriate styles of conduct on the part of the representatives is facilitated by the choice and play of appropriate metaphors. For Ankersmit, in the development of an appropriate style of conduct for a representative the metaphor of a 'maintenance man' or woman is more facilitating for self-growth than an architect. While the architect thinks that he or she is designing a building of which he or she is the creator, a maintenance person has a much more modest understanding of his or her role and does not look at his or her effort as creating a building out of nothing, rather continuing a work to which many others have contributed. Such a metaphor of 'maintenance man' can provide new self-understanding to actors in the fields of both politics and development, where we do not have any dearth of actors, institutions or world-views which attribute to themselves the role of the original creator, the architect, the god. But such a self-understanding of ourselves as architects often leads to arrogance and dominance. In this context, there is modesty in the metaphor of the 'maintenance person', which is further facilitated by the choice of the metaphor of the captain of a ship. It is not enough for a captain to have only an *a priori* plan; he or she must know how to negotiate between *a priori* plans and the contingent situations on the ground. Such a capacity for negotiation, which is facilitated by one's choice of an appropriate metaphor, such as captain or 'maintenance person', is crucial for the development of appropriate styles of conduct on the part of the actors in the fields of politics and development. In developing his outline of aesthetic politics, an outline which has enormous significance for reconstituting the field of development as a field of artistic rather than mimetic representation, which, in turn, calls for the cultivation of an appropriate lifestyle on the part of the actors of development, Ankersmit writes: "[...] when asking himself or herself how best to represent the represented, the representative should ask what political style would best suit the electorate. And this question requires an essentially creative answer on the part of the representative, in the sense that there exists no style in the electorate that is quietly waiting to be copied" (Ankersmit 1996: 54). For Ankersmit, "aesthetics will provide us with a most fruitful point of departure if we desire to improve our political self-knowledge" and in this self-knowledge autonomy of actors, units and institutions has crucial significance. In fact, nurturing the autonomous spaces of self, institutions and society itself as spaces of creative self-fashioning and development of creative styles of action becomes an aesthetic activity par excellence. Of course, autonomy here is not meant in a defensive sense of preserving the established structures rather than transforming them in accordance with the transformative imagination of actors and a democratic public discursive formation of will.

Aesthetic politics in Ankersmit are not geared to a will to wield power but inspired by a will to gain political self-knowledge and the will to develop oneself as a 'maintenance man', which can be linked to developing oneself as a servant of people, the divine and society as part of the new ethics, aesthetics and spirituality of servanthood (cf. Giri 2002). In contrast to the tyranny of unity in certain strands of German aesthetics, such as Schiller's, Ankersmit's aesthetics celebrates and works "within an irrevocably broken world" (Ankersmit 1996: 53) but the brokenness of the world is not an excuse to abandon one's responsibility. This is facilitated by further creative elaborations of an aesthetic mode of engagement by Taylor (1991) and Benhabib (1996), where aesthetics is characterised by both quest of authenticity and striving for establishing non-domineering relationships with others (also see Scarry 1999; Welsch 1997). In the words of Benhabib:

> The overcoming of the compulsive logic of modernism can only be a matter of giving back to the non-identical, the suppressed, and the dominated their right to be. We can invoke the other but we cannot name it. Like the God of the Jewish tradition who must not be named but evoked, the utopian transcendence of the compulsive logic of Enlightenment and modernism cannot be named but awakened in memory. The evocation of this memory, the 'rethinking of nature in the subject' is the achievement of the aesthetic (Benhabib 1996: 333).

Our engagement with various new ways of understanding the work of aesthetics has important lessons for us in thinking about and relating to the field of development. First, aesthetics as critical and creative memory work helps us to transform development as a multi-dimensional memory work in our lives, which also involves working and meditating with both our roots and routes as well as their complex arts of cross-fertilisation (Giri 2016c).[5] Second, aesthetics, as sensitivity to configurations of togetherness without reducing it to an *a priori* plan or teleology of order, can help us to look at the field of development as a field of togetherness – a movement for the generation of the commons in our world of individualism and fragmentation (Reid/Taylor 2010). But this togetherness is not a product of an ordered plan, nor is it teleologically geared to the production of order. A preoccupation with order has led to dangerous consequences in the field of development, where leaders have deliberately tried to sweep conflict, ambiguity and contradictions under the carpet. It has also led to a denial of the work of contingencies in developmental dynamics. Aesthetics as openness to the contingent also helps us overcome the creed of certainty, better prepare ourselves for appreciating the work of uncertainty in the developmental world, and fashion an appropriate mode of action and management which reflects such a concern. For instance, Lyla Mehta, Melissa Leach and her colleagues at the Institute of Development Studies, Sussex urge us to explore new directions in natural resource management which take the uncertainty of people's lives – ecological uncertainty, livelihood uncertainty, and knowledge uncertainties – seriously, and in this engagement an aesthetic awareness compared with a positivist preoccupation with regulation can help us too (Mehta et al. 1999). Finally, aesthetics as artistic representation rather than mimetic representation can enable us first to understand the mimetic nature of most of development interventions and then encourage us to cultivate various alternative ways of coming out of this closed mimetic world in which creative and critical memory work can help us. One aspect of the mimetic character of the contemporary world of development interventions is that the representatives of development are self-confident that they can represent the interest of the donor agencies on the one hand and beneficiaries on the other in a transparent and unproblematic manner. But such an assumption condemns them to a world of self-created continuity, while the

---

[5]As Alka Wali writes about her work with artists in New York city and Chicago: "We discovered that the desire to practise art led people to cross deep social boundaries of gender, class, and even, at times race. We discovered that the serious dedication to the crafts led people to overcome their fear of each other, to develop trust and engagement in many ways that were not possible in their work place or home place" (Wali 2015: 183). To make sense of this work of art, Wali presents us with the perspective of relational aesthetics cultivated by Bourriaud, an art critic. Here what Wali writes deserves our careful consideration: Bourriaud, an art critic and curator, attempts to characterise trends in conceptual arts that emerged in the 1990s, suggesting that certain artists are positioning art as a form of social activism in ways that emphasise social interaction and its context. Relational aesthetics is defined by art that is more participatory, collaborative and activist. Following Foucault and Guattari, Bourriaud posits that this type of art works at the 'micro-political' level, focusing on individual or localised transformation rather than striving for grander-scale social movements (Wali 2015: 185).

field of development is characterised by a lack of fit between intentions and outcome. And with aesthetic sensibility, once the representatives realise the practical and moral untenability of such a mimetic world, they can engage themselves with various modes of aesthetic ethics and politics which enable them to articulate the interests of donors and beneficiaries in a more responsible manner.

But while these are some of the potentials for renewing development practice with an engagement with aesthetics, unfortunately there are some fundamental limits, too. One of these relates to a narrow valorisation of care of the self in an aesthetic engagement, a valorisation which does not take seriously and is even blissfully oblivious to its responsibility to others. In fact, this problem lies at the core of Foucauldian ethics as care of the self. As Gardiner helps us realise: "In Foucault's ontology of the subjects, there are only scattered and essentially gratuitous references to our relations with others, little real acknowledgment of the centrality of non-repressive solidarity and dialogue for human existence. One must not have the care for others precede the care of the self, he [Foucault] bluntly declares at one point" (Gardiner 1996: 38). Critical reflections on Foucault's own scripting of life also points to a preoccupation with sado-masochism in his life which points to the limits of his aesthetic ethics (Miller 1993: 327). In this context, aesthetic ethics in itself cannot help us come out of our current impasse in the field of development; instead we need to engage ourselves with development as the embodiment of responsibility. Development as responsibility involves both ethics and aesthetics in a transformational way. Development as responsibility includes striving for beauty, dignity and dialogues across borders in which art plays an important role (see Giri/Quarles van Ufford 2015; Giri 2016d).

7.3   Broadening and Deepening the Vision and Practice of Art, Aesthetics and Human Development

Aesthetic ethics in Foucault and aesthetic politics in Ankersmit urge us to realise the broader and deeper vision of art and aesthetics. Art is not only what we draw on a piece of paper but the quality of life we live and create in our lives and society. It is the poetry that we write in our bodies, social as well as self. As Chitta Ranjan Das (2008) argues, there are two streams of aesthetic consciousness – *Anna* and *Ananda* (food and bliss) – and human development lies in establishing bridges between the two. This brings us to a dialogue between Gandhi and Tagore vis-à-vis these two wings of aesthetic consciousness. To the comment of Tagore that Gandhi does not see the beauty of birds flying early in the morning, Gandhi replied that he sees the beauty but he also understands the pain of those birds who are not able to flap their wings because they did not have anything to eat the night before.[6] Aesthetic

---

[6]The vision and pathways of multi-*topial* hermeneutics builds upon the idea of *diatopical* hermeneutics proposed by Raimundo Panikkar. Building upon the seminal work of Raimundo

development is thus a bridge between these two wings of human consciousness – food and freedom – and facilitates border-crossing movements across many domains which are considered isolated from and opposed to each other, such as economic and artistic, material and spiritual.

We can here engage with several experiments in broadening the vision and practice of aesthetic development. In the famous Bauhaus movement in the Weimar Republic in pre-Nazi Germany, artists and architects were inspired by a vision of social beauty for many and not only for the elites. They built aesthetically rich houses for the working class. There is also an emerging new aesthetics of experimentation and participation in Europe, India and many parts of the world. The small country of Liechtenstein has a laboratory for artists called BKK Labor, where around twenty artists work together. According to an artist who works in such a shared space whom I met during a visit in 2008: "As an artist I am used to working alone but in this place I work in a space shared by other artists. They come and comment on my work, which also softens the edges of my own ego." Artistic development in such places not only means movement from gross to subtle but a transformation of one's ego through the social practice of working, sharing and experimenting together. In the same gallery I met a female photographer who collaborated with two middle-aged men in taking photographs of their nude bodies to reflect upon the fragility of the body through the flow of time and its attendant anxiety. Barbara Muller, this creative artist, who through artwork overcomes the marked bodies of gender, told me about a fellow artist in Malta named Pierre Portelli. I visited Portelli in Malta in 2008, and during our conversation he told me how, in his artistic production, he seeks to narrow the gap between viewers and

---

Panikkar, Boaventuara de Sousa Santos (2014: 92) elaborates *diatopical* hermeneutics thus: "The aim of *diatopical* hermeneutics is to maximize the awareness of the reciprocal incompleteness of cultures by engaging in a dialogue, as it were, with one foot in one culture and the other in another – hence its *diatopical* character. *Diatopical* hermeneutics is an exercise in reciprocity among cultures that consists in transforming the premises of argumentation in a given culture into intelligible and credible arguments in another."

Santos here talks about putting one's feet in cultures which resonates with my idea of footwork, footwork in landscapes of self, culture and society as part of creative research (Giri 2012). Hermeneutics does not mean only reading texts and cultures as texts but also foot-walking with texts and cultures as foot walks, and foot works resonating with what Heidegger calls a hermeneutics of *facticity* (cf. Mehta 2004). Santos talks about *diatopical* hermeneutics, but this need not be confined to our feet only in two cultures; it needs to move beyond two cultures and embrace many cultures. Spiritual traditions can also help us realise that though we have physically two feet, we can have many feet. In the Vedas it is considered that the Divine has a million feet, and similarly we can realise that humans also have a million feet, and with our million feet we can engage ourselves with not only creative foot work but also heart work (*herzwerk*, as it is called in German) in our acts of gathering knowledge and caring for self and the world. Supplementing Santos's *diatopical* hermeneutics, we can cultivate *multi-topial* hermeneutics, which is accompanied by a multi-valued logic of autonomy and interpenetration going beyond either-or logic. Art and aesthetics play an important role in both multi-topial hermeneutics and multi-valued logic, as they help us to take gentle and careful artistic steps in difficult journeys across terrains and domains and make connections across fields usually construed as isolated and separate (see Giri 2016b).

artists by inviting the viewers to take part in the process of creating art. Portelli deliberately keeps his artwork incomplete so that the viewers may take part in it and join the ongoing process of completion. Portelli also produces contemporary art in which objects of art, such as bread, gather fungus over time, showing the transitory nature of life. Art thus contributes to an understanding of the transitory nature of all things, which can embue it with a deeply spiritual dimension.

Art is thus related to transcendence and critique of self, culture and society. The avant-garde artistic and cultural movements in modernity also presented such a move of deepening and broadening art and its relationship to self, culture, society and the world. Art here was not just tied to reproduction of the status quo but also explored its transcendence and transformation (see Strydom 1984, 1994). Piet Strydom, who has reflected upon the avant-garde movements, also urges us to realise the wider and deeper significance of the aesthetic in human development, which resonates with similar emphases by Clammer (2017). As Clammer tells us in his insightful essay, "Art and Social Transformation: Challenges to the Discourse and Practice of Human Development":

> The sterile question of 'what is art' […] is better replaced by the question of what art *does*. Apart from its role as a natural expression of human creativity (children almost always spontaneously produce art), it clearly has four major functions. The first of these is to create new imaginative spaces […] The second is to reflect, record or symbolise, often in indirect but nevertheless unmistakable forms, the fundamental existential issues built into being human – suffering, mortality, death, belief, embodiment, sexuality, strangeness, curiosity, fear, our relationship to nature and our desire to represent in some physical form our current and cultural perceptions of the world around us and its varied inhabitants, and our ideas of divinity.

The third is the real but again indirect relationship between ethics and aesthetics, between truth and beauty. In contemporary analytical social science, the two are, of course, unrelated. So called 'development', for example, is 'successful' if it brings about economic and material growth, even at the expense of immense ugliness, destruction of natural beauty and devastated landscapes and cityscapes, all issues thought to be peripheral to real 'progress'. Even as the emerging field of eco-psychology has clearly showed that prolonged lack of exposure to nature is a source of stress, neurosis and violence, so too lack of exposure to beauty is exhausting and causes similar mental and behavioural problems, and the extent to which art therapy is now prescribed as a remedy for such ills clearly points to the role of art as an essential part of the human psyche's make-up, which in turn has ethical implications: to impose ugliness and lack of form on any natural environment or humanscape is to do violence, not only symbolic violence, but also to create the conditions for many forms of behavioural disorder, crime and alienation.

All these transformational discourses also point to the transformation in the discourse and practice of art now, which links it to establishing relationships with beauty, dignity and dialogue in society in a world marred by monologue, ugliness and violence of many kinds. They also bring us in many different related ways to seekers and practitioners such as Tagore, Gandhi and Sri Aurobindo. Tagore urges us to realise the crucial significance of art and aesthetics to the gift and functioning of life and the way art helps to improve imagination and generate creative webs of relationships. In Tagore, art and spirituality are also linked, as both help us cultivate our inner life in creative relational ways – neither solipsistic nor collectivist.[7] In Gandhi, our way of life, which is based upon a mode of renunciation, and not just endless gratification of desire, becomes an art which is crucial for development – self as well as social.[8]

Sri Aurobindo (1962) challenges us to realise the limits of the ethical, which can degenerate into a regime of control which resonates with the Foucauldian critique of ethics. Aesthetics needs to transform this rigidity into flows of ease, spontaneity, joy and beauty. But Sri Aurobindo here also challenges limited understanding of aesthetics when it is bound to senses only. In modernity, there is a predominance of

---

[7]As distinct from Max Weber, for Hannah Arendt, power is the ability to work in concert rather than exercise one's will over others.

[8]Memory work involves both work and meditation with memory as well as our roots and routes of life. The following poem by the author explores these entangled pathways of critique, creativity and transformations:

Roots and Routes: Memory Works and Meditations

Roots and Routes

Routes within Roots

Roots with Routes

Multiple Roots and Multiple Routes

Criss-crossing With Love

Care and *Karuna*

Criss-crossing and Cross-firing

Root work and Route Work

Footwork and Memory Work

Weaving threads

Amidst threats

Dancing in front of terror

Dancing with terrorists

Meditating with threat

Meditating with threads

Meditating with Roots and Routes

Root Meditation

Route Meditation

Memory Work as Meditating with Earth

Dancing with Soul, Cultures and Cosmos

[9 a.m., 13 February 2015, UNPAR Guest House, Bandung].

the senses, as we are within a predominantly sensate civilisation, as Pritrim Sorokin, the great cultural critic and sociologist, told us a long time ago. Sri Aurobindo thus makes a distinction between aesthetics and *aestheses*, the former limited within a sensualist perspective and the latter realising the limits of it and exploring different modes of transcendence of senses, self and society. In *aestheses*, there is a profound connection between art and spiritual quest (see Bidwaiker 2012).[9]

7.4 Art, Education and Human Development

Sri Aurobindo, Tagore, Gandhi and many seekers and thinkers challenge us to realise the link between art, education and human development. Art plays an important role in educational initiatives started by these pioneers in their own ways. Art is dancingly significant in the vision and practice of integral education which draws inspiration from Sri Aurobindo and his spiritual companion, Mira Richards, who is also known as The Mother.[10] Integral education seeks to go beyond the one-sided mechanical and mentalist emphasis in modern education and integrate all the dimensions of our self and society, such as physical, vital, mental, psychic and spiritual (Giri 2008). Art plays an important role in such a journey of education and human development. Art also plays an important role in the educational pathways

---

[9]In his dialogue with Tagore, Gandhi writes in his article, "The Great Sentinel", published in *Young India* of 13 October 1921: "True to his poetical instinct the Poet lives for the morrow and would have us do likewise. He presents to our admiring gaze the beautiful picture of the birds early in the morning singing hymns of praise as they soar into the sky. These birds had their day's food and soared with rested wings in whose veins new blood has flown during the previous nights. But I have had the pain of watching birds who, for want of strength, could not be coaxed even into a flutter of their wings. The human bird under Indian sky gets up weaker than when he pretended to retire" (in Bhattacharya 1997: 91).

Just as the above presents a glimpse of Gandhi's mind, there are many aspects of Gandhi's approach to beauty which can inspire us to think, meditate and walk further with him and the calling of beauty in life. The following glimpse from Gandhi's visit to the Paris exhibition of 1890 can be instructive. As Hassan (1980: 52) writes: "He appreciated the wonderful construction of Notre Dame and the elaborate decoration of interior with its beautiful sculptures. There was much fashion and frivolity about the streets but inside churches, he found a different atmosphere as he saw people kneeling and praying before the image of the Virgin [...] On the other hand, he found no beauty in the Eiffel Tower and like Tolstoy before him disparaged it: 'It was the toy of the exhibition. So long as we are children we are attracted by toys.'"

[10]Here what Margaret Chatterjee writes deserves our careful consideration: "When Rabindranath Tagore writes of the spiritual, especially in his Herbert Lectures [...] he expressed his dissatisfaction with 'the solitary enjoyment of the infinite in meditation.' He quoted approvingly Kabir's opinion that to say the Supreme Reality dwells in the inner realm of the Spirit 'shames out the outer world of matter.' But how can the two pilgrimages be combined, the within and without? Tagore's answer is clear – through artistic activity. The harmony of relationship created by the poet and musician can be mirrored in the nature of each individual, for each man is endowed with a perpetual surplus of powers which transcends the desultory facts about him (Chatterjee 2009: 107; also see Miri 2016).

of Tagore. In a different way, art and music play an important role in the educational pathways of Gandhi, which is known as basic education. Art also plays an important role in educational initiatives, such as the Steiner Waldorf schools which draw inspiration from Rudolf Steiner. In Steiner'sWaldorf schools there is a connection between arts and craft as well as dances like eurhythmy, which creates movements for artistic self-creation and co-creation of learners.

In all these, art, education and human development blossom and dance together, which has a wider significance. In creative experimental schools such as Bifrost in Denmark, rainbow art plays an important role. Art has also been an important part of many initiatives in education in the modern world, and now there is a need for a broader base for the social foundation of arts education as well as a need to include it in the curriculum as part of our inter-linked wider practices and thoughts.[11] John Dewey, who is an inspiration behind many strands of progressive education in the modern world, also challenged us to understand the link between art, education and human development. Dewey talked about developing an aesthetic ecology of public intelligence (Reid/Taylor 2006). In this the aesthetic, ecology and our intelligence come together as a creative process of human development. This contributes to the creation of an integral being – not only mentalist or intellectualist, but a folded ontology of many-layered self and society, not just the flat ontology of modernity (Taylor 2016). Aesthetics helps us to generate such configurations of folding and unfolding beings and society.

All these point to profound challenges to the existing models and practices of both education and development. Here there is a need to transform both education and development as multi-dimensional visions and experiments in learning – co-learning and collaborative learning. Building upon the seminal work of Piet Strydom, we can realise both education and development as manifold initiatives and movements in collective learning and triple contingency learning (Strydom 2009). Triple contingency learning goes beyond the double contingency of just the self and the other and embraces the third as an inauguration of working, meditating and dancing with the multiple contingencies of self, other, culture, society and the world. Triple contingency learning in education and development has both an artistic and spiritual dimension, as it is facilitated by the aesthetic sensibility to embrace many and the spiritual works of going beyond the logic of closure of self and the other. Triple contingency learning is thus linked to both multi-valued logic and multi-*topial* hermeneutics (Giri 2016b). Multi-*topial* hermeneutics, as discussed earlier in in this essay in a footnote, involves walking and meditating across different cultural, social and spiritual terrains which is also an artistic and spiritual process. It involves foot work, or what David Henry Thoreau (1947) calls walking like a camel whereby we ruminate while walking. In such meditative co-walking, we open the very themes of life and discourse, such as art, spirituality and human

---

[11]Renunciation in Gandhi can bring us to Foucault's path of self-restraint in life as a work of art, though in his own life Foucault may not have followed this closely, as he enjoyed bodily pleasure without limits, as is evident from his visits to many bath houses in San Francisco when he was teaching at Berkeley.

development, to cross-cultural dialogues and personal and transpersonal realisations in which our co-walking becomes a passage and hermeneutics of a new revelation and gathering of meaning.[12] Multi-valued logic and multi-topial hermeneutics

---

[12]For Sri Aurobindo (1973: 40), aestheses can awaken us – even the soul in us – to something yet deeper and more fundamental than mere pleasure and enjoyment, to some form of the spirit's delight in existence, *Ananda*. According to Sri Aurobindo (1973: 44), "There is not only physical beauty in the world – there is moral, intellectual and spiritual beauty too. There are not only aesthetic values but life values, mind values and soul values that enter into art. Beyond the ideals and idea forces even there are other presences more inner and inmost realities, a soul behind things and beings, the spirit and its powers, which could be subject matter of an art still more rich and deep and abundant in its interest than any of these." Walking and meditating with Sri Aurobindo and some of his co-walkers, the following poem of the author explores different possibilities of art, collaboration and transcendence:

Half-Birth Day

This is my half-birth day

This is my friend's birthday

We are friends

of soul, art and the world

We create art in the beach

A public art of aesthetics and *aestheses*

Aesthetics touching the visible

Aestheses embracing the deeper

We create murals in the streets

Not only in our drawing rooms

We create fusion of flags and music

A new art of border-crossing

Art becomes a call for transformation

Calling friends to break out of

Routines of repetition and reproduction

To discover the spring within and around

To sing again

We call people to their streets and souls

We become clean

We become green

We create beauty in our lips and cosmos

We have faith in each other

Faith in Nature, Human and Divine

When we sing

The donkey and divine

creates a dance of transpositionality as it frees us from prisons of fixed positions and standpoints (see Giri 2016e). Such movements are crucial for human development, which helps go beyond prisons and prisms of dominant and established models and practices of human development. Art and spirituality play important roles in triple contingency learning, multi-valued logic, multi-*topial* hermeneutics and the accompanying movements of a new imagination and practice which are crucial for thinking and practising development in a different way, as they help us realise that another development is possible.

7.5 The Broadening and Deepening of Vision and Practice of Spirituality: The Dance of Practical Spirituality and Human Development

The broadening and deepening of vision and practice of art resonates with broadening and deepening of both spirituality and society. If, in the discourse of development, there is now a move to include visions and movements like self-development, ethical development and aesthetic development, in the discourse of spirituality there is now a move to make spirituality relational and practical movement beyond the frames of individual excellence, salvation and isolated meditation – a multi-dimensional initiative in self and social transformation. There is a practical and social turn in spirituality whereby many movements of spirituality wish to address concrete problems in society, such as poverty, shelter and suffering – physical as well as spiritual. This gives birth to the reality and movement of practical spirituality (Giri 2013, 2016). Practical spirituality emphasises continued

---

Come to listen

This is our joint birth day

Of co-birthing and co-breathing

Surrender and co-creation

[For and with Kirti and Lelya, Tasmai Art Gallery, 6.10 p.m., 17 July 2014.]

Sri Aurobindo's pointer to aestheses as movement of beauty with and beyond sensuality is also reflected in the following passage on art and beauty by Andrew Harvey, a deep seeker of both art and spirituality: "Fyodor Dostoevsky wrote in the *Brothers Karamazov*: 'The world will be saved by beauty.' What Dostoevsky meant by beauty was not mere aesthetic beauty but an illumined and initiatory radiance of a vision of holiness. This radiance is art's highest and noblest function to represent, and when through art's holy magic, the heart is awakened to a vision of the sacrality of all creation, beauty can become the fuel for a passion to transform the world" (Harvey: 60).

practice, not only euphoric movement of realisation, enthusiasm and miraculous experience. As Robert Wuthnow tells us, drawing on his work with the spiritual quest of the artists: "Many artists speak of their work as a form of meditation. For some the sheer rhythm of the daily routine brings them closer to the essence of their being. Writing all morning or practising for the next musical performance requires mental and emotional toughness [...] For spiritual dabbers the insight that these artists provide is that persistence and hard work may still be the best way to attain spiritual growth" (Wuthnow 2001: 10). Like art, practical spirituality accepts the brokenness of the world and does not want to assert any totalising unity or total-itarian absorption (cf. Bellah 1970). At the same time, practical spirituality is a striving for wholeness in the midst of our inescapable brokenness and the frag-mentation of this world. This wholeness is emergent as it is manifested in the work of the artists. Artists strive to paint landscapes of emergent wholeness in the midst of fragmentation and brokenness. Artists incorporate their "experimental approach into one's spiritual quest" (Wuthnow 2001: 276).

An artist is a *bricoleur*, creating beauty and images of emergent wholeness out of many fragments. There is an artistic dimension to our striving to establish con-nections and communications across fragments. Practical spirituality follows a new logic – a multi-valued logic of autonomy and interpenetration. This is different from the dualistic logic of either/or, and it seeks to find out and weave threads of connections among different fragments and disjunctions. This involves both ontology and epistemology in which art plays an important role. This art of establishing connections across isolated fields and domains creates a new *yoga* of human development, as *yoga* also means the ethics, aesthetics and spirituality of establishing connections in the midst of disjunctions, disruptions and violence of many kinds.[13] It is also part of the warmth of being connected in the midst of the cold logic of isolation and alienation which gives rise to a new *tantra* of human development in which the fire of mutual creative warmth becomes a vehicle of self and social transformation. Art and spirituality play an important role in the vision and experiments with human development as *yoga* and *tantra* of development. The yoga of human and social development involves creating fields of mutual learning and connectedness. The *tantra* of development involves creating vibrant spaces of conviviality where all concerned would enjoy being together and grow in each other's warmth of relationship. The broadening of art and spirituality is also accompanied by deepening of the discourse and realisation of the social where 'social' no longer means only structures but also spaces of self and mutual realisation.

---

[13]The Mother herself was a creative artist and she was part of circle of creative artists and circles in Paris around the turn of the nineteenth century. She knew such famous artists and sculptors, such as Auguste Rodin.

7.6 Art, Spirituality and Human Development: An Invitation for New Paths of Thinking and Meditative Verbs of Realisations

Development has both an artistic and a spiritual dimension. Conventional visions and practices of human development have not paid enough attention to these. This essay is an invitation for us to realise these links and manifold processes of self, cultural, social and world transformations. Art and spirituality help us realise that development is not only a noun but also a verb and as a verb it is not only activistic but also meditative. Development as multi-dimensional processes of transformations involves meditative verbs of co-realisations of self, culture, society, Nature, Divine and Cosmos as it also struggles with conflicts and contradictions with confrontation and compassion (Giri 2012, 2013). Such a re-envisioning and reconstitution of development is crucial for building sustainable community systems where art and spirituality help us to transform existing communities and build new ones.

References

Arendt, Hannah, 1958: *The Human Condition* (Chicago: University of Chicago Press).

Agamben, Giorgio, 1999: *TheMan Without Contents* (Stanford: Stanford U. Press).

Ankersmit, Frank R., 1996: *Aesthetic Politics: Political Philosophy Beyond Fact and Value* (Stanford: Stanford Press).

Bellah, Robert N., 1970: *Beyond Belief* (New York: Harper & Row).

Benhabib, Seyla, 1996: "Critical Theory and Postmodernism: On the Interplay of Ethics, Aesthetics and Utopia in Critical Theory", in: Rasmussen, David M. (Ed.): *The Handbook of Critical Theory* (Cambridge, MA: Blackwell): 327–339.

Bidwaiker, Shruti, 2012: *Vision, Experience and Experiment in Sri Aurobindo's Poetry and Poetics* (PhD thesis, Department of English, Central University of Pondicherry).

Bhattacharya, Sabyasachi, 1997: *The Mahatma and the Poet: Letters and Debates Between Gandhi and Tagore 1915–1941* (New Delhi: National Book Trust).

Boughton, Doug; Mason, Rachel (Eds.), 1999: *Beyond Multicultural Art Education: International Perspectives* (Germany: Waxman).

Cavell, Stanley, 1988: *Conditions Handsome and Unhandsome* (Chicago: University of Chicago Press).

Chambers, Robert, 1997: *Whose Reality Counts? Putting the Last First* (London: Alternative Technology).

Chatterjee, Margaret, 2009: *Inter-Religious Communication: A Gandhian Perspective* (New Delhi: Promilla & Co).

Clammer, John, 2017: "Art and Social Transformations: Challenges to the Discourse and Practice of Human Development", in: Giri, Ananta Kumar (Ed.): *Cultivating Pathways of Creative Research: New Horizons of Transformative Practice and Collaborative Imagination* (Delhi: Primus Books).

Eisenstadt, S.N., 2002: *Political Theory In Search of the Political* (Jerusalem: Manuscript).

Eisenstadt, S.N., 2007, 2008: *Political Theory in Search of the Political* (Liverpool: Liverpool University Press).

Foucault, Michel, 1988: "An Aesthetics of Existence", in: Foucault, M.: *Politics, Philosophy, Culture: Interviews and Other Writings 1977–1984*, translated by A. Sheridan and edited by L. D. Kritzman, (New York; London: Routledge): 47–53.

Foucault, Michel, 2005: *The Hermeneutics of the Subject: Lectures at the College de France, 1981–82* (New York: Palgrave).

Gardiner, Michel, 1996: "Foucault, Ethics and Dialogue", in: *History of the Human Sciences*, 9(3): 27–46.

Giri, Ananta Kumar, 2002: "The Calling of an Ethics of Servanthood", in: Giri, Ananta Kumar: *Conversations and Transformations: Towards a New Ethics of Self and Society* (Lanham, MD: Lexington Books).

Giri, Ananta Kumar (Ed.), 2009: *The Modern Prince and the Modern Sage: Transforming Power and Freedom* (Delhi: Sage).

Giri, Ananta Kumar, 2012: *Sociology and Beyond: Windows and Horizons* (Jaipur: Rawat Publications).

Giri, Ananta Kumar, 2013: "The Calling of Practical Spirituality", in: Giri, Ananta Kumar: *Knowledge and Human Liberation: Towards Planetary Realizations* (London: Anthem Press).

Giri, Ananta Kumar, 2016a: "With and Beyond Epistemologies from the South: Ontological Epistemology, Multi-topial Hermeneutics and the Contemporary Challenges of Planetary Realizations", Paper.

Giri, Ananta Kumar, 2016b: "Cross-Fertilizing Roots and Routes: Ethnicity, Socio-Cultural Regeneration and Planetary Realizations", Paper.

Giri, Ananta Kumar, 2016c: *The Calling of Global Responsibility: New Initiatives in Justice, Dialogues and Planetary Co-Realizations.* Madras: Madras Institute of Development Studies: Report to Indian Council of Social Science Research.

Giri, Ananta Kumar, 2016d: "Transforming the Subjective and the Objective: Transpositional Subject objectivity", Paper.

Giri, Ananta Kumar, 2017a: "Introduction", in: Giri, Ananta Kumar (Ed.): *Cultivating Pathways of Creative Research: New Horizons of Transformative Theory and Practice and the Work of Collaborative Imaginatio* (Delhi: Primus Books).

Giri, Ananta Kumar, 2017b (Ed.): *Practical Spirituality and Human Development* (Basingstoke: Palgrave Macmillan).

Hassan, Z., 1980: *Gandhi and Ruskin* (Delhi: Shree Publications).

Harvey, Andrew: "Afterword", in: Harvey, Andrew: *Goddess of the Celestial Gallery* (San Rafael: Mandala Publishing House).

Mason, Rachel: "Multicultural Art Education and Global Reform", in: *Beyond Multicultural Art Education: International Perspectives,* 3–17 (New York: Waxmann).

Mehta, Lyla, et al., 1999: *Institutions and Uncertainty: New Directions in Natural Resource Management* (Falmer, Sussex: Institute of Development Studies, University of Brighton: Discussion Paper 372).

Miller, James, 1993: *The Passion of Michel Foucault* (New York: Simon & Schuster).

Miri, Mrinal (Ed.), 2015: *The Idea of Surplus: Tagore and the Humanities* (Delhi: Routledge).

Mohanty, J.N., 2002: *Explorations in Philosophy: Western Philosophy* (Delhi: Oxford University Press).

Nussbaum, Martha, 1990: *Love's Knowledge: Essays on Philosophy and Literature* (New York: Oxford University Press).

Nussbaum, Martha, 1997: "Kant and Stoic Cosmopolitanism", in: *Journal of Political Philosophy*, 5(1): 1–25.

Nussbaum, Martha, 2006: *Frontiers of Justice: Disability, Nationality, Species Membership* (Cambridge MA: Harvard University Press).

Osborne, Thomas, 1997: "The Aesthetic Problematic", in: *Economy and Society*, 26(1): 126–147.

Rahenema, Majid, 1997: "Towards Post-Development: Searching for Signposts, A New Language and New Paradigm", in: Rehenema, Majid; Bawtree, (Eds.): *The Post-Development Reader*, 377–403 (London: Zed Publications).

Reid, Herbert; Taylor, Betsy, 2006: "Globalization, Democracy and the Aesthetic Ecology of Emergent Publics for A Sustainable World: Working from John Dewey", in: *Asian Journal of Social Sciences*, 34(1): 22–46.

Reid, Herbert; Taylor, Betsy, 2010: *Recovering the Commons: Democracy, Place, and Global Justice* (Urbana Champaign, IL: University of Illinois Press).

Safranski, Rüdiger, 2005: *How Much Globalization Can We Bear?* (Cambridge: Polity Press).

Santos, Boaventuara de Sousa, 2014: *Epistemologies from the South: Justice Against Epistemicide* (Boulder, Co: Paradigm Publishers).

Scarry, Elaine, 1999: *On Beauty and Being Just* (Princeton: Princeton UniversityPress).

Sen, Amartya, 1999: *Development as Freedom* (Oxford: Oxford University Press).

Sri Aurobindo, 1962: "The Aesthetic and Ethical Culture", in: Sri Aurobindo: *The Human Cycle. The Ideal of Human Unity. War and Self-Determination* (Pondicherry: Sri Aurobindo Ashram).

Sri Aurobindo, 1973: *Collected Works of Sri Aurobindo* (Pondicherry: Sri Aurobindo Ashram).

Strydom, Piet, 1984: Theory of the Avant-Garde (Cork: Dept of Sociology, University College Cork: A Course Text).

Strydom, Piet, 1994: "The Ambivalence of the Avant-Garde Movement in Late Twentieth-Century Social Movement Perspective." University College Cork, Centre for European Research, Ireland: Working Paper No. 1.

Strydom, Piet, 2009: *New Horizons of Critical Theory: Triple Contingency and Collective Learning* (Delhi: Shipra).

Taylor, Charles, 1991: *The Ethics of Authenticity* (Cambridge, MA: Harvard University Press).

Taylor, Betsy, 2016: "Body place commons: reclaiming professional practice, reclaiming democracy", in: Taylor, Betsy: *Pathways of Creative Research: Rethinking Theories and Methods and the Calling of an Ontological Epistemology of Participation* (Delhi: Primus Books).

Thoreau, Henry David, 1947: "Walking", in: *Portable Thoreau* (New York: Penguin).

Vatsayan, Kapila, 2011: "Arts in Education and Society Today", In: Vatsyan, Kapila (Ed.): *Transmissions and Transformations: Learning Through the Arts in Asia,* 1–15 (Delhi: Primus).

Wali, Alka, 2015: "Listening with Passion: A Journey with Engagement and Exchange", in: Sajnek, Roger (Ed.): *Mutuality: Anthropology's Changing Terms of Engagement* (Philadelphia: University of Pennsylvania Press).

Welsch, Wolfgang, 1997: *Undoing Aesthetics* (London: Sage).

Wuthnow, Robert, 2001: *Creative Spirituality: The Way of the Artis* (Princeton: Princeton University Press).

Chapter 8
A Treatise on Interpretation, Viewpoint and Perspectives on Trust

Abhranil Gupta

Abstract 'Trust' is being studied from multiple angles by many disciplines, and hence many definitions of trust exist (Corazzini, Psychological Reports, 40:75–80, 1977). Numerous research projects have been done and continue to be done on trust by researchers all around the world. Here we argue trust to be of a multidimensional nature, portrayed at various levels: individual, organisational and societal or economic. There is no consensus on whether trust is an abstract quality, a concept, a decision, some neuropeptide in the brain, or all of them or none of them. As we dig further into the trust literature, we find more and more angles from which trust is being and has been studied, but all reflect each researcher's own individual approach, with no literature having a full unanimous view on trust. Here we try to collate the studies on trust.

Keywords Trust · Social trust · Economic trust · Social capital

Trust is the basis of all relationships and is also one of the most important aspects or factors in transitioning from the present situation to a sustainable community. A community that is based upon trust is beneficial to both the individual in question and society as a whole. The basic components of trust that make a successful relationship are addressed below. This is a sort of sketch that tries to acquaint the reader with the various degrees of trust and answers the question of how trust can be of any help to a sustainable society at large.

8.1 Introduction

Before going further into the literature of trust and the basis for it, it is necessary to understand the mechanisms by which trust takes a central position in the research into the management of economic transactions. There are many multidisciplinary

Abhranil Gupta, Ph.D. scholar, XIMB, Bhubaneshwar, India; Email: abhranil@stu.ximb.ac.in

© Springer Nature Switzerland AG 2019   87
A. K. J. R. Nayak (ed.), *Transition Strategies for Sustainable Community Systems*,
The Anthropocene: Politik—Economics—Society—Science 26,
https://doi.org/10.1007/978-3-030-00356-2_8

fields of study at play. In a post-bureaucratic work environment, decentralisation of production methods, organisational structure and other important dimensions play a pivotal role in defining the reasons why trust takes the central position in the field of business management today.

8.2 Trust in the Classical Era

In the classic era of bureaucratic organisation the presence of trust was negligible. In these types of organisation, we argue that the rules set in the formal structure of the organisation were the mainstay of the organisation and that the protocols present then had the last laugh in the organisation. Trust was somewhat undermined in the then scenario.

Things have changed since the time of bureaucracy (Weber 2013). There have been signals of bureaucracy (Reed 2005) and structural changes with the changes in the relationships, which can be observed in the intra- and inter-organisational level that evolved with the modernisation of organisations. The level of uncertainty that has crept into organisations is now unprecedented, which calls for trust in an organisation to become a matter of major concern. In the twentieth century trust arguably became the most sought-after mode of coordination mechanism in organisational relationships (Bachmann 2001; McEvily et al. 2003a, b). Authority or monetary incentive also play a role in the coordination of business relationships (Bradach/Eccles 1989).

8.3 Social Capital

Social capital is demonstrated to provide economic pay-off. Social capital is the resource of society based upon trust. In the seminal work of Hume (2003), trust is often regarded as the psycho-social contract that naturally comes into the mind of the beholder. When we have no regard for co-operative behaviour we lose both faith and profit. Hume describes this in a small yet riveting story of two farmers.

One farmer notices that his colleague's farm is ripe one day, and on the same day he foresees that the crops on his own farm will be ready by the next. Here, if the former helps at his colleague's farm today and his colleague aids him on the next day, it is beneficial to both, but if one helps and the other doesn't, only the individual who gets the help benefits. In such a scenario, if neither helps the other, neither benefits. The case of trust is such that one has to come forward and benefit the other in faith that in the coming days his help will be reciprocated. It is this faith, or psycho-social contract, that is called trust as we know it today. Hume's solution to these kinds of problem is a philosophical approach to human nature. His treatise helps us understand the philosophical underpinnings of human nature. Faith in human nature and understanding of the nature of the person on which trust is made is the key for its predisposition. Generally, trust in another person germinates and grows on the basis of previous conversations and exchanges, social or otherwise. In other words, it is self-reinforced within an environment of trust during a social situation. Otherwise Hume's description of trust stems from the nobility of the persons involved in the situation. One solution to the problem of trust is based upon social contract, a kind of treaty between the parties involved which acts as the overseer of trust. This is rejected by Hume. Hume conceived the view of trust as a word that has moral and social consequences. This moral view of trust is what sets Hume apart from Hobbes.

Although Hume has done away with the proposal of social contract, he offers solutions to the problem of free-riding. In a society where risk and uncertainty is rising, this is a concern for all. Social capital concerned with different entities has two things in common; it consists of certain social structures and it facilitates certain actions of the social actors. Recent articles by other authors put the elements of social capital – such as trust and the other characteristics that make up social relationships – as features of social organisation. This view of social capital is a view that the social virtues of love, co-operation and trust are the few basic ingredients of society that thrive and sustain organisations for long.

Very few early sociologists have viewed trust as a cornerstone of modern society. With the advent of uncertainty and the random nature of relationships at individual, organisational and cross level, trust and belief bring order to society. Order within society is an emergent phenomenon that arises from the individual preferences and decisions that social agents take in a micro environment which emerge in the macro behaviour of society.

The political take on liberalism must consider trust as an important issue, as the basis of liberalism is founded upon social justice. Social justice and trust are two parts of the same coin. Hobbes views unlimited authority as the crucial factor in the case of interpersonal trust in the era of uncertainty and the risk of free riders. Locke focuses on the maintenance of a governance that is worthy of the trust bestowed upon it by civil society, whereas Hume envisions a naturally trustworthy society where interpersonal trust is high. Although Hobbes and Hume have political stances on trust, their shared perception is that society thrives on information about the people to be trusted; knowledge of their previous behaviour and reputation is more important for bestowing trust than anything else. Though their analysis is profound and their logical deductions infallible, our contention about trust in modern society is a little different. As previously mentioned, trust is a self-reinforcing phenomenon, and it thrives on the information we receive from various sources. Today sources of information about people and perceptions of them are amply available in many domains. From only word of mouth in the early years to the social media platforms of today, information about other people is everywhere. This information explosion is essential on the one hand, as it dynamically changes, allowing astute decisions to be made regarding the level of trust it is sensible to place in a particular person; but, on the other hand, it is very bad, since there is no place for privacy these days.

The trust in today's scenario is particularly a modern problem. Every era has its own set of problems and solutions. Trust is one such phenomenon, which remoulds itself in every era and has its own set of problems and solutions. As pointed out earlier, information and misinformation pose their own fallout in the problem of trust in modern society. If we call the post-industrial era modern society, then the post-information era may be referred to as the ultra-modern society of today. In this ultra-modern era we can and do have information about anything and everything. Trust on an interpersonal level is therefore not only a mediated word of mouth, as earlier mentioned, but based on a lot more information we can receive from other sources available.

Social trust is a collective confidence of individuals in the norms of society. The societal norms of a community are based on the principles upon which the trust is bestowed. An important aspect of the community that makes it sustainable is nothing but trust. Let alone sustainability, the community or society cannot exist if trust is lower amongst the actors of the community or society. Like non-market forces in business coordination, communitarian forces also affect the efficiency of the market economy. Social trust in the economy also has important effect on the market.

8.4 Trust Dimensions

Trust is a contextual phenomenon. In a family trust stems from compassion for others and love. In traditional societies, children trust their parent in all aspects while they are growing up, and in old age parents trust their children to ensure their well-being and provide a living. A similar scenario prevails in organisations, where trust in their peers, superiors and subordinates is an important factor for people who work for a living. Trust in every sphere of life stems from love for other individuals and the feeling of connectedness. This connectedness is something we develop continually. This feeling brings trust in others as we perceive that at a certain level we are all are connected by many facets of life itself, such as religion, community, kin and kith and many other factors. Religion may be the first construct of humankind which tried to create order in society and build trust in the followers of the religion concerned. I distinguish trust in three forms:

1. Interpersonal
2. Community
3. Economic

8.4.1 Interpersonal Trust

The most basic level of trust is manifested at the level of a newborn. This, according to the notes of the developmentalist Erick Erickson, is the 'first take' of one's ego. At the most basic level trust is thought to manifest when an individual has the benevolence to accept the vulnerability of being cheated or not experiencing co-operative behaviour from the person on whom trust has been bestowed. Trust is the individual's prediction of the future state of affairs in a given circumstance. Much confusion arises from the assumption that it is straightforward to extend an individual phenomenon like trust from an interpersonal level to an organisational level. However, extrapolating trust from an individual to an organisational level is by no means a simple linear progression. A community or society also develops a culture. This culture is also a product of systems of norms and rules set by the community. This culture of trust needs to be built upon and it is built on the interactive process of the individual agents in an interactive process. Though this chapter will not delve into the mechanisms of building trust per se, it will demonstrate how it is an interactive process.

Leadership is a good area for research on trust at the individual or personal level. Leaders who appeal to their followers are able to evoke trust (Kirkpatrick/Locke 1996; Podsakoff et al. 1990). Trust is also an important aspect of being chosen to be a leader (Fleishman/Harris 1962) and of leader-member exchange theory (Schriesheim et al. 1999). Bass (1990) and Hogan et al. (1994) have demonstrated that to be an effective leader, dissemination of trust is essential. Studies on trust in leadership are mainly based on two qualitative-based research areas – relationship-based and character-based perspectives (Dirks/Ferrin 2002]. Trust is an essential component in relationships, whether in business or social contexts. Although trust literature does not broadly differentiate between trust and interpersonal trust, it needs to be mentioned here that interpersonal trust is distinct from 'Trust' as a concept. Interpersonal trust is built on the lines of reciprocity of trust and word of mouth. In order to define interpersonal trust, many authors have shared different views on it. Six (2007) defines interpersonal trust as a psychological state of accept the vulnerability of being in a position where the actions of others might not match one's expectations. Similarly, McAllister (1995) defines it as confidence in another and willingness to place faith in another. Interpersonal trust is succinct from other forms of trust and more than a psychological state or the confidence one individual has in another. In his study on interpersonal trust, Gellar (1999) argues that confidence in others and their abilities is akin to what we call trust.

A character-based relationship focuses on the character of the leader and its impact on followers. Consequently, placing trust in the leader can result in vulnerability (Mayer et al. 1995). For instance, leadership decisions may affect conditions at work (incentives, promotions, redundancy), which in turn may affect the individual's perception of the leader's ability and the extent to which the leader can be trusted. From the perspective of relationship-based theory, the focus is on the chance of receiving some benefits from the leadership (Clark/Mills 1979; Fiske 1992). Instances of research using this route include studies on the characteristics of the supervisor and their followers' perceptions of it (Cunningham/MacGregor 2000; Oldham 1975) and research on leaders' behaviour (Jones et al. 1975).

Trust is more important than anything in all spheres of social life. It is the binding component in friendships (Gibbons 2004), it facilitates bargaining and negotiations (Olekalns/Smith 2005), reduces transaction costs in inter-firm exchanges (Bharadwaj/Matsuno 2006) and is crucial in international political conflicts (Kelman 2005; Lorenz 1991). It is argued that trust also facilitates investment as it assures the parties of no foul play in the transactions. In the above instances, all the authors have tried to bring out one important aspect of human behaviour: confidence in another individual's future behaviour. This trait in humans makes social and business transactions easy. Information and its transmission is only possible when there is trust between the social players. Accuracy of information also plays a vital part. Information shared by the concerned parties in a trustworthy environment reinforces the trust that already exists.

One of the interesting aspects of research into trust is that it is quantified by survey questions only. Mathematical modelling techniques to quantify trust are tried only in online communities and reputation systems. We don't find any research detailing the mathematical equations upon which we can model trust. Modelling and simulating trust in artificial societies created in the computing environment is a gap in literature.

With regard to interpersonal trust, then, trust is believed to be a kind of confidence measure in the future behaviour of others. This confidence is a type of reinforcing circuit that is strengthened further every time events and experience demonstrate that the trust placed in a particular individual is justified.

8.4.2 Community Trust and Social Trust

Trust is viewed as a kind of lubricating element that facilitates easy economic exchange between individuals (Arrow 1985). Mistrust leads to stringency amongst individuals in operations with individuals. This stringency of operations can be eased off by trust amongst individuals in an economic exchange or other ways of exchange (Bromiley/Cummings 1995; John 1984). Research into strategy has shown that trust is a resource which can secure an enduring reputation as a market leader (Barney/Hansen 1995). While trust is viewed as a strategic resource by some researchers, others are of the opinion that it as an approach to administer and harmonise commercial movement (Bradach/Eccles 1989; McEvily et al. 2003a, b; Powell 1990). Whether we view trust as a resource to sustain economic activities or a scheme to govern groups, one thing is obvious from the discussion here: that it is a crucial element that, though intangible, leads to tangible results. Whatever may be the actual reason, this line of enquiry leads us to reap results in the form of benefits to groups working together. These paybacks are adequately proved by studies ranging from empirical ones (Ostrom/Walker 2003) to real world studies in organisational settings (Fukuyama 1995; Kramer/Cook 2004; Lane/Bachmann 1998; Sztompka 1999), though getting the paybacks are more difficult (Brothers 1995; Janoff-Bulman 1992; Kanter/Mirvis 1989; Seligman 1997). Trust, as said

earlier, acts as a lubricating agent to facilitate the process of governance or transaction. Studies have shown the benefits, but those are under experimental settings
which are not often achieved in the real world. Experimental settings try to restrict
some angles of real life and let go of other angles to interpret and quantify the
variables hidden. Sometimes we may try to obtain real-life situations but we may
not be able to attain the fullest reality in any experiment.

Community, as previously mentioned, is a concretisation of society. Community
can be as large as a nation to as small as a village. Whether small or large, complex
or simple, a community is sustained on the basis of the norms that are set in the
community. Generally, these are based upon social values and culture. Let us first
understand what values are and what social value is.

In sociology values are the concepts people have about society and social order.
Order in society is brought about when there is trust between individuals. Values in
society are relative, as what is valuable for one individual might not be valuable or
worth striving for, for another.

Although the terms 'social trust' and 'social capital' are often used inter-
changeably, they refer to two distinct concepts in academic literature about trust.
Social trust is one of the main constituents of social capital, but they are not the
same thing. Social trust is faith in people, whether strangers or otherwise.
Community systems built on social trust make the community more compassionate
and sustainable. When most people meet a stranger they generally reserve judgement about whether to trust them or not. But instances occur when we have to trust
someone we have never met before.

Social capital – the social values that bind together the people in a community or
society – is an important asset to society. Social capital produces tangible results in
economic interaction, social interaction and is a type of lubricating agent in all types
of transaction. After a discourse on trust, social trust and capital we might start to
ask whether all transactions are done by contract of pen, paper and law or whether
there is a level of trust involved. Even if contracts are based on legal terms, do we
not trust in the law of the land? If contracts were to be based upon trust would it be
better for the economy?

Community, therefore, is more than just a conglomeration of people. Values play
an important role in the conglomeration of people. Social values are the reason why
people stay together. They are the binding force in the background, and their loss
will only break relationships. The science of relationships upon which this society
stands is understanding. Loss of understanding leads to loneliness and a drive to
walk alone. In today's uncertain environment, fast life leads social values to take
the backstage.

The withering of social capital is alarmingly high in communities both simple
and complex. Some authors claim that Social capital is a symbolic quality which is
essential for economic transactions and the public good. This leads to a natural
conclusion that it is a sort of facilitating agent.

8.4.3   Economic Trust

Technology is set to be the clear winner in the new generation of business today. But economists worldwide are arguing that the latest entrant in the new economics is trust.

Trust is often identified as the currency upon which the new economy is built. Trust is the basis of what is otherwise known as a sharing economy. While in traditional economic models the means of wealth production are owned by the State or an individual, in a shared economy they are shared. In a shared economy, there is less underutilisation of the resources that lie within our assets. Technology has allowed us to use the resources at an optimum level so as we can make the most of the resources we own. Although technology today lets us share on a much bigger scale, this is not a new phenomenon. From age immemorial we have been sharing; we cultivated in co-operatives, we hunted in packs, and so on and so forth.

The market is the instrument for the coordination of prices in a place of trade. For an economy to work, a market is an essential system, otherwise prices will be set by non-economic criteria. Another of the criteria is Trust, as it gives a social order, a norm or protocol to work with in a real market scenario. Though there are preconditions for market mechanisms to work, it is essential for the market to be safeguarded from instances of mistrust leading to scams. Researchers in sociology (Granovetter 1985; Fligstein 2001) and institutions (Williamson 1975) have studied this in great detail and found that Trust and Mistrust are both very much valid and operational in real-life market scenarios. Micro-economics presume some protocols that, if maintained in market conditions, mean Trust has a very low role to play; but if those protocols are not maintained, things such as Trust play a major role in social and economic exchanges in markets. Information and its asymmetry play a vital role in making or breaking market relations. Such studies are done by researchers in the domain of game theory and the economics of information (Arrow 1985; Stiglitz 1987).

Trust is an intrinsic element of sharing. Although competition is prevalent in many of spheres of life, trust, sharing and co-operation are also facets of human behaviour. Sharing is one of the primitive behaviours of humans. Trust is intrinsically a personal phenomenon. Trust in an organisation or institution depends on whom we trust in the said organisation. In other words, there is a face of trust. We tend to have trust in an individual who represents the organisation.

It is argued that natural resources like gasoline and water (consumable) are used at an alarming rate, older people are living longer than before and population is on the rise. At this juncture we really need to think about a sustainable community. It is proclaimed that for a community to be sustainable, the scarce resources that we have need to be used efficiently. This is only possible when we learn to use technology to our benefit and share the resources we have. This is the basic rule of trust and a sharing economy.

Societies of today live in an environment of uncertainty, pluralism of views and ideologies and government interventions of many sorts. One can view this as diversity, but this chapter takes a different approach. Here it is viewed as coherent distinct modules of societal systems working in synchronisation with each other wherein each has a system of its own. Society is certainly a complex of various people with various micro motives leading to the macro behaviour of society. Complexity perceives observable phenomena from a different perspective. It is the dynamic relationship between the low-level systems that gives rise to the macro phenomena of the society. Communities are just a simplification of the complexities of daily observed non-linear phenomena. Systems are developed to make communities simpler and sustainable. But as technologies become more developed, people tend to become short-sighted and to adopt less sustainable behaviours. Communities based on trust and compassion are observed to be more sustainable, decentralised in operation and more innovative in their approach.

8.5   Discussion

From the old days to the modern day, the phenomenon of trust has led to friendship, while a breach of it has brought wars. From the wars that we fought to the friendships we earned we have gained insights into what it is to be trustworthy. Trustworthy relationships evolve, and trust itself evolves to be sustainable and less complex. It is argued that in an interaction between two individuals, if one shows trust in the other for whatever reason, that individual enters into an unspoken relationship which exposes the individual to the vulnerability of being cheated by the opportunistic behaviour of the other. An individual showing trust in another individual commits his/her resources in the hope that this commitment will be reciprocated. While this is the case when trust is sustained, it may not be the case if trust is not displayed by the other individual. Reciprocation of the belief set by one individual is the only way to maintain a trustworthy relationship.

References

Arrow, Kenneth J., 1985: "The economics of agency", in: Pratt, J.W.; Zeckhauser, R.J. (Eds.): *Principals and Agents. The Structure of Business* (Boston, MA: Harvard Business School Press), 37–51.

Bacharach, M.; Gambetta, D., 2001: "Trust in signs", in: Cook, K.S. (Ed.): *Trust in Society* (New York: Russell Sage Foundation), 148–184.

Bachmann, R., 2001: "Trust, power and control in trans-organizational relations", in: *Organization Studies*, 22(2): 337–365.

Bass, B.M., 1990: *Bass and Stogdill's Handbook of Leadership: Theory, Research, and Managerial Applications* (New York: Free Press).

Bharadwaj, N.; Matsuno, K., 2006: "Investigating the antecedents and outcomes of customer firm transaction cost savings in a supply chain relationship", in: *Journal of Business Research, 59* (1): 62–72.

Bradach, J.; Eccles, R.G., 1989: "Price, authority and trust: From ideal types to plural forms", in: *Annual Review of Sociology*, 15: 97–118.

Bromiley, P.; Cummings, L.L., 1995: "Transaction costs in organizations with trust", in: Bies, R.; Sheppard, B.; Lewicki, R. (Eds.), *Research on Negotiation in Organizations* (Greenwich, CT: JAI Press).

Brothers, D., 1995: *Falling Backwards: An Exploration of Trust and Self-Experience* (New York: Norton).

Clark, M.; Mills, J., 1979: "Interpersonal attraction in exchange and communal relationships", in: *Journal of Personality and Social Psychology*, 37: 12–24.

Corazzini, R., 1977: "Trust as a complex multi-dimensional construct", in: *Psychological Reports*, 40: 75–80.

Cunningham, J.B., & MacGregor, J. (2000). "Trust and the design of work complementary constructs in satisfaction and performance", in: *Human Relations*, 53(12): 1,575–1,591.

Dirks, K.T.; Ferrin, D.L., 2002: "Trust in leadership: Meta-analytic findings and implications for organizational research", in: *Journal of Applied Psychology*, 87: 611–628.

Fiske, A.P., 1992: "The four elementary forms of sociality: Framework for a unified theory of social relations", in: *Psychological Bulletin*, 99: 689–723.

Fligstein, N., 2001: *The Architecture of Markets. An Economic Sociology of Twenty-First Century Capitalist Societies* (Princeton, NJ: Princeton University Press).

Fukuyama, F., 1995: *Trust: The Social Virtues and the Creation of Prosperity* (New York: Free Press).

Gibbons, D.E. 2004: "Friendship and advice networks in the context of changing professional values", in: *Administrative Science Quarterly*, 49(2): 238–262.

Granovetter, M., 1985: "Economic action and social structure. The problem of embeddedness", in: *American Journal of Sociology*, 91: 481–510.

Hogan, R.; Curphy, G.; Hogan, J., 1994: "What we know about leadership: Effectiveness and personality", in: *American Psychologist*, 49: 493–504.

Hume, D. (2003): *A Treatise of Human Nature* (New Chelmsford, MA: Courier Corporation).

Janoff-Bulman, R., 1992: *Shattered Assumptions: Towards a New Psychology of Trauma* (New York: Free Press).

John, G., 1984: "An empirical investigation of some antecedents of opportunism in a marketing channel", in: *Journal of Marketing Research*, 21: 278–289.

Jones, A.; James, L.; Bruni, J., 1975: "Perceived leadership behavior and employee confidence in the leader as moderated by job involvement", in: *Journal of Applied Psychology*, 60: 146–149.

Kanter, D.L.; Mirvis, P.H., 1989: *The Cynical Americans: Living and Working in an Age of Discontent and Disillusion* (San Francisco, CA: Jossey-Bass).

Kelman, H.C.; 2005: "Building trust among enemies: The central challenge for international conflict resolution", in: *International Journal of Intercultural Relations*, 29(6): 639–650.

Kirkpatrick, S.A.; Locke, E.A., 1996: "Direct and indirect effects of three core charismatic leadership components on performance and attitudes", in: *Journal of Applied Psychology*, 81: 36–51.

Kramer, R.M.; Cook, K.S., 2004: *Trust and Distrust in Organizations: Approaches and Dilemmas* (New York: Russell Sage Foundation).

Lane, C., Bachmann, R., 1998: *Trust Within and Between Organizations: Conceptual Issues and Empirical Applications* (New York: Oxford University Press).

Lorenz, E.H., 1991. "Neither friends nor strangers: Informal networks of subcontracting in French industry". *Markets, Hierarchies and Networks: The Coordination of Social Life*: 183–191.

Mayer, R.C.; Davis, J.H.; Schoorman, F.D., 1995: "An integrative model of organizational trust", in: *Academy of Management Review*, 20: 709–734.

McAllister, D.J., 1995: "Affect-and cognition-based trust as foundations for interpersonal cooperation in organizations", in: *Academy of Management Journal*, 38(1): 24–59.

McEvily, B.; Perrone, V.; Zaheer, A., 2003a: "Trust as an organizing principle", in: *Organization Science*, 14: 91–103.

McEvily, B.; Perrone, V.; Zaheer, A., 2003b: "Trust as an organizing principle", in: *Organization Science*, 14: 91–103.

Oldham, G.R., 1975: "The impact of supervisory characteristics on goal acceptance", in: *Academy of Management Journal*, 18: 461–475.

Olekalns, M.; Smith, P. L., 2005: "Moments in time: Metacognition, trust, and outcomes in dyadic negotiations", in: *Personality and Social Psychology Bulletin*, 31(12): 1,696–1,707. (Accessed December 2015.)

Ostrom, E.: Walker, J., 2003: *Trust and Reciprocity: Interdisciplinary Lessons* (Oxford: Oxford University Press).

Podsakoff, P.; MacKenzie, S.; Moorman, R.; Fetter, R., 1990: "Transformational leader behaviors and their effects on followers' trust in leader, satisfaction, and organizational citizenship behaviors", in: *Leadership Quarterly*, 1: 107–142.

Powell, W.W., 1990: "Neither market nor hierarchy: Network forms of organization", in: Cummings, L.L.; Staw, B. (Eds.): *Research in Organizational Behavior* (Greenwich, CT: JAI Press), vol. 12: 295–336.

Reed, M., 2005: "Beyond the iron cage? Bureaucracy and democracy in the knowledge economy and society", in: Du Gay, P. (Ed.): *The Values of Bureaucracy* (Oxford: Oxford University Press): 115–140.

Schriesheim, C.; Castro, S.; Cogliser, C., 1999: "Leader-member exchange (LMX) research: A comprehensive review of theory, measurement, and data-analytic procedures", in: *Leadership Quarterly*, 10: 63–113.

Seligman, A.B., 1997: *The Problem of Trust* (New Haven, CT: Princeton University Press).

Six, F.E., 2007: "Building interpersonal trust within organizations: a relational signalling perspective", in: *Journal of Management and Governance*, 11(3): 285–309.

Stiglitz, J.E., 1987: "Principal and agent", in: Eatwell, J. et al. (Eds.): *Allocation, Information and Markets* (New York: W.W. Norton): 241–253.

Sztompka, P., 1999: *Trust: A Sociological Theory* (Cambridge, UK: Cambridge University Press).

Weber, M., 2013: *From Max Weber: Essays in Sociology* (Abingdon: Routledge).

Williamson, O., 1975: *Markets and Hierarchies* (New York: Free Press).

Part II
Institutions

The nature and type of institutions at the district, state, national and global levels can either facilitate or destroy sustainable principles adopted by the other four inner layers, namely community governance, community enterprise systems, production systems and culture of relationships at the core. Even if there are inconsistencies in the institutions at the higher levels from district to global levels, the coherence of institutions within a district can greatly facilitate sustainability in the four inner layers within a district, especially at the community (GP) level.

This section is about the five key factors of institutions: *norms and conventions, rules and regulations, principles of justice, interaction intensity* and *institutional loading*. However, the chapters in this section largely provide insights into approaches to the development interventions of the government, as a key institution.

Chapter 9
Strategy-Driven Institutional Convergence: A Policy Vista for Integrated Agriculture Development

Hemnath Rao Hanumankar

Abstract In the quest to usher in the second revolution in Indian agriculture, several options are being explored along the value chain of agricultural development. In this context, it is imperative to recognise the need to align the development strategy in agriculture with institutional support structures. While there is much talk and much has been written about programme and project-level convergence for facilitating the synergistic growth of agriculture and rural development, hardly any attention has been devoted in either academic or policy literature to the need to develop convergent institutional structures driven by a coherent strategy which follows the principle of primacy of strategy over structure. This chapter seeks to signal a policy perspective that stimulates a rethink of traditional institutional systems and the need for greater integration across the sectoral value chain, including non-arable farming activities. Based on the author's involvement in developing guidelines for converging schemes of the Ministries of Agriculture and Rural Development of the Government of India, the chapter proposes the constitution of a *State Agricultural Development Agency* (SADA) as a way forward for strategy-driven institutional integration to ensure desired outcomes from agricultural development programmes. The integrative approach to institutional convergence suggested here has the potential to ease access for farming communities to extension and public services.

Keywords Convergence · Synergy · Farmers · Strategy versus structure Sectoral value chain

Hemnath Rao Hanumankar, Senior Professor and Dean, Development Management Institute (DMI) Patna, Bihar, India; Email: hrhanumankar@dmi.ac.in

The author is grateful to Professor Surya Bhushan for his valuable inputs to the first draft. The usual disclaimer applies.

A. K. J. R. Nayak (ed.), *Transition Strategies for Sustainable Community Systems*,
The Anthropocene: Politik—Economics—Society—Science 26,
https://doi.org/10.1007/978-3-030-00356-2_9

9.1   Convergence Advocacy for Agriculture Development

The Vision 2020 document of the *Indian Agricultural Research Institute* (IARI) emphasises that a policy framework for resolving agricultural distress must spell out "a clear long term vision where inter sectoral linkages are explicit" (IARI 1999). Further, the World Bank's Development Policy Review, notes that "experience with watershed programmes in India points to a number of institutional, design and implementation issues that are hindering the programmes' full effectiveness" (World Bank 2006: 138). Institutional constraints include (a) a multiplicity of programmes financed and delivered through different central agencies; (b) weaknesses in state government capacity for watershed management planning, monitoring, and evaluation; (c) weak coordination among different government agencies charged with planning and delivery; (d) lack of coordination between decentralisation policy and local authority capacity for delivering watershed programmes.

Farmers and their associations have also been emphatic about the need for convergence as the Confederation of Indian Farmers Associations has been advocating. The model of convergence demanded by the Confederation for single window delivery of all farm support services may not appear implementable in the short term. The Confederation's attempt reflects, however, a serious concern for the convergence of all development schemes targeted at farmers, farm labour, artisans, rural youth, women and others so that the energies of multiple actors in the agricultural and rural development sector are not lost in identification of target groups and each scheme does not remain disconnected from the other, as is so often observed during field studies. In the wake of the farmer suicides that rocked the southern states of India in 2004–05, the State Government of Andhra Pradesh set up a Cabinet Sub-Committee to examine the causes and remedies. In its report the Committee recommended incorporating the Integrated Farming Systems Approach with activities from secondary and tertiary sectors as a longer term measure to end suicides among the farming community.

The most telling advocacy for the convergence of development schemes and programmes for agriculture and rural development came from the *National Commission for Farmers* (NCF), chaired by Dr. M.S. Swaminathan, in the very first report of the Commission that was meant to be the wake-up call to arrest agrarian distress in the country (Swaminathan 2006). The National Commission for Farmers, constituted three decades after the National Commission on Agriculture, drew the nation's attention to major concerns which included a minimum net income for farmers, mainstreaming the human and gender dimensions, attention to sustainable livelihoods, fostering youth participation in farming and post-harvest activities, and highlighted the livelihood security of farmers. Among the basket of choices recommended by the Commission for enhancing the competitiveness of Indian agriculture was "Promoting the active involvement of *Panchayati Raj Institutions* (PRIs) and local bodies to foster water-shed/command area communities and making the water-shed or the irrigation command area the point of convergence and integration of various Technology Missions relating to oilseeds, pulses, horticulture, cotton, milk, etc."

Convergence and Synergy among the numerous Technology Missions was considered imperative by the NCF, both for enhanced utility and deeper impact. All the Missions could be integrated under an umbrella set-up which could be termed the "National Federation of Farm Technology Missions". In a subsequent report of the Commission, the "need for a Rural Non-Farm Livelihood Initiative for rural areas" was also highlighted. "The initiative could have as its core a more market-orientated and professionalised *Khadi and Village Industries Commission* (KVIC) and a restructured and financially strengthened *Small Farmers Agribusiness Consortium* (SFAC) to bring all rural non-farm employment programmes together in order to generate convergence and synergy among them." The thrust on convergence was also evident in the design of the *National Agricultural Innovation Project* (NAIP), which was mandated to transform the *National Agricultural Research System* (NARS) into the *National Agricultural Innovation System* (NAIS) to support poverty alleviation.

9.2 Ground Realities of Programme: Schematic Convergence

As the foregoing overview of literature shows, convergence of development schemes implemented by various ministries and departments of state and central governments offers advantages that are widely accepted. Yet, in practice, convergence is not easy to achieve, due primarily to entrenched bureaucracies with vested interests and strong territorial imperatives. Many a time, guidelines for schematic convergence issued from national and state-level responsibility centres do not translate into project-based convergence at the grass-roots level, simply because the subordinate bureaucracies do not demonstrate the commitment to make convergence happen or the communities are not sufficiently informed or empowered to demand convergent delivery of public services, which leaves a yawning gap between the convergence envisioned by senior policy-makers and the field-level realities. Everything else, including the targeted synergies between development programmes and projects, remains elusive.

In the context of rural development, the *Integrated Rural Development Programme* (IRDP) launched initially in 1978 and merged later into the *Swarnjayanti Gram SwarozgarYojana* (SGSY) in 1999 was the first national programme that sought to projectise multiple streams of rural economic activities for the benefit of the rural poor. In retrospect, development pundits concur that the lack of convergence with rural credit institutions and technical, as well as administrative, wings of the project implementation agencies was among the major causes for not fully realising the project objectives. This view is heard very often with regard to other major development schemes and more stridently during the last decade in the course of the implementation of the *Mahatma Gandhi National Rural Employment Guarantee Scheme* (MGNREGS) implemented within a rights-based framework, guaranteeing one hundred days of wage employment to rural households willing to do manual work. A well-researched set of guidelines was centrally drafted and shared with state governments to converge MGNREGS works with all development schemes, such as the National Horticulture Scheme, Watershed Management Scheme, Village Road Scheme, Integrated Fisheries Development Programme and other, but progress has not been satisfactory.

Convergence is widely believed to offer the following development advantages:

- Synergies in learning and knowledge-sharing
- Leveraging scarce physical and financial resources
- Increased transparency in decision-making and process design
- Objectivity and clarity in targeting programme beneficiaries
- Reduced time and cost slippages in programme implementation
- Breaking the self-seeking nexus between contractors and stakeholders.

The above and other advantages are generally appreciated by participants during training sessions and conferences, as the author has experienced in the classroom environment while offering courses in development management to diverse groups of civil servants and rural development professionals. At the District Collectors' Workshop on Rural Development and Urban Renewal, held in April 2005, several themed chapters and presentations highlighted the fact that the battle for sustainable development in the twenty-first century will be won or lost depending on how we managed to consolidate and converge the multiplicity of schemes through well-defined operational guidelines. Even if the bureaucratic guidelines are sufficiently elaborate and empowering to facilitate inter-departmental convergence for synergies in programme implementation, farmers and rural communities are left more confused about the responsibility centres for service delivery and, curiously enough, poor service quality is blamed on compulsions for convergence, driven by the higher administrative echelons.

In the past, convergence may have been blocked by vested interests, whether it was for sharing a co-operative channel of distribution between two marketing co-operatives or even sharing cold storage facilities between two horticulture co-operatives. However, in the present environment, convergence is a way forward for development, and the *Mahatma Gandhi National Rural Employment Guarantee Scheme* (MGNREGS) notified on 7 September 2005 emerged as a strategic entry point for the convergence of agriculture and allied activities with the wage employment programme, in a symbiotic spirit. A series of guidelines were developed for converging the MGNREGS with development projects implemented by various other ministries of the *Government of India* (GoI). Studies commissioned by the Ministry of Rural Development in the GoI through the *professional institutional network* (PIN) had provided penetrating insights into how programmes and projects pursued in a mission mode with well-defined milestones and active co-ordination among all state and central agencies with a stake in agricultural development, yielded very positive outcomes.

9.3   Strategising for Agriculture Development

Beyond making strides in agricultural production with regard to food grains as well as commercial crops during the last couple of decades, the country is keen to leverage the productivity and sustainability gains in agriculture to help enhance the income levels of farmers. In recent months the GoI has been emphasising the need for strategic interventions to double the real income at the level of individual farmers by 2022. In keeping with the spirit of the globalisation of agri-business markets around the world, the competitiveness of Indian agricultural produce deserves serious attention, even as we push the frontiers of production. If farm production is not in tune with the shifting patterns of market demand, valuable opportunities for value addition and value realisation by farmers can be lost to competing countries. This means that agricultural research and development activities need to be realigned with the emerging global market scenario and, in the process of transferring such market-led research to the agricultural production system, the traditional extension machinery will have to gear up to new challenges.

The lab to land approach to extension based on farmers' visits to demonstration plots and the distribution of minikits has contributed richly to empowering the farmers with knowledge of new farming techniques, efficient use of inputs and crop husbandry practices for enhancing farm production. In the unfolding scenario, the extension machinery will have to do more to leverage the new information and communication technologies to feed the farmers/producers with market intelligence and information that can guide them through the entire value chain of production planning, choice of products, harvesting schedules, delivery channels, market prices and changing needs of the marketplace. Undoubtedly, the agriculture extension system in the country is the world's largest organised extension education system, linking the Union Departments of Agriculture and Farmers' Welfare as well as the National Agriculture Research System to the individual farmers and farming units at grass-roots level.

Further, given the fact that over 2,000 institutions are functioning under various ministries of the GoI, State Governments, private sector, NGOs etc., but all contributing to agricultural extension in varying degrees, there is an urgent need to promote networking between all these institutions to capture their knowledge and functional synergies for sharing, learning and networking for innovation. There is an urgent need to reinvent and recapacitate the giant wheel of agriculture extension through comprehensive review and restructuring of the apex, regional, state and district level extension institutions, including the 652 Agriculture *Technology Management Agencies* (ATMAs) and the 680-odd *Krishi Vigyan Kendras* (KVKs) across the country. An overhaul of their financial, human and physical resource systems, cannot wait further any more.

9.4 Exploring and Exploiting the Value Chain

To impart increased gravitas to Indian agriculture and to combat the current agrarian crisis, it is necessary to ensure that no point in the value chain across the major agricultural crops/commodities is allowed to go unexplored and unexploited in terms of economic value. While the value chain serves as a conceptual tool to strategise and operationalise secondary processing initiatives, the outcomes in terms of increased returns from farming, employment opportunities, checking migration and, above all, improving the sagging portion of agriculture's share in the country's GDP, cannot be exaggerated. As gains from secondary processing become evident, price buoyancy of the primary produce will motivate farmers to strive for higher yields and productivity gains. In the current situation of distress enveloping farmers in most geographical parts of the country and segments of agriculture, the value chain approach to crop-planning, production and processing seems to be the way forward for sustainable growth.

Globally, the industrial models of value chain-based business strategies have appealed to the imagination of agribusinesses and farming groups, such as the soybean industry, corn and cotton growers, who have been deeply influenced and motivated to deploy value chain-based approaches for enhancing returns from farm production. The results have been quite positive, as certain industries, such as wineries, tea and coffee processors, have realised, which is prompting the farmers to adopt value-chain-based strategies across all crop and animal husbandry produce. The value chain approach to planning crop production and secondary processing is very nascent to the Indian context. In fact, many African and Latin American economies seem far ahead of India on the learning curve.

In other words, if the emergent strategy for agricultural growth in India is to be viewed as doubling farm incomes with the realignment of the agricultural research and extension systems along the sector value chain, state governments will have to be supported and partnered with both knowledge and financial resources. Typically, the state governments articulate their agriculture development goals as follows:

- To increase farm incomes in a sustainable and environmentally friendly manner.
- To improve the efficiency of agricultural production by improving yields, reducing costs, developing export markets, building infrastructure and investing in research, development and extension etc.
- To develop the state as a leader in agriculture/horticulture/fisheries/animal husbandry etc.
- To focus on crop diversification, based on understanding agro-climatic zones, so as to ensure higher farm incomes.

- Ensure adequate and timely supply of quality inputs – seeds, pesticides, and fertilizers – to the farmer at his doorstep.
- Bring all eligible farmers into the banking net by restructuring existing credit delivery institutions.
- Institutionalise integrated pest management for forecasting and managing outbreaks of pests and diseases.
- Promote private sector participation in agriculture through contract farming, buy-back arrangements, private investment in storage facilities and marketing.
- Balanced regional development in agriculture within the states.

The above goals provide an indication of the strategic orientation of the states, and the goals do have the underpinning to translate the strategy of doubling farm incomes into an actionable agenda. The institutional support and the decision-making structures, however, are either taken for granted or left to the discretion of individual departments in the states Therefore, while the farming systems approach is generally accepted and highlighted as the premise for developing agriculture and enhancing farm incomes, in actual practice, the implementing departments do not even care to understand the complementarity of their schemes and programmes, leaving a wide gap between strategic intent and performance. To overcome this gap, an overarching integrated institutional system, to be titled the *State Agricultural Development Authority* (SADA), could be created in each state for integrated development and management of agricultural resources in the State, in the spirit of a strategy-driven convergent institutional support system.

9.5 State Agriculture Development Authority

The SADA, as a convergent umbrella institution, could be empowered to plan and close the decision-making loop in all financial and administrative matters related to the state's agricultural development across the sectoral value chain with the same authenticity that decisions of the State Cabinet carry (Fig. 9.1).

Serving both as a policy-making and project implementation body, the new institutional mechanism should be responsible and accountable for allocation of resources, appointment and transfer of personnel and reorientating, where necessary, the existing institutions to make way for successful implementation of the projects. The SADA could serve as the single point agency for formulating and implementing projects, allocating resources from the state exchequer as well as from central and multilateral agencies, and for co-ordinating the work of all related departments/institutions across the sector to achieve its strategic and operating goals (Hanumankar 2008). The organisational structure for the proposed SADA is depicted in Fig. 9.2.

The SADA would help integrate the efforts of numerous departments, organisations and projects involved in agricultural development in the State at the apex level. Besides integration at the policy level, the Executive Board, comprising three

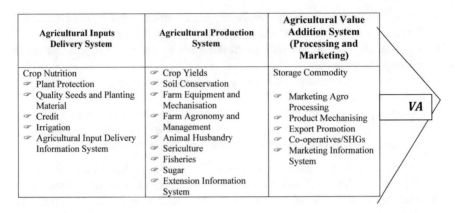

Agricultural Inputs Delivery System	Agricultural Production System	Agricultural Value Addition System (Processing and Marketing)	
Crop Nutrition ☞ Plant Protection ☞ Quality Seeds and Planting Material ☞ Credit ☞ Irrigation ☞ Agricultural Input Delivery Information System	☞ Crop Yields ☞ Soil Conservation ☞ Farm Equipment and Mechanisation ☞ Farm Agronomy and Management ☞ Animal Husbandry ☞ Sericulture ☞ Fisheries ☞ Sugar ☞ Extension Information System	Storage Commodity ☞ Marketing Agro Processing ☞ Product Mechanising ☞ Export Promotion ☞ Co-operatives/SHGs ☞ Marketing Information System	VA

Fig. 9.1 Sector value chain for agriculture. *Source* The author

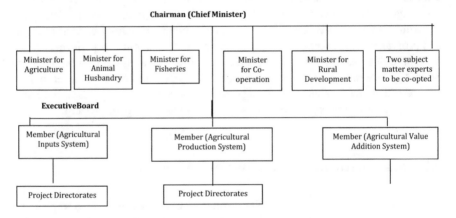

Fig. 9.2 Suggested structure for SADA governing board. *Source* The author

members (preferably of the rank of Principal Secretary to the State Government) would help bring a process orientation to the sector along three distinct streams of activity-agricultural inputs: delivery, agricultural production and value addition process. While details of the administrative linkages between the members of the executive board and the field formation would need to be discussed and defined in the context of each state, depending on the level and stage of agricultural development, select Project Directorates could be set up under the supervision of each member to address areas of immediate concern to farmers.

All the staffing needs of SADA should be met through transfers or redeployment of existing officers of the state governments in the agricultural sector, except where specialist positions need to be created with competencies not currently available within the states. In effect, the SADA should operate with a matrix structure in the initial years, with cross linkages between it and the field commissionerates/

directorates. As the process and projects mature, a situation could be envisaged where a more monolithic structure could emerge, with the field organisation becoming an integral part of the SADA. The entire process involves a complex process of change which would need the facilitating intervention of a reputed management development institution that has strong agricultural orientation in its academic portfolio of activities. Such an agency could be designated the *Design, Implementation, Monitoring and Evaluation* (DIME) partner for the state government, with responsibility for making the transition towards institutional convergence happen in a smooth and non-disruptive way, steering clear of all politics.

9.6 Design, Implementation, Monitoring and Evaluation (DIME) Partner

As a partner institution with the state government facilitating the above process of change, the DIME partner would be responsible for identifying and sourcing expertise from various state, central and, where necessary, international agencies, in pursuit of programmes and projects for integrated development of agriculture in the state. In addition, formal and informal co-ordination with existing institutions of the *Government of India* (GOI) and states like the *Indian Council of Agriculture Research* (ICAR), *National Institute of Agriculture Extension Management* (MANAGE), *National Bank for Agriculture and Rural Development* (NABARD), *National Co-operative Development Corporation* (NCDC), *National Dairy Development Board* (NDDB), *National Horticulture Board* (NHB), *National Fisheries Development Board* (NFDB), *Agriculture Products Export Development Authority* (APEDA), *State Agriculture Universities* (SAU's), State Seeds and Agro Industries Corporations and various other institutions linking up to the sector value chain, would be facilitated by the partner institution through the entire cycle of *Design, Implementation, Monitoring and Evaluation* (DIME) of projects initiated by SADA.

While working closely with State Governments in designing and implementing projects across the three distinct sub-sectors identified above in the sectoral value chain, the DIME partner could also carry out independent studies and surveys to capture the voice of the farmer and the responses of the farming community to the progress and problems of various development projects. The concerns expressed by the farmers and other stakeholders would be debated and discussed in public fora with the DIME partner's facilitation to make way for corrections, modifications etc. in the course and content of the projects. Where necessary, the DIME partner would also assist the SADA in designing and delivering capacity development programmes by way of training and development interventions, on and off the field, particularly for the extension, agricultural marketing and rural development professionals. Since Social Audit is fast emerging as a tool for extracting accountability of project functionaries and governance structures involved in agriculture and rural development programmes, the DIME partner could also facilitate social audit of the convergence between multiple projects and streams of services delivered for the benefit of farmers, strengthening the SADA in the process.

9.7 Conclusion

The above discussion highlights the need for a more integrated institutional support system for desired outcomes from agricultural development programmes. The traditional silo-like structures dissipate synergies across the sectoral value chain, which impedes effective implementation of strategies, however well formulated. The constitution of a State Agricultural Development Agency (SADA) could help bring about institutional convergence driven by a cogent strategy that is widely understood and shared among all actors involved in agricultural development, both arable and non-arable. A broad organisation structure has been proposed for SADA to underpin institutional convergence and serve as an over-arching system of promoting a shared understanding of the challenges and prospects for integrated agricultural development, going deep down to the farmers and the rural communities who are currently hostage to bureaucratic barriers and their territorial trappings. The role of suitable machinery for the Design, Implementation, Monitoring and Evaluation (DIME) of agricultural development projects has also been flagged as an independent observer, analyst, adviser and auditor for the SADA to enhance its service quality and outreach to rural communities, particularly farmers.

References

Hanumankar, H.R., 2008: "Bridging Strategy and Structure for Agricultural Development", in: *Journal of Rural Management*, ICFAI University, 1(1): 55–59.
IARI, 1999: *Vision 2020* (New Delhi: PUSA, Indian Agriculture Research Institute).
Swaminathan, M.S., 2006: *National Commission on Farmers* (New Delhi: Government of India, Ministry of Agriculture).
World Bank, 2006: *India Inclusive Growth and Service Delivery: Building on India's Success*. Development Policy Review of India Country Management Unit, South Asia Region (Washington, D.C.: World Bank).

Chapter 10
Convergence and Flexibility for Last-Mile Delivery of Citizen-Centric Services: A Case Study from Nabarangpur, Odisha

Usha Padhee

Abstract Since independence, India has been implementing several government schemes to target poverty alleviation among the marginalised communities. However, the impact of these programmes has been limited and could not address the requirement of the communities completely. The planning and design of the schemes are more challenging due to the context and need of the diversified communities. Lack of flexibility and convergence while implementing the scheme adversely influences the effectiveness of the scheme. Any efforts by practitioners/duty-bearers to bring convergence and flexibility yield better results with regard to sustaining the communities. A case study from the tribal district of *Nabarangpur* may help in understanding the complexities involved in a real-life situation and how synergy can leverage the strength of each scheme.

Keywords Government schemes · Poverty alleviation · Marginalised communities · Flexibility · Convergence · Practitioners/duty-bearers Case study · Tribal district

In a democracy the obligation of a welfare state is to enhance the quality of life of its citizens. A plethora of schemes and programmes is executed to achieve these objectives. How effectively they are implemented depends on the policy, programme design and execution. An ideal programme envisages the involvement of all stakeholders to ensure the desired outcome. However, development activities often stand in isolation, defeating the objectives.

The landscape of the developed sector becomes more complicated due to different sets of institutions responsible for achieving different objectives. In larger picture, all may be working towards achieving the same goal, both at state or national level, but in a fragmented manner. Summing up of all the activities of individual schemes may not lead to the main outcome and there could be issues of transaction cost and

Usha Padhee, Joint Secretary, Ministry of Civil Aviation, Government of India and Doctoral Scholar, XIMB, Bhubaneswar, State, India; Email: ushapadhee1996@yahoo.com

© Springer Nature Switzerland AG 2019   111
A. K. J. R. Nayak (ed.), *Transition Strategies for Sustainable Community Systems*,
The Anthropocene: Politik—Economics—Society—Science 26,
https://doi.org/10.1007/978-3-030-00356-2_10

effectiveness. But well-structured convergence of programmes with the same activities may lead to better results. It is established by many studies that convergence will bring synergies between different government programmes in terms of planning, design, process and execution (UNDP 2007). Convergence, if it is need-based, will facilitate sustainable development. So bringing the programmes together to converge the efforts will help to achieve synergy and optimal use of resources. There have been discourses in parlance of policy-making to identify the size, scope and level of the programmes to achieve the desired goals. Against this backdrop, the optimal convergence of developmental activities to achieve desirable goals would not only be useful to the policy-makers, but would help the institutions to understand their accountabilities better. However, good intentions, large programmes and projects, and lots of financial resources are not enough to ensure that the development objectives will be achieved. The quality of those plans, programmes and projects, and how well resources are used, are also critical factors for success (UNDP 2009).

While discussing convergence, overvaluing the concept may be avoided and polemical views regarding technical aspects and limitations need to be considered. Factors like competency and commitment may deter the convergence. Different departments with different sets of personnel implement various developmental programmes in the present-day governance. To have focus and to maintain the competencies required for planning and execution, departments are likely to maintain seclusion. There is also manifold growth of programmes designed in different times and contexts. So although the concept of convergence sounds logical, it can be difficult to adopt in some circumstances.

'Best is enemy of good', it is often quoted. Likewise, minimum convergence has to trade off with timeliness or efficiency. To achieve complete convergence, it is necessary to make additional efforts and to depend on other institutions. This can be compared to the bounded rationality theory of Herbert Simon (1976), where choices are based on logic but with limited information. For example, in a poverty alleviation programme, the concerned agency may not have access to all the information needed for interventions. There may be a lack of awareness of different programmes with the same objectives implemented at different levels or by different agencies. So a framework may be necessary to understand the optimal level and limitations of convergence, without which the core elements of efficiency and economy may be the casualties.

Complementing the convergence is the flexibility in administration to facilitate delivery of public services. Flexibility as a strategy can happen at two levels: (a) own-utilisation of workforce and resources with fungibility and (b) flexibility in the regulatory/implementation frameworks. Both levels are interdependent and the latter is crucial to achieve flexibility in a sustainable manner. Flexibility enables the organisation to enhance its capacity to undertake a wide range of tasks. Flexibility also provides speed and agility as competencies to the service delivery (Marquiss 2012).

It is not only in developing countries but also in developed countries (where the trust level is high) that rigid rules create administrative and regulatory barriers. These barriers can be actual and sometimes perceptional. Many institutions would like to stick to the rule books because of disquiet about the consequences of failure to do so. To address the issue of flexibility in implementation, there has to be a legal framework to allow flexibility for the utilisation of the resources and manpower within the implementation framework. The spirit of the scheme is to be seen and not the restrictive guidelines. But this requires very high quality institutions to bear the responsibility for optimum utilisation of the resources. For example, during any disaster, all the resources available at district level are under the command of the designated authority (District Magistrate). An enabling regulation has given this flexibility, which improves the efficiency of the public delivery of services at the time of the emergency. Flexibility also fosters innovations which will provide a fillip to the efficient service delivery. For example, the flexi funds provided to various self-government agencies have triggered many innovative processes as well as programmes (13th Finance Committee Report 2014). However, mechanisms of quick deterrent action in cases of individual aberrations need to be strengthened to avoid misutilisation of this strategy.

Convergence and flexibility can be effective strategies to optimise the resources and achieve the objectives within the time frame. The following case study illustrates the failure of public delivery of services though all welfare schemes existed in their respective verticals but never converged. Lack of flexibility in the existing schemes created barriers to achieving the desired goods.

Nabarangpur, a tribal district, is considered one of the most backward districts in Odisha in terms of Human Development Indices as well as income. It has a dismal female literacy rate and the populace of the district mostly depend on the bounties of nature. A young boy from this district attempted to commit suicide due to non-completion of a form which would have enabled him to the All India Secondary School Examination at the end of 10 years of schooling. This tragic incident hit the headlines in the local media and local authorities initiated an immediate enquiry. The enquiry report revealed following facts:

Makra (name changed to protect identity) comes from a poor family and his parents are daily wage earners. He has two brothers, both younger than him, and three sisters, all of them Class 8 dropouts and involved in daily labour. Makra was in Class 10 but his school attendance was never regular. An enquiry revealed that the student had not received a scholarship as he had failed to open an account at the local bank. The district unit of tribal education was authorised to disburse the scholarship after the completion of due formalities. The technical support team visited the home of the student but reported the lack of interest shown by the family. The Project officials also enclosed the statement of the Sarpanch of the Gram Panchayat to substantiate their claims. At the same time, the district unit also reported that during the 2013–14 sessions, Makra attended the school on only 47 days out of 214. Similarly, during 2014–15 (i.e. in Class 10) he attended only 30 days out of 125 working days. It was also reported that the boy regularly helped his father on their small agricultural plot. When his mother was engaged in the collection of Kendu leaves, Makra also supported his mother. In one instance Makra received wages in the form of a cheque for Rs. 600 from the Forest Department for collecting Kendu leaves. The Forest Department could only disburse cheques but no cash as per their guidelines. Makra could not cash the cheque as he did not have a bank account. Makra had approached the headmaster of the school to help him cash the cheque, but in vain. The cheque had remained uncashed at the time of the incident. At the end of the ordeal, Makra, in a state of extreme grief and distress, consumed poison for not filling up the examination form. With an attendance rate of below 25%, whether Makra would have been allowed to take the exams or not is another question.

The above story depicts how an incident can expose the failure of the governance systems at all levels. A poor family of eight people struggling to make ends meet was not linked effectively to the welfare programmes. They had no support from any institution to apply for assistance. Makra did not go to school on most of the working days. Because he was regularly absent, the school had no concern or capacity to counsel the family or the student. No state functionaries were aware of the problem and no one followed it up. Scholarships were not disbursed, and the blame for this failure was placed on the family for not opening an account at the local bank. Makra's irregular attendance throughout his schooling divulges the status of his circumstances. Poverty at home forced Makra either to labour alongside his father on their small agricultural plot or go to the forest to collect *Kendu* leaves with his mother. If the plethora of poverty alleviation programmes had targeted this family effectively, there would have been no pressure on the boy to miss school. Forest officials did not recognise the child labour component and allowed an adolescent boy to work and issued the cheque without providing any facilitation or assistance. When Makra wanted to open an account, there could have been a system to open a no-frill account at school. If the school had accepted the cheque or provided an extension for paying the fees, the student could have been saved from duress.

Revisiting the above incident within the framework of convergence and flexibility, the family's problems could have been addressed through the poverty alleviation programmes which provide various supports for survival, including the resources for consumptional needs. An effort to explain how a life cycle approach could be adopted along with a sustainability model for the welfare of the family is illustrated (Fig. 10.1).

If the family had the support of convergent efforts from all corners, the state of affairs would have been different. Let's imagine there was one institution – a self-help group (SHG), Gram Panchayat or any other formal or informal organisation – which worked in tandem with all the service providers to ensure delivery of services to all people in their jurisdiction, thereby becoming a one-stop shop where financial and personnel resources converge to optimise the benefits. In Makra's circumstances, if the Gram Panchayat had intervened and provided the financial resources and fees to the school, the needs of the family could also have been addressed. As an institution, it could have negotiated with financial institutions such as banks to have better outreach.

Community institutions can act as single-window service-providers with all the inputs required to ensure sustainable livelihoods for struggling households. This can come in five broad categories of Physical, Natural, Financial, Human and Social Capital. The service-providing organisation itself can be the social capital.

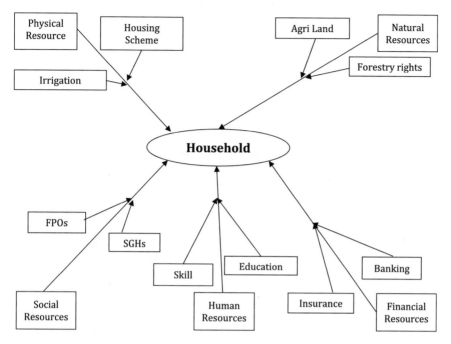

Fig. 10.1 Life cycle welfare model of a household. *Source* Author's conceptualisation

The other types of capital may be delivered as different components of the State Welfare Schemes. The functions, functionaries and funds can be transferred to these institutions for achieving the services. The complementing verticals can be converged and a holistic approach can be adopted.

But the challenges are many when we want to bring convergence at the level of one institution. Considering the need for expertise, limitations in the extension services might exist. Also, the transaction cost would determine whether convergence efforts would be sustainable. The availability of resources would determine how effectively the institution could function. Too small an institution would not be able to manage due to higher transaction costs. Too big an institution might lose focus and people might have difficulty in accessing the institution for the required services (Nayak 2009). Based on practical experience, there is a need to strengthen a body which has the constitutional backing to bring such services together. The elected body needs to be broad-based and could invite representatives from all other informal organisations to achieve convergence and flexibility. An ideal institution would be an inclusive democratic entity at the level of Gram Panchayat (Fig. 10.2).

Convergence and flexibility can happen by proper planning and design. Institutions, which naturally allow convergence and flexibility, can sustain to provide quality service-delivery to the public. Just enhancing the resources may not yield good results unless the livelihood-related elements are addressed through individual and household-level activities. Every household is unique and converged

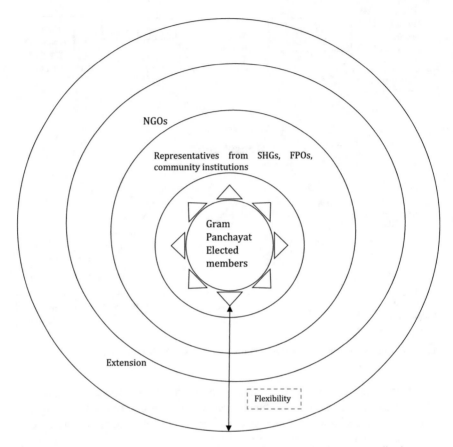

Fig. 10.2 Convergence strategy in a Gram Panchayat. *Source* Author's conceptualisation

efforts with flexibility will lead to better service delivery. It is ultimately hoped that governance in India can be simplified so as to bring to its citizens a single-window delivery structure for channelling all types of government services at local level in the most efficient manner possible.

References

Department of Rural Development, 2008: *Report of the Task Force on Convergence – India Environment* (New Delhi: Department of Rural Development, Government of India).
Future of Work Institute, 2012: *The Benefits of Flexible Working Arrangements: A Future of Work Report* (London: Future of Work Institute).
Government of Andhra Pradesh, 2003: Circular for 'Guidelines about formation and strengthening of village organizations', 9911/SERP/PM/Circular on Velugu implementation/2012/18, 30 August 2003 (Hyderabad: Government of Andhra Pradesh).

Government of India, 10th, 11th and 12th Plan documents (New Delhi: Government of India December 2002, December 2007, December 2012) Guidelines for convergence of NREGS with programmes of Ministry of Agriculture for enhancing productivity (New Delhi: Government of India, MoRD, October 2009).

Inter-sectoral Convergence-National Rural Health Mission: at: nrhm.gov.in/mediamenu/presentations/otherr-presentation/intersectoral-convergence.html (24 June 2016).

Nayak, Amar K.J.R., 2009: *Optimizing Asymmetries for Sustainability: Design Issues of Producers' Organisation* (New Delhi: GoI, Planning Commission, January).

Simon, Herbert A., 1976: *From Substantive to Procedural Rationality* (Cambridge: Cambridge University Press).

UNDP, 2007: *Annual Report. Making Globalization Work for Everyone* (New York: UNDP).

UNDP, 2009: *Reaching Last Beneficiary: Resource Convergence Mantra Model* (New York: UNDP).

Part III
Production: Sustainable Agricultural Systems

Agriculture, the primary production activity, has greatly been impacted by the climate changes and is increasingly becoming unsustainable across the world. It appears that in the course of our taking agriculture forward for greater productivity through intensive external inputs, we have made small farms unviable, jeopardised food safety and security, increased environmental degradation and risk, making agriculture unsustainable.

This part discusses the principles of *seed, soil, moisture, diversity* and *ecology* in line with principles of *agro-ecology* that can enable a small farmer's agricultural field to become sustainable and the farm viable in the short run and sustainable in the long run. This section has the largest number of chapters. While some of them deal with these issues directly, others deal with the related issues of significance, performance measurement methods and policy-related issues.

Chapter 11
In-Situ Water Conservation on One-Hectare Farmland: An Action Research

Amar K. J. R. Nayak, Tanmay Khuntia and Birendra Suna

Abstract This chapter narrates the technicality of in-situ water conservation on a one-hectare farm and its impact on its field ecology. It discusses 'in-situ water conservation' through rainwater harvesting on a farmer's field by the use of micro-locks and trenches. Its arguments are based on an action research project where rainwater was conserved by increasing the water-holding capacity of the field through developing a sustainable agriculture system. The amount of water that could be captured by adapting these techniques has improved soil porosity eight times and the water-holding capacity accordingly. In-situ water conservation techniques played a significant role in developing groundwater by capturing rainfall on the farm.

Keywords Moisture · In-situ water conservation (IWC)
Rainwater harvesting (RWH) · Trench · Micro-locks

11.1 Introduction

Numerous nations are confronting the issue of diminishing per capita accessibility of fresh water and that of water efficiency. *In-situ water conservation* (IWC) through *rainwater harvesting* (RWH) might be a critical way to alleviate the trial of water shortages. This article focuses on an experiment which can significantly contribute towards *sustainable development goal* (SDG) number 2: no hunger, and SDG number 6: clean water and sanitation. Clean potable water for domestic use, adequate fresh water for agricultural production, balancing the demand and supply for various purposes, optimisation in procurement and utili-

Amar K. J. R. Nayak, Professor of Strategic Management, XIMB, Bhubaneswar, Odisha, India; Email: amar@ximb.ac.in

Tanmay Khuntia, Senior Research Fellow, XIMB, Bhubaneswar, Odisha, India; Email: tanmaykhuntia@gmail.com

Birendra Suna, Assistant Professor, Department of Social Work, NISWASS, Bhubaneswar, Odisha, India; Email: sona.birendra@gmail.com

© Springer Nature Switzerland AG 2019
A. K. J. R. Nayak (ed.), *Transition Strategies for Sustainable Community Systems*,
The Anthropocene: Politik—Economics—Society—Science 26,
https://doi.org/10.1007/978-3-030-00356-2_11

sation is inevitable for a sustainable community system. As water is the lifeline for sustenance of any community, indigenous knowledge of in-situ water conservation might be a crucial way to relieve the challenge of water shortages.

This experiment on in-situ water conservation could significantly and unswervingly contribute towards sustainable development goal (SDG) 2: End hunger, achieve food security and improved nutrition and promote sustainable agriculture, and SDG 6: Ensure access to water and sanitation for all. Apart from these two goals, IWC also affects other SDGs, such as goal 1: End poverty in all its forms everywhere; goal 3: Ensure healthy lives and promote well-being for all at all ages; goal 11: Make cities inclusive, safe, resilient and sustainable; goal 12: Ensure sustainable consumption and production patterns; goal 13: Take urgent action to combat climate change and its impacts; goal 14: Conserve and sustainably use the oceans, seas and marine resources; goal 15: Sustainably manage forests, combat desertification, halt and reverse land degradation, halt biodiversity loss (FAO 2016). Clean potable water for domestic use, adequate fresh water for agriculture production, balancing the demand and supply for various purposes, and optimisation in procurement and utilisation are essential for a sustainable community system. As water is the lifeline for the sustenance of any community, indigenous knowledge of in-situ water conservation might be an crucial way to relieve the trial of water shortages.

Agriculture alone in the present day consumes about 89% of groundwater (GoIMWR 2014) in the Indian context, while its farm production uses three to four times more water to produce one unit of major food crops than the US, Brazil and China. Groundwater today accounts for 62.4% of the net irrigation need of India (Kant 2016). Here high volume of irrigation is believed to be inevitable for agricultural production. Public discourse is that over-exploitation of groundwater is a matter of worry for many states of India. There is over-exploitation of groundwater in states like Punjab, Delhi, Haryana and Rajasthan, and the current rate of withdrawal may lead to the complete exhaustion of groundwater in a decade.

Preservationist thinkers like the water man of India, Shri Rajendra Singh, and the environmentalist and water conservationist Shri Anupam Mishra (1948–2016) highlighted the importance of water and introduced traditional knowledge about preserving water into mainstream discourse. Influenced by renowned practitioners and farmers, an intervention was carried out to conserve rainwater in the natural setting of the farm in the experiment. This chapter cites evidence from the action research project on one hectare (2.47 acres) of land where each drop of rainwaterfall on the ground is conserved in-situ. It describes the technical layout of the farm that facilitated in-situ water conservation (IWC). It emphasises 'in-situ water conservation' through rainwater harvesting (RWH) in the crop field by the use of trenches and micro-locks. It argues plants and crops essentially need moisture for their growth, and a small amount of additional water supply may suffice. It also asserts that in-situ water conservation, along with other factors – soil fertility, seed quality, crop diversity, and ecology – leads to sustainable agricultural systems.

11.2 Existing Literature on Soil Moisture

Considering the importance of water conservation, several initiatives were observed across the globe and included in the literature review on IWC. A few of the key papers are cited below.

In-situ moisture conservation methods help to conserve maximum moisture in the root zone even after a few months of the rainy season. This helps the plants to overcome the water stress conditions of the dry season. Whether cultivation occurs during scarcity or abundance of moisture and water, plants are forced to react to reach maximum efficiency. Plant species utilise the conserved moisture to survive and protect soil erosion (Anju/Koppad 2013).

In-situ rainwater harvesting practices improve hydrological indicators, such as percolation and groundwater recharge. As a result, soil remains lively and porous, which fosters enriching nutrients. This results in better biomass production and, in return, the yield is higher. Higher biomass supports a favourable number of plants and animals, but the native species might be replaced by crops so the overall landscape might change. Farmers applying in-situ rainwater harvesting practices encounter low risk of food security and greater income (Vohland/Barry 2009).

The world's dry sub-humid and semi-arid regions are critical places where agriculture suffers a lot due to water related constraints. So investment to facilitate water in those areas is very important. Water harvesting systems are a key strategy for supplemental irrigation to minimise the risk of crop failure in the dry spells. "Rainfall may be regarded as the entry point for the procurement of fresh water, thus incorporating green water resources (sustaining rain-fed agriculture and terrestrial ecosystems) and blue water resources (local runoff). The divide between rain-fed and irrigated agriculture needs to be reconsidered in favour of a governance, investment, and management paradigm (Rockström et al. 2009).

Soil moisture is regarded as the most limiting variable in farm production. For instance, in Tigray region the loss of rainwater because of overflow and, in addition, the soil loss has been seen as a basic issue. The contour system (locally called *terwah*), which is a customary furrowing strategy, consists of making a 1.5–2 m contour system for lasting raised beds. The contour furrows are at an interval of 60–70 cm mediations. The system was tested and assessed as a fruitful practice that could help considerably in enhancing the successful use of in-situ water and soil preservation (Gebreegziabher et al. 2009).

In-situ moisture conservation by the use of ridge and furrow resulted in higher moisture storage than deep and shallow (DAS) tillage. This experiment was conducted over sixty days on a plot where the soil is sandy loam to loamy sand in nature, with below average water retention capacity and poor fertility. The moisture conservation experiment and the application of mulching on the plot resulted in high soil moisture content, which leads to better crop production than before (Arora/Bhatt 2004).

A few tests have been done on the efficacy of ridge and furrow strategy using different materials, such as a plastic-secured ridge and rock mulched furrow, rock sand, and mulching, when investigating in-situ water conservation in China. This has been tried for various crops in different places. The above rainwater preservation enhanced both the crop yield and the effectiveness of water use in semi-arid areas. The uncovered ridge gathers the high-power precipitation, though the distinctive mix of mulch might be exceptionally helpful for the low-force precipitation. The technique brought about a record high in crop production (Li et al. 2000, 2001; Tian et al. 2003; Wang et al. 2009).

It is essential to understand the needs and aims of farmers and to recognise that they possess innovative ideas about moisture and soil conservation by various means, since reducing moisture affects productivity. The soil and water conservation methods are therefore best designed in collaboration with farmers. This makes it easier for farmers to accept and implement the designed programme. Farmers also need to know the cost of implementing the project. Their willingness to participate is of prime importance in in-situ conservation (Kerr/Sanghi 1992)

Different moisture conservation techniques, such as mulching, weeding and tillage, are being used to harvest runoff water, which improves the yield of both seasonal and perennial crops. The study witnessed the improvement in growth of trees and crops even in the Indian desert because of the rainwater harvesting and in-situ conservation practice. The ridge and furrow method was again found to be one of the most effective methods of moisture storage in the establishment of new forests. The moisture conservation technique has proved to be a boon for the arid ecosystem in terms of forest creation and crop production with higher socio-economic stability (Gupta 1995).

Good rainfall alone does not automatically signify better crop production, as rainfall is not a valid measure for the retention of soil moisture, which is a prime factor in plant growth. Instead of going for weather forecasting and rainfall prediction, soil moisture observation can be a better decision-making tool for farmers. In conventional farming the relationship between rainfall and subsequent moisture retention is unpredictable. Generally, the water evaporates or flows away through surface runoff if it is not conserved in time. The moisture level goes down because of tillage and surface aeration (Horrigan et al. 2002).

In an action research project on in-situ moisture conservation the crop yield became better due to the availability of moisture at various stages of the crop's life cycle. Rainwater was harvested in-situ by use of a compartmental bunding and ridge-furrow which locked water in the field by curbing runoff. As a result, the water percolated into the soil. Part of that water remained in the ground in the form of moisture and the rest went inside and got converted into groundwater. Using the farm manure and other materials from organic/natural sources improved the soil health, which developed the porosity and infiltration rate. Hence in-situ soil and water conservation became effective (Patil/Sheelavantar 2004).

In several action research projects, it has been observed that in-situ water/moisture conservation by use of various methods improved the crop production in comparison to conventional methods in different geographical areas in India and across the globe. The development of physical barriers to check the runoff water in the farm by use of gravel mulch, micro-locks, trenches, ridges-furrows, farm ponds, planting of grass and trees, bore-well recharge unit, mulching with plastic and plant residue played a key role in infiltration of water inside soil and in checking evaporation (Muthamilsel van et al. 2006; Vita et al. 2007; Rockstrom et al. 2010).

It is reasonably evident from the literature that in-situ water conservation improves the land and benefits the farmer's yield. There are different methods of in-situ water conservation practised by farmers across the world. IWC positively impacts the soil moisture, which improves the ecology and productivity as a result. A specific type of intervention is being implemented on the action research site to observe the impact and cost benefit on a hectare of land.

11.3 Practitioners' Perspectives

Many of the farmers who advocate natural farming, organic farming and other forms of inorganic-chemical farming believe crops need moisture for their growth, but do not need to be flooded with water. They have experimented and experienced that present-day use of water in agriculture through irrigation uses more water than required. The unethical practices in farming promoted by corporate giants, such as mono-cropping, the use of chemical fertilizers and pesticides, and the use of hybrid and genetically modified (GM) seeds, are leading to a world where the principles of agriculture are going against ecology. As a result, large quantities of groundwater are being depleted. However, if farmers opt for mixed cropping, use indigenous seeds, and keep diversity, ecology, and soil fertility in place, the current use of water will be halved. So water stress on farmers will come down, according to leading organic farmers.

Natural farming practitioner Subash Sharma of Yevatmal district, Maharashtra says, "We can conserve around 1 crore litres of fresh water by the use of micro-locks and trenches for a rainfall of 1,000 mm on one hectare of land. Approximately 80 lakhs litres of water can be conserved even after evaporation. For multiple cropping on one hectare of land a farmer needs 40 lakhs litres of water throughout the year if we can manage the moisture properly.[1] So the level of groundwater recharge in the aquifer will keep on increasing year by year."

Natueco farmer Deepak Suchde of Malpani trust, Madhya Pradesh, who works on a horticulture-based mixed cropping method, says "If we can develop the firm scientifically with a proper management of ecology, diversity, sunlight, moisture, seed, and soil, there is no need to provide water from external sources to the crops. Moisture alone can take care of the farm."

Gandhian farmer Bhaskar Hiraraj Save of Gujarat, who received a place in the *Limca Book of World Records* (1993) for generating the highest production and profit in the world with his Organic Farming System, emphasised the "need for moisture for plant growth instead of the conventional practice of irrigation."

---

[1] 1 lakh = 100,000.

Shri Natabara Sarangi of Odisha, whose dedication and commitment have enabled him to preserve nearly 700 varieties of indigenous paddy seeds, says none of the paddy seeds requires a huge amount of water, just proper moisture management. Plentiful amounts of rice can be produced without chemical pesticides or synthetic fertilizers.

11.4   Action Research Observation and Findings

The action research project is situated at the *Centre for Development Education and Communication* (CEDEC) in collaboration with the *National Institute of Social Work and Social Sciences* (NISWASS), Bhubaneswar, Odisha, India. The plot used for the pilot project is one hectare (2.47 acres). The length and width of the farm are 425 feet and 250 feet respectively. It is surrounded by a concrete wall. For the balance of ecology inside the farm 11,000 sq. feet (roughly 10%) of the total land is used for the plantation of perennial plants. Now almost a thousand perennial plants surround the cropland. There is a cowshed with the capacity to hold eight cows, a farmers' residence, a drying yard, a godown (store-house), a dug well, plus space for fodder preparation, a compost pit and storing the farm equipment on 0.5 acres of the total land. Around 3% of the total land, roughly 3,500 sq. feet, is used for making the trenches. The space left for cultivation is around 1.8 acres (78,000 sq. feet). This cultivated area is surrounded by the trench, followed by a living fence (trees and shrubs on three sides). Different varieties, like mango, chiku, custard apple, neem, jackfruit, coconut, gooseberry, acacia, guava, timber and flower-bearing trees, have been planted to enrich the natural ecosystem so that there will be flowers throughout the year and birds and bees will avail themselves of the eatables. This is helpful for pollination and conserves the balance in the ecosystem (Fig. 11.1).

Land levelling and balancing of surface water flow: Land levelling with a tractor-mounted leveller was done when the land was prepared for the first time. This was a one-time investment and use of heavy machinery for long-term usefulness and the minimisation of slope in the crop field for water conservation and cropping. Land levelling was done with the help of a simple tool (zero level contour markers) which is made of wood and attached to a narrow pipe (0.02 cm) with a clamp and measuring scales. The instrument is used to measure the ups and downs in a crop field so that micro-locks and contours can be made accordingly. It is very useful for better conservation of rainwater and optimum utilisation of water in a crop field.

Micro-locks: Micro-locks are the locking of the space in the ridges and furrows in a field to keep the rainwater locked at a certain height for a certain time period. Micro-locks were made on the whole farm so that rainwater could recharge there and then. By this method of micro-locking, rainwater remains in contact with the soil surface for a longer period and penetrates underground rather than freely flowing away from the field. The overflown water has been stored in the trenches

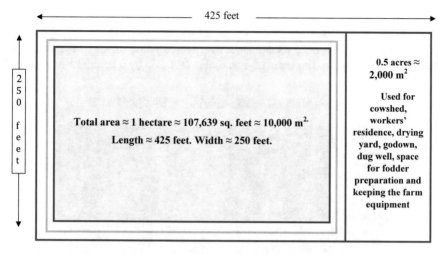

Fig. 11.1 Basic technique used for in-situ water conservation. *Source* The authors

adjacent to the main cropland. The micro-locks are made by using a simple instrument (little harrow/*daud*) driven by bullocks. The size of the micro-locks varies from 4 feet to 12 feet depending on the slope of the land.

Trenches: Trenches are a one-time investment on a farm. On the action research site 3–4% of the farmland is occupied by them. The farmland is generally protected by a boundary wall/wire fence followed by a living fence inside, which consists of diverse varieties of trees and shrubs. There are around 35–40 feet between the fence and the trench. The trenches surround the rest of the cultivable land. Ideally trenches are 6 feet deep and 4 feet wide with a length of 80 feet. A locking system of 10 feet long by 4 feet wide and 1.5 feet deep is constructed at an interval of 80 feet. It is patched by stones and gravel at regular intervals. It takes 10–12 h for an excavator to make a trench in one hectare of land, and two tractors need to be engaged to shift the excavated soil. Altogether this costs around Rs. 15,000/- to Rs. 17,000/-. The trenches need maintenance every three years, which costs around Rs. 5,000/- to Rs. 7,000/-. The trenches could also be made by a manual labour force.

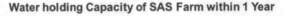

Water holding Capacity of SAS Farm within 1 Year

Year 1   Year 2

11.5   Estimation of In-Situ Water Conservation

As per a simple calculation, 1 cm of rainfall on 1 hectare (10,000 m^2) of land would amount to 1 lakh litre of water (1 cm * 100 m * 100 m = 1 * 10,000 * 10,000 cm^3 = 10^8 cm^3 = 10^5 L). For example, if there is a rainfall of 100 cm per year then approximately 1 crore litres of pure fresh water reach the ground. In the majority of the cropland in India, this amount of water flows away as runoff water and is of little use to the cropland. In spite of getting a huge supply of water in the form of rainfall, Indian farmers face severe water scarcity because of current practices in agriculture and lack of rainwater harvesting systems. The major challenge is how this huge amount of precious water can be converted to groundwater and properly utilised for crops.

Recording process of actual rainfall data: We measure the rainfall data by using a simple measuring beaker featuring a measuring scale. The beaker is placed in the open field. We keep measuring the rainfall data at regular intervals. The recording of the data is as mentioned in the Table 11.1.

Table 11.1 Rainfall data

Month	Average rainfall in the project area (mm)	State average (mm)
June 2016	155.8	159
July 2016	113.0	277
August 2016	260.7	330.6
September 2016	285.4	274.7
October 2016	88.5	88
November 2016	15	166
Total	918.4	1128.7

Source Author's observed data

Increase in Soil Porosity of SAS Farm within 1 Year

Control Plot   Experimental Plot

11.6 Result

During the monsoon season six months (June–November 2016) 92 cm of rainfall occurred and 92 lakhs litres of water were locked via micro-locks and trenches. A major part of that rainfall was converted to groundwater and used by the crops.

- Moisture in the cropland and trenches remained for a longer time period. Even fifteen days after the rainfall, moisture was found at the surface of the trench. It helped the crops to use less water from external sources.
- The water column of the well present in the farm increased from a depth of 18 feet in May 2016 to 21 feet in May 2017 even after utilisation for the cultivation. It was also observed that the water column of the farm well was 25 feet in June–

July 2016, which increased significantly to 33 feet in June–July 2017, maybe because of groundwater recharge.

- Soil erosion was checked because of the micro-locks and trenches present in the action research project.

11.7   Water Drawn from the Farm Pond/Well

Utilisation of water differs on the basis of the type of crop produced and how many times in a year the land is used for cropping. If we assume that the land is engaged for cropping 365 days a year, the maximum amount of water a cropland needs is as

Table 11.2 Water extracted from the farm well

Date	Tinie (min.)	Volume of water (L)
18.10.2016	25	3750
22.10.2016	20	3000
25.10.2016	70	10,500
30.10.2016	40	6000
2.11.2016	30	4500
10.11.2016	80	12,000
14.11.2016	45	6750
20.11.2016	40	6000
25.11.2016	40	6000
29.11.2016	40	6000
03.12.2016	30	4500
6.12.2016	90	13,500
8.12.2016	120	18,000
10.12.2016	120	18,000
12.12.2016	120	18,000
14.12.2016	120	18,000
15.12.2016	120	18,000
16.12.2016	120	18,000
17.12.2016	120	18,000
20.12.2016	40	6000
26.12.2016	30	4500
28.12.2016	120	18,000
29.12.2016	120	18,000
30.12.2016	100	15,000
04.01.2017	20	3000
04.01.2017	20	3000
Total	1840	276,000

Source Data collected by the authors

follows. One hectare of sugar cane needs around 64 lakhs litres of water a year. Arhar uses around 16 lakhs litres of water for one hectare of land (as per the opinion of the farmer, Mr. Subash Sharma). So in mono-cropping and by use of flood irrigation the consumption of water is huge. But in mixed cropping conditions the use of water reduces by almost 50%. The use of water in a multi-cropping scenario is 40 lakhs litres per annum.

Size of the dug well in the action research farm: The dug well is a square-shaped opening. The length and breadth are 7.7 feet. The depth of the well is 37 feet from the earth surface. The volume of the well is approximately 2,200 cube feet. Water extracted from the farm well is shown in Table 11.2.

The given data is recorded as observed in the farmland. The calculation has been made on the basis of the following method. It has been observed that the water level is reduced by 5–5.5 feet after extracting water for an hour. The length, breath and depth of the water reduced is 7.7 feet, 7.7 feet and 5–5.5 feet respectively. The volume of the space created after one hour's extraction of water is 315 cubic feet, which is roughly equal to 9,000 L (one cubic foot equals 28.3168 L) per hour.

11.8 The Impact of In-Situ Water Conservation for a Sustainable Agriculture System (SAS) along with other Factors of SAS

Soil Health: A strong microbial ecosystem signifies better soil health, which means the soil is more alive. Living soil absorbs rainwater very quickly because of the porousness and connecting space inside the soil. That reduces the overflow of water and increases the water-holding capacity. However, because of the current-day practice of chemical farming and the abundant use of inorganic pesticides, fertilizers and tillage through machines, the microbial ecosystem inside the soil is heavily damaged. To improve the soil health substantially, on-farm biomass and green manures are used. On-farm green manure is produced on the farm in a systematic way using a mix of Dhanicha, Barbati, Til, Urad and Jowar that grow very fast with the first rain, and these are cut and used to mulch the soil within about 6–8 weeks. The technique is designed to sequester carbon from the air and fix it into the soil. This is expected to increase the organic carbon in the soil. Arhar and Maize is cultivated at the same time to fix atmospheric nitrogen in the soil and make the soil conducive to life. This method was implemented in the action research project to improve soil health.

Diversity: The farm adopted diversity of cropping. There are as many as 200 cropping cycles. The farm needs to follow the practice of mixed cropping to provide diversity on the farm and enhance the use of the water available in nature. This leads to surplus water on the farm. Increasing diversity also reduces the pest levels on the farm. In the process, it also avoids the use of any external pesticides and fertilizers on the farm, which increases the life of the soil, and aids rainwater harvesting. Diversity was maintained on the project site not only by practising mixed cropping but also by planting perennial plants and keeping livestock and an apiary.

Ecology: Ideally, 10% of the farm should be used for creating a farm ecology consisting of various plants and trees to provide ecological balance to the life system on the farm. On the outer side of the trench, a natural fence with a variety of plant and tree species enriches the ecology. Fruit trees need to be planted such that the farm has some flowering tree at any time of the year – a necessary condition for the bees and flies that support pollination in the farm. Additionally, the mini-forest around the farm can reduce temperatures by around 3–4 degrees centigrade and also protects the farm from high-velocity winds and heatwaves. All this helps to manage climate at a micro-farm level, which indirectly aids the retention of soil moisture and reduces evaporation. The action research farm is already in the process of building up a rich environment.

11.9   Limitations

- Though we have experienced very good results, the experimentation of this action research project is only eighteen months old. So it needs some more time for validation.
- It is initially being tested only in one cropland in a rain-fed urban context; it needs to be tested in several places with different physical environments to assess its suitability for generalisation.
- It needs to be tested in different agro-climatic and topographical conditions for better assessment of its impact.

11.10   Summary

In-situ water conservation is an important tool to conserve rainwater on the farm and make moisture available to the micro-organisms and plants on the farm throughout the year. For in-situ water conservation, two things are important: *trenches* and *micro-locks*.

By using this method, nearly one lakh litre of water can be conserved underground for every centimetre of rainfall on one hectare of land. In this process, the above techniques will provide moisture to the soil for about 6 months. At the same time, because of groundwater recharge, nearby water bodies such as wells and ponds will recover their health. For the remaining 6 months of the year, an open dug well on the farm can provide sufficient water to the farm. However, micro-locks and trenches alone are not sufficient for the conservation of all the rainwater that falls on the ground. Other biosystems, such as better soil health, indigenous seeds, diversity and ecology of the cropland, are essential for optimum utilisation of in-situ water conservation and groundwater recharge.

The in-situ water conservation techniques of this action research have already been replicated by two farmers and furthermore by the Government in three different areas. This conservation model is greatly appreciated by the Government, farmers and policy-makers at different levels. Despite the fact that it is at an early stage, the effect is clearly noticeable. Replication of this model in three different districts is being undertaken to validate the model of water conservation and SAS in different topographical and agro-climatic conditions.

References

Anju, S.V.; Koppad, A.G., 2013: "Influence of in Situ Soil Moisture Conservation Measures on Growth and Productivity of Acacia Auriculiformis", in: *Karnataka Journal of Agricultural Science*, 26(1): 170–71.

Arora, Sanjay; Bhatt, Rajan, 2004: "Impact of Improved Soil and In-Situ Water Conservation Practices on Productivity in Rainfed Foothill Region of North-West India".

FAO, 2016: "Key to Achieving the 2030 Agenda for Sustainable Development", in: *Food and Agriculture*, 32.

Gebreegziabher, Tewodros; Nyssen, Jan; Govaerts, Bram; Getnet, Fekadu; Behailu, Mintesinot; Haile, Mitiku; Deckers, Jozef, 2009: "Contour Furrows for in Situ Soil and Water Conservation, Tigray, Northern Ethiopia", in: *Soil and Tillage Research*, 103(2): 257–64, https://doi.org/10.1016/j.still.2008.05.021 (10 August 2016).

Gupta G. N., 1995: "Rain-Water Management for Tree Planting in the Indian Desert", in: *Journal of Arid Environments*, 31(2): 219–35, https://doi.org/10.1006/jare.1995.0062 (24 September 2016).

Horrigan, Leo; Lawrence, Robert S.; Polly Walker, 2002: "How Sustainable Agriculture Can Address the Environmental and Human Health Harms of Industrial Agriculture", in: *Environmental Health Perspectives*, 110(5): 445–56, https://doi.org/10.1289/ehp.02110445 (15 October 2016).

Kant, A., 2016: "India's Great Drying Out", in: *The Times of India* (21 September): 10.

Kerr, John; Sanghi., N.K., 1992: "Indigenous Soil and Water Conservation in India's Semi-Arid Tropics", in: *Working paper by International Institute for Environment and Development*.

Li, Xiao-Yan; Gong, Jia-Dong; Wei, Xing-Hu, 2000: "In-Situ Rainwater Harvesting and Gravel Mulch Combination for Corn Production in the Dry Semi-Arid Region of China", in: Journal of Arid Environments (2000) 46: 371–382, (October), https://doi.org/10.1006/jare.2000.0705 (24 April 2018).

Li, Xiao Y.; Gong, Jia Dong; Zhao Gao, Qian; F Li, Eng R., 2001: "Incorporation of Ridge and Furrow Method of Rainfall Harvesting with Mulching for Crop Production under Semiarid Conditions", in: *Agricultural Water Management*, 50(3): 173–83, https://doi.org/10.1016/s0378-3774(01)00105-6 (12 June 2018).

Ministry of Water Resources, River Development and Ganga Rejuvenation, 2015: *Annual Report 2013–14* (India); at: http://mowr.gov.in/sites/default/files/AR_/CWC_2013-14_2.pdf (10 August 2017).

Muthamilsel van, M.; Manian, R.; Kathirvel, K., 2006: "In-Situ Moisture Conservation Techniques in Dryfarming – A Review", in: *Agricultural Reviews*, 27(1): 67–72.

Patil, S.L.; Sheelavantar, M.N., 2004: "Effect of Cultural Practices on Soil Properties, Moisture Conservation and Grain Yield of Winter Sorghum (Sorghum bicolar L. Moench)", in: *Agricultural Water Management*, 64 (February 2004), https://doi.org/10.1016/s0378-3774(03)00178-1 (3 February 2018).

Rockstrom, Johan; Karlberg, Louise; Wani, Suhas P.; Barron, Jennie; Hatibu, Nuhu; Oweis, Theib; Bruggeman, Adriana; Farahani, Jalali; Qiang, Zhu, 2010: "Managing Water in Rainfed Agriculture – The Need for a Paradigm Shift", in: *Agricultural Water Management*, 97, https://doi.org/10.1016/j.agwat.2009.09.009 (18 November 2016).

Rockström, Johan; Steffen, Will; Noone, Kevin; Persson, Åsa; Chapin, F. Stuart; Lambin, Eric; Lenton, Timothy M., et al. 2009: "Planetary Boundaries: Exploring the Safe Operating Space for Humanity", in: *Ecology and Society*, 14(2), https://doi.org/10.1038/461472a (6 December 2016).

Tian, Yuan; Su, Derong; Li, Fengmin; Li, Xiaoling, 2003: "Effect of Rainwater Harvesting with Ridge and Furrow on Yield of Potato in Semiarid Areas", in: *Field Crops Research*, 84(3): 385–91, https://doi.org/10.1016/s0378-4290(03)00118-7 (8 November 2016).

Vita, P. De; Paolo, E. Di; Fecondo, G.; Fonzo, N, Di; Pisante, M., 2007: "No-Tillage and Conventional Tillage Effects on Durum Wheat Yield, Grain Quality and Soil Moisture Content in Southern Italy", in: *Soil and Tillage Research*, 92: 69–78, https://doi.org/10.1016/j.still.2006.01.012 (4 January 2016).

Vohland, Katrin; Barry, Boubacar, 2009: "A Review of in Situ Rainwater Harvesting (RWH) Practices Modifying Landscape Functions in African Drylands", in: *Agriculture, Ecosystems and Environment*, 131(3–4): 119–27, https://doi.org/10.1016/j.agee.2009.01.010 (23 December 2016).

Wang, Yajun; Xie, Zhongkui; Malhi, Sukhdev S.; Vera, Cecil L.; Zhang, Yubao; Wang, Jinniu, 2009: "Effects of Rainfall Harvesting and Mulching Technologies on Water Use Efficiency and Crop Yield in the Semi-Arid Loess Plateau, China", in: *Agricultural Water Management*, 96(3): 374–82, https://doi.org/10.1016/j.agwat.2008.09.012 (17 October 2017).

Chapter 12
Intensification of *Bodi*-Based Farming for Sustainable Livelihood Assurance

Chandrakanta B. Prasan and Joshua N. Daniel

Abstract It is an agricultural tradition in Gadchiroli district of Maharashtra to make large ponds within paddy fields. These water bodies, locally known as *bodis*, serve the purpose of storing rainwater. Besides irrigating the paddy crop, the water in the *bodi* may be used for fish culture. A *bodi*-based farming system having greater diversification with ducks, poultry, fishery, fruit trees, vegetable cultivation and vermicompost-making was designed and analysed for production, income and employment. The study was carried out in two farms having *bodis* of 0.40 and 0.28 ha, respectively, in BAIF-MITTRA's Etapalli project villages of Tumargunda and Gurupalli in Gadchiroli district of Maharashtra during 2012–13. Preliminary results indicated that the integrated *bodi*-based farming system can generate an income of more than Rs. 60,000. The main income, as expected, was in the rainy season, but there were occasional returns during the rest of the year as well.

Keywords *Bodi*-based farming · Livelihood · Resource use · Income generation Integrated approach

12.1 Introduction

Gadchiroli district, situated in the south-eastern corner of Maharashtra, is populated mainly by tribal communities and is relatively less developed than most other districts of the state. About 92 per cent has forest cover and the topography is comprised of flat land interspaced with scattered hills. People of the district are predominantly engaged in agriculture, and the main crop is rice. A traditional practice of the area is to make ponds, locally referred to as *bodis*, within the paddy

Chandrakanta B. Prasan, Senior Project Officer, BAIF Development Research Foundation, Pune, Maharashtra, India; Email: ckbaradaprasan@gmail.com

Joshua N. Daniel, Programme Advisor, BAIF Development Research Foundation, Pune, Maharashtra, India; Email: jndaniel@baif.org.in

© Springer Nature Switzerland AG 2019   135
A. K. J. R. Nayak (ed.), *Transition Strategies for Sustainable Community Systems*,
The Anthropocene: Politik—Economics—Society—Science 26,
https://doi.org/10.1007/978-3-030-00356-2_12

fields. While supplementing water for the rice crop, *bodi*s are also used for fish culture. They differ in size and period of water retention. Typically, a *bodi* fills up during the rainy season and water level gradually declines during September to November. Depending on the depth of the structure, some water may remain during the winter months of December to February, and most *bodi*s dry up during the summer. The average availability of water is generally for a period of about five to six months.

There is tremendous scope for increasing the use of *bodi* for multiple income generation activities. Systems with several farm enterprises operate in areas where water is not a limiting factor, but such farming approaches are uncommon in drier areas because of the uncertainty associated with water availability. Therefore, a *bodi*-based farming system was designed and its functioning and income generation potential were assessed in two farms in Gadchiroli district of Maharashtra.

12.2   Traditional Resource Use in a *Bodi*-Based Farming System

The central resource in this farm production system is the *bodi*, a large pond of irregular shape which collects and stores rainwater within the paddy field. Like any other water-holding structure, it has an inlet to regulate inflow and an outlet to discharge excess water. A bund constructed along the ridge of the *bodi* helps prevent erosion and store water. The size and depth of the *bodi* depends on farm size, soil type, the farmer's water requirement and the cost of excavation and possible uses of excavated earth. Irrigation is by carrying the water in the *bodi* manually to the fields or by pumping water into furrows. The selection of a site for a *bodi* is critical in maximising its storage capacity. Farmers cultivate paddy in the lowland, while the upland is devoted to vegetable and grain crops. The *bodi* is located in a corner or one side of the paddy field so that it does not inconvenience farm operations while making it easier for the distribution of water to the rest of the crop land. The rainy season crop from June to October, usually lowland rice, is irrigated with water in the *bodi* if the rains cease before crop maturity. If there is surplus water after the rice crop is harvested, the farmer will be able to grow a post-rainy season crop and sometimes even a summer crop. Some enterprising farmers use the *bodi* for raising fish in addition to irrigating the fields. It is also a source of drinking water for farm animals. Although the *bodi* is an attempt to enhance the use of rainwater, it is evident that the majority of farmers do not exploit the full potential of the harvested water. The appearance of a *bodi* during the different seasons of the year is shown in Fig. 12.1a–d.

(a)   **(b)**

(c)   **(d)**

Fig. 12.1 Appearance of a *bodi* during different seasons of the year: **a** summer; **b** rainy season; **c** post-rainy season; **d** winter. *Source* Photo by author, Chandrakanta Barada Prasan

12.3 Potential for Intensification of Resource Use in a *Bodi*-Based System

Intensive production in the *bodi*-based farming system encompasses the principle of creating inter-relationships among all components of the farm enterprises in such a way that maximum efficiency in resource use is achieved. This is possible when waste of one component becomes an input for another. For example, poultry droppings can become fish feed and fish droppings can be collected at the time of desilting the *bodi* and used in vermicompost-making or directly as manure. Thus, the waste from the water body becomes an input for crop production and crop wastes in turn are used as feed for livestock and poultry. If the nutritional requirement of any component is not fully met by the linkages it has, supplementation in the form of feed or fertiliser is necessary. Hence the system is designed to have linkages that would use resources efficiently while minimising the requirement for external inputs.

The system designed for *bodi* intensification envisaged several components of farm production. It had ducks and poultry whose droppings have value as feed and manure for fish and vegetables, respectively. The embankment of the *bodi* can be used as support for creeper types of vegetables and tuber crops, and the space above it for vegetable cultivation. The outside embankment of *bodi* can be utilised for fruit trees. Simultaneously, some aquatic cash crops can be grown in *bodis* which may also have value as fish feed as well as other economic benefits. Proper stocking rate with area and depth of water bodies is necessary to optimise fish production. The overall aim is the integration of the inter-dependencies among the components of a *bodi*-based farming system.

The *bodi*-based farming system is an opportunity for farmers to utilise their resources judiciously and effectively. With the objective to assess the structure and function of the system, a study was carried out to understand the production and income of different components with benefits and costs. Additionally, it examined the utilisation of labour, mainly family labour, to undertake the various activities required by the system.

12.4   Implementation of the Model

12.4.1   Poultry and Ducks

Bodi interventions were started in the month of March with the construction of sheds for poultry and ducks. The poultry shed was erected on the deeper side of the pond with two poles inside the pond and the other two on the bund of the *bodi* (Fig. 12.2). The shed for ducks was erected in the middle of the dyke in the *bodi*.

Fig. 12.2 Sheds constructed for ducks and poultry in *bodi*-based farming. *Source* Photo by the author, Chandrakanta Barada Prasan

The sheds for ducks and poultry were separated by a distance of 1.0 m. The floor of the sheds was made with bamboo planks that were spaced 1.5 cm apart to allow droppings and feed wastage to fall through. The distance between the shed and the highest water surface of the *bodi* was 60 cm to prevent water reaching the floor of the shed during the rainy season. Adult birds, ten ducks and forty poultry birds were provided for each *bodi*-holder. After one and a half months, ducks were allowed into the *bodi*.

Ducks received most of their nutrition from the *bodi*, but supplementary feed was provided by preparing a meal consisting of rice bran (3.0 kg), wheat (35.5 kg), soybean (9.0 kg), fish meal (2.5 kg) and mineral mixture (0.25 kg). Poultry was fed with commercially available finisher feed (mash). Each bird was fed 150 gm twice daily. Clean and disinfected water was also provided in the shed. Scheduled vaccination was done when the chicks were four weeks old. Proper hygienic measures were taken regularly.

12.4.2 Vegetable and Paddy Cultivation

The space above the *bodi* embankment was used for vegetable cultivation by making small plots on and around the bund. This vegetable production was an additional innovation over traditional *bodi* management. Brinjal, tomato and yam were taken in plots, and cowpea, cluster bean, country bean, bitter gourd, yam and climbing vegetables were planted in rows. Trellises were prepared for creepers by covering the edge of the *bodi* bank with bamboo and wood. The downstream side of the *bodi* was used for paddy cultivation.

12.4.3 Fishery

Once the *bodi* filled up with water with the onset of the monsoon season in June, it was prepared for stocking fingerlings. After removing the sediments and other unwanted material, lime was applied to the wall of the *bodi* at the rate of 110 kg (60 kg in 1.0 acre and 50 kg in 0.75 acre). After twenty-one days of liming, fish species such as *Rohu*, *Catla*, Grass Carp, Common Carp and Silver Carp were released at a stocking density of 5.0 kg in 1.0 acre. The fish were mostly dependent on the droppings and feed wastage falling from the duck and poultry sheds above for their feed. Additionally, as nutrients for fish, fresh cow dung with paddy husk was provided at the rate of 50 kg in one acre up to three months. At monthly intervals, netting was done.

12.4.4   Fruit Tree Plantation

The space adjacent to the outer side dyke of the *bodi* was used for planting five saplings each of custard apple, lime, jackfruit and drumstick at a spacing of 3.5 m.

12.4.5   Vermicompost

Vermicompost-making was another activity introduced, as it can recycle waste biomass and serve in crop and fish nutrition. A mixed layer of chopped leaf litter, grasses and cow dung in the proportion of 3:1 was used as raw material for vermicomposting. The substrate material was covered with gunny bags and left for pre-treatment for a week. Thereafter, earthworms were introduced and the surface was covered with dry soil. After sixty days, the vermicompost was applied to vegetable plots.

12.4.6   Economic Benefits of the Bodi-Based System

The diversity in farm activities ensures the distribution of income so that a reasonable amount of income is earned during most of the months except the summer period. The months of October and November, coinciding with the festival season, are the most favourable in this regard. Therefore, the model of the *bodi*-based farming system is likely to be readily accepted by farmers in this region. The economics of income (Table 12.1) shows the possibility of doubling income over a period of two years.

The economic analysis of the *bodi*-based farming system revealed that the gross income of Rs. 44,645 was obtained by investing Rs. 31,631 (Table 12.1) and, additionally, the farmer obtained paddy production with the same resources. This system provided employment of 240 man-days with an attractive benefit-cost ratio. The net income from the *bodi*-based farming system was Rs. 6368. It was also evident that the risk of economic loss in this system is very low.

This integrated approach was aimed at maximising resource use as well as reducing input costs on feed while increasing fish production compared to the traditional method adopted in *bodi*s. It has the potential to provide food for rural poor families throughout the year. The utilisation of family labour during the study period of eight months also showed its employment potential. The overall results revealed that integrated *bodi*-based farming with ducks, poultry, fish, fruit and vegetable cultivation was an excellent approach for sustainable production, resource use maximisation, income generation and employment for resource-poor rural households. The concept of the *bodi*-based farming system through agro-enterprise convergence revolves round the interactive use and efficient

Table 12.1 Economics of *bodi*-based farming system in the first year

No.	Components	Investment (Rs.)	Gross Return (Rs.)	Net Return (Rs.)	Remarks
1	Fishery	4,490	9,720	5,230	
2	Duck farming	7,892	8,700	808	Includes one-time investment of Rs. 4,176 for construction of sheds for poultry and ducks
3	Poultry farming	13,282	8,800	*	Includes one-time investment of Rs. 6,254 for poultry equipment and transportation
4	Vegetable cultivation	345	675	330	
5	Horticulture	2,020	*	*	
6	Vermicompost	3,602	3,150 (in 1st Year)	*	Rs. 1,477 vermibed is an investment for three years
7	Paddy	*	13,600	*	Paddy production cost within the *bodis*
	Grand total	31,631	44,645	6,368	

Source The authors

utilisation of land, labour and other available resources. Based on their experience over the previous three years, farmers made certain modifications to the original model. Thus it is necessary for a multi-component system of this nature to have the flexibility to incorporate changes as it evolves.

12.4.7 Ensuring Sustainability

A challenge often faced while introducing complex food production systems is ensuring their sustainability. In the *bodi*-based farming system, there are certain critical aspects that have to be carefully managed to make sure the annual cycle of activities continues undisturbed. As the *bodis* dry up during the summer, fish has to be harvested as the water level declines. This makes it necessary to introduce fingerlings every year, requiring a dependable and affordable source nearby. Farmers may be able to raise poultry chicks and ducklings at home without having to depend on external sources.

In subsequent years, it was observed that farmers had made certain modifications to the original model. One of them was to shelter poultry and ducks in the household itself during the dry months, when their visit to the farm is not frequent. As ducks require a wet place, farmers made a ditch in the homestead for the summer season. Thus, depending on the local conditions and experience, farmers can play a key role in helping to evolve the model, which can become efficient in resource use as well as high in productivity. Unlike normal cultivated land, *bodis* and the associated structures have to be regularly maintained and repaired to prevent breakages. Desiltation of *bodis* will help deepen the structure to hold more water, while the sediments removed can be incorporated into the crop land to enhance fertility. The fruit trees will start bearing only after about four years and they will eventually become a regular source of income.

12.5   Concluding Remarks

The concept of the *bodi*-based farming system through agro-enterprise convergence revolves around the interactive use and efficient utilisation of land, labour and other available resources. Intensification of farming with poultry, ducks, fish, vegetables and other crops can play a significant role in increasing manifold production, income, and the nutrition and employment opportunities of rural populations. Four different components – fish, ducks, poultry and vegetable production on bunds – were employed in the current investigation. The results indicated that the intensification of *bodi*-based farming practice in a sustainable manner with poultry and duck rearing year-round resulted in encouraging production over traditional management. The average fish production by the intensification strategy was also higher than the traditional management. It reveals that *bodi*-based farming with poultry, ducks, fish and vegetables is sustainable and will lead to improved diversification of food production and income generation on resource-poor farm households. Intensification with these components is not only technically feasible, but is also economically viable.

References

Alam, M.R.; Islam, F.; Molla, M.S.H.; Hossain, M.A.; Hoque, M., 2001: "Pond based integrated farming systems with fish, poultry and vegetables", in: *Annual Report 2001–2002* (Pabna: Bangladesh Agril. Res. Institute, On-Farm Res. Divn.): 126–128. Csavas, I., 1991: "*Regional review on livestock-fish production systems in Asia*", Paper presented at International Workshop on Integrated Livestock-Fish Production Systems, 16–20 December 1991, University of Malaysia, Kuala Lumpur (Bangkok: FAO/RAPA).

Delmendo, M.N., 1980: "A review of integrated livestock-fowl-fish farming systems", in: Pullin, R.S.V.; Shehadeh, Z.H. (Eds.): *Integrated Agriculture Aquaculture Farming Systems*, ICLARM Conf. Proc. 4 (Manila, Philippines): 59–71.

Devendra, C., 1991: "Integrated animal-fish-mixed cropping systems", FAO/IPT Workshop on Integrated Livestock-Fish Production Systems, 16–20 December 1991, Institute of Advanced Studies, University of Malaya, Kuala Lumpur, Malaysia (Singapore: International Development Research Centre, Division of Agriculture, Food and Nutrition Sciences); at: http://www.fao.org/docrep/004/ac155E/AC155E10.htm (15 June 2014).

Edwards, P., 1990: "An alternative excreta re-use strategy for aquaculture: The production of high protein animal feed", in: Edwards, P.; Pullin, R.S.V. (Eds.): *Proceedings of the International Seminar on Waste Water Reclamation and Re Use for Aquaculture* (Calcutta, India, December 1988): 209–221.

Edwards, P.; Demaine, H., 1997: *Rural Aquaculture: Overview and Framework for Country Reviews* (Bangkok: RAP/FAO).

Edwards, P.; Demaine, H.; Innes-Taylor, N.; Turonguang, D., 1996: "Sustainable aquaculture for small-scale farmers: Need for a balanced model", in: *Outlook on Agriculture*, 25(1): 19–26.

FAO, 1990: *Women in Agricultural Development: Gender Issues in Rural Food Security in Developing Countries* (Rome: FAO).

FAO; ICLARM; IIRR, 2001: *Integrated agriculture-aquaculture: a primer*, FAO Fisheries Technical Paper No. 407 (Rome: FAO).

Hasan, M.R.; Hecht, T.; De Silva, S.S.; Tacon, A.G.J. (Eds), 2007: *Study and analysis of feeds and fertilizers for sustainable aquaculture development*. FAO Fisheries Technical Paper. No. 497 (Rome: FAO).

Iqbal, S., 1999: *Duckweed Aquaculture: Potentials, Possibilities and Limitations for combined Waste Water Treatment and Animal Feed Production in Developing Countries*. SUNDEC Report no. 6/99, Switzerland.

Kumar, S.; Jain, D.K., 2005: "Are linkages between crops and livestock important for the sustainability of the farming system?", in: *Asian Economic Review*, 47(1): 90–101.

Lightfoot, C., 1990: "Integration of aquaculture and agriculture: A route to sustainable farming systems", in: *The ICLARM Quarterly*, 13(1): 9–12.

Little, D.C.; Innes-Taylor, N.L.; Turongruang, D.; Komolmarl, S., 1991: "Large seed for small-scale aquaculture", in: *Aquabyte*, 4(2): 2–3.

Man, L.H., 1991: "Duck-Fish integration in Vietnam" (Hanoi, Vietnam: Vietnam Union of Poultry Enterprises); at: http://www.fao.org/DOCREP/004/AC155E/AC155E15.htm (16 December 1991).

Men, B.X.; Su, V.V., 1990: "'A' molasses in diets for growing ducks", in: *Livestock Research for Rural Development*, 2(3), article #30 (30 December 1990); at: http://www.lrrd.org/lrrd2/3/vietnam1.htm (8 March 2006).

Mohamad Hanif, M.J.; Tajuddin, Z.A.; Doyah, O.S.; Mohamad, M., 1990: "Maximising farm production output through fish, prawn, chicken, duck and crop farming", *Proceedings of the 13th Malaysian Society of Animal Protection Annual Conference* (Malacca, Malaysia, Kuala Lumpur).

Mukherjee, T.K.; Geeta S.; Rohani A.; Phang S.M., 1991: *Study on Integrated Duck-Fish and Goat-Fish Production Systems* (Kuala Lumpur: University of Malaya, Institute of Advanced Studies).

Nuruzzaman, A.K.M., 1991: *Integrated Fish Farming System Holds Promise in Bangladesh* (Dhaka, Bangladesh: 5/H Eastern Housing Apt.).

Pingel, H.; Tieu H.V., 2005: *Duck Production* (Vietnam: The Agricultural Publishing House).

Sahoo, B.; Lenka, A.; Nedunchezhiyan, M., 2012: "Sustainable livelihood support through enterprise convergence in pond based farming system", in: *Odisha Review* (April 2012).

Samantray, B.R.; Ku, B.D.; Kurukshetra, C.P.K.: "Community based fish farming a new facet of employment for rural women", in: *India*: 41 (February 2013).

Skillicorn, P.; Spira, W.; Journey, W., 1993: *Duckweed Aquaculture: A New Aquatic Farming System for Developing Countries* (Washington DC: The World Bank); at: http://www.p2pays.org/ref/09/08875.htm (14 October 2016).

Tajuddin, Z.A., 1980: "Integrated peking duck-fish farming", in: *Annual Report* (Kuala Lumpur, Malaysia: MARDI, Freshwater Fisheries Branch).

Vincke, M.J., 1991: *Integrated Farming of Fish and Livestock: Present Status and Future Development* (Rome: FAO).

Vincke, M., 1988: "Developing productive systems under village conditions", in: King, H.R.; Ibrahim, K.H. (Eds.): "Village Level Aquaculture Development in Africa", *Proceedings of the Commonwealth Consultative Workshop on Village Level Aquaculture Development in Africa*, Freetown, Sierra Leone, 14–20 February 1985 (London: Commonwealth Secretariat): 39–57.

Chapter 13
Farm-Diversity Management for Sustainable Production Systems

Arun K. Sharma

Abstract Diversity at farm level spreads the risk of uncertainties of climate and market, aids efficient use of available resources and additionally increases farm output by enhancing soil health, pollination and minimising pests and diseases. Thus it has multiple benefits but needs intelligent integration of all the components of diversity at all the input and output levels. Intensive monoculture with chemicals has been the major cause of the failure of industrial agriculture, and enhancing farm diversity may be the best solution for the restoration of such a production system. A major part of world food comes from smallholder farming, and diversity assures sustainable livelihoods for small-scale growers. Designing diversity for better productivity with sustainability is discussed in this chapter.

Keywords Farm diversity · Designing farm diversity · Resource diversity Benefits of farm diversity

Component diversity in a biological production system is the major contributor to system sustainability. This diversity spreads the risk of abiotic and biotic stresses and, in addition to that, components of biodiversity in agricultural landscapes maintain ecosystem services, such as pollination, biological pest control, soil and water conservation and nutrient cycling. This diversity is present in all natural production systems and is fundamental to the life and livelihood of any community. Diversity fulfils the varied needs of humans and other fauna dependent on this system. The analogue of this natural system was prevalent in all agricultural systems until the advent of chemical-based intensive (conventional) agriculture in the early nineteenth century. In intensive agriculture, monoculture models of crop production and animal husbandry have become popular. However, its unsustainability is becoming more and more apparent in recent decades due to various ill effects on soil and human health, loss of biodiversity and decreasing productivity with increasing cost and risk.

Arun K. Sharma, Senior Scientist, CAZRI, Jodhpur, Rajasthan, India; Email: arun. k_sharma@yahoo.co.in

© Springer Nature Switzerland AG 2019   145
A. K. J. R. Nayak (ed.), *Transition Strategies for Sustainable Community Systems*,
The Anthropocene: Politik—Economics—Society—Science 26,
https://doi.org/10.1007/978-3-030-00356-2_13

In recent decades there has been growing awareness and recognition in different sectors of society that the conservation and sustainable use of biodiversity is the key to human well-being. This inclusion of all biological components in a system for efficient resource recycling, reduced dependency on external inputs and reducing climatic uncertainties means various eco-friendly farming systems – e.g. organic farming, ecological farming, bio-farming – are being promoted not only to restore soil and human health, but also to sustain the system under variations of social-climatic-economic conditions.

13.1 Eco-farming Verus Conventional (Chemical) Farming

Ecofarming (EF) is commonly understood as the minimum use of external inputs. However, in actuality it is the ideological differences with conventional farming that make EF friendly, both to society and the environment. These differences are given below (Sharma 2001) (Table 13.1).

In this table it is clearly stated that diversity is one of the major components of EF systems. Since organic farming is based on the use of internal resources and is considered to be the most sustainable system, the phrase 'organic farming' is often used in texts as a substitute for EF.

13.1.1 Farm Diversity and Ecofarming

Diversity is nature's gift to provide several products and services from one system. Most of the practices or technologies in organic farming are based on harnessing nature's potential from all sources by maintaining cyclic use of resources. Diversification makes this cyclic utilisation possible – e.g. a crop is eaten by humans and animals, and their excreta is used for manuring the crop. This complementary relationship between components results in sustainable production.

Table 13.1 Ideological differences between ecofarming and conventional farming

Eco farming	Conventional (chemical) farming
Holistic approach	Reductionist approach
Decentralise production	Centralise production
Harmony with *Nature* (Harness the benefit)	Domination over *Nature* (Exploit for *profit*)
Diversity	Specialisation
Input optimisation (save more)	Output maximisation (spend more)

Source Sharma (2001)

Farm diversity: This is the variability among living organisms associated with the cultivation of crops and rearing of animals, and the ecological complexes of which those species are part; this includes diversity within and between species, and of ecosystems (Ploeg et al. 2009).

13.2 Benefits of Farm Diversity

- *Development of favourable soil quality*: Soil organisms ensure the maintenance and development of good soil structure, with positive effects on water regulation, erosion, leaching, etc. In addition, a healthy soil ecosystem ensures the release of nutrients and better disease resistance, both of which are essential for good crop development. Farm diversity is also a good way to recycle nutrients.
- *Crop pollination*: Crops are mainly pollinated by honeybees, which are introduced for this purpose by beekeepers or in collaboration with them. Wild agro biodiversity also plays an important supplementary role. A significant portion of fruit production is dependent on pollination by insects.
- *Biological pest control*: Natural enemies of plague insects can help to keep them below the damage threshold, so that the use of crop protection products is limited or not necessary. Populations of disease-causing micro-organisms can also be minimised with crop rotation.
- *Conservation and other services*: Natural vegetation can help in erosion control and relieve heat stress in livestock. It also provides a habitat for other useful organisms, such a natural enemies of plague species, and it determines the appearance of the landscape.
- *Enhancing ability to adapt*: It is vitally important to maintain genetic diversity, in order for agricultural crops and livestock to be able to adapt – naturally or with human intervention – to future needs and challenges of the climate and market.

13.3 Management of Farm Diversity

There are broadly two types of technology/methods for maintaining diversity. They are:

A. Technologies for enrichment of diversity – biofertilizer, agro-forestry, mixed farming.
B. Technologies for conserving diversity – seeds of traditional varieties or landraces, mulching, conservation villages etc.

From a management point of view, diversification can be divided into two main groups with their sub groups:

- At Input level:
  - Diversified nutrient management.
  - Diversified resource conservation.
  - Diversified protection system.

- At Production Level: Diversified production system.
  - Crop diversification.
  - Agro-forestry.
  - Mixed farming.

No single model of diversity in organic farming can be prescribed for all agro-ecological conditions; models even differ from farmer to farmer as their resources and priorities are different (Lin 2011). Therefore in this article various components of a diversified system are described with their method of application and how to make them complimentary to the system.

13.4 Diversification at the Input Level

- *Diversified nutrient management*: These are components which suppliers input mainly for their nutrients. Some of them are discussed here and others in the related chapters. In the management of organic farming soil comes first in relation to nutrition management. There are several resources which are used for nutrition management in soil. These components are:
  - *Micro Organisms*: These include mainly nitrogen-fixing and phosphorous solubilising micro-organisms and micro-organisms which increase the process of composting animal and crop waste. These micro-organisms actually play a major role in nutrient availability from natural sources, e.g. air, soil reserves. Besides nutrients, these micro-organisms release several bio-chemicals which act as plant growth promoters and help to boost plant resilience to pests and drought.
  - *Compost*: This is a well decomposed organic material made from diversified agro waste. Besides being a nutrient provider, this material improves soil structure, water and nutrient retention capacity, and serves as a food source to all the micro organisms working in soil. Using compost and vermicompost (by earthworms) is the best way to maintain a cyclic system of nutrient supply.

- *Diversified resource conservation*: Resource conservation is to be done at multiple levels e.g. conservation of rainwater (Rockström et al. 2010), conservation of soil and conservation of genetic diversity. This is also possible by:

In-situ rainwater conservation by mulching: Mulch is a decomposable material (grass, hay, leaves, twigs, plant residue, and uprooted weeds) for covering the soil surface so that soil is not exposed to direct sun or the beating action of raindrops. The process is called mulching.

Benefits of mulching:

- Reduces moisture loss from soil and checks soil erosion caused by rainfall.
- Improves the micro-climate in soils (temperature, moisture), which helps micro-organisms in the soil work more efficiently, resulting in the continuous availability of nutrients to the crops being grown.
- Reduces weed growth.
- Provides nutrient-rich humus on decomposition.

The above ways of mulching are highly beneficial vis-à-vis cost-effective methods of conserving resources in Ecofriendly or organic farming.

Method of application: Mulch can be applied using two methods.

- Before sowing crops.
- After sowing crops.

Mulching before sowing of crops: This is a major component in conservation tillage. This is being practised successfully in the humid tropics of Latin America and Africa due to abundantly available organic residue in humid tropics. These residues are spread on crop fields to make a permanent cover, then crops are sown in lines with special equipment which opens the furrow in this mulch and sows seeds.

Mulching after sowing: In semi-arid or arid areas with a shortage of organic residue, the crop is sown in lines, and after 20–25 days a layer of organic residue 3–5 cm thick is spread in the inter-row space of crops. Another way is to manually weed crops after 15–20 days of sowing, allow these uprooted weeds to dry for 3–5 days, and then spread them in the inter-row space. In this method, weeding and mulching follow a relay and thus minimise the requirement for organic residue from outside the field. A similar practice can be followed during the second and third weeding. Inter-culture has the additional advantage of soil moisture conservation and better aerations.

In this method two precautions should be taken.

- Partial drying of weeds for 3–5 days to avoid re-establishment.
- Either do not use weeds if seeds have developed or remove seed portions.

Cover crop or live mulch: Many leguminous annuals grow and spread near (20–40 cm) the soil surface. Because of this characteristic, these crops are grown in the inter-row spaces between tall crops like maize, sorghum and pearl millet. This intercropping of spreading legumes has additional advantages which make the residue a particularly valuable mulch:

- Legumes shed nitrogen-rich leaves which are easily decomposable and make natural compost.
- Legumes fix nitrogen in the soil.
- This type of intercropping gives additional yield and minimises the risk of drought as legumes exhibit drought-resistant properties.

This cover crop or mulching is applicable in both humid and arid regions. The only change is the legume species. Cover crops for high rainfall areas include mucuna, lablub and cowpea, while cover crops for low rainfall areas include cluster beans, green gram etc. Most of the cover crops have a shorter growing period than the main crop and therefore require separate sowing, generally 10–15 days after sowing the main crop.

Soil conservation: Soil conservation can be done by using agrodiversity such as growing grasses on high slopes, bushes on medium slopes and trees on mild slopes to reduce the soil loss from water or winds. Also, bund planting of trees/bushes of economic use reduces the soil erosion from crop fields and additionally provides some income to the farmer.

Genetic diversification: A greater variety of crops with a wider genetic base need to be used in ecofarming. Hybrid varieties have a very narrow genetic base, which makes them suitable for specific high input environments, whereas varieties developed from local landraces with a wider genetic base which is better adapted to that eco-region minimise the cost of nutrients, irrigation and pest management. Developing variety at local level reduces the cost of seed, which farmers otherwise have to buy every year. The development and use of locally improved varieties in ecofarming is just one step in a series of practices to develop, produce and store seeds which possess diversity in terms of one or multiple characteristics, e.g. colour, taste, aroma, yield, processing quality, tolerance to drought, resistance to pests, dominance over weeds, responsive to organic manure and any other characteristics desirable due to market or social demand.

Diversified protection system: Protection in a farming system can be achieved through integrated use of various plant extracts, micro-organisms, birds, friendly insects, repellents or trap crops etc. The use of diversified protection methods not only increases the efficiency of pest control but also reduces the risk of reccurrence.

13.5 Diversification at the Production Level

From a management point of view, diversification in production systems can broadly be divided into three groups. However, to derive maximum benefit from diversity these three need to be integrated at each organic unit.

- Crop diversification
- Agroforestry
- Mixed farming.

Crop diversification: Instead of monoculture of a nutrient-extrusive crop like maize, sorghum, wheat, rice, etc., some other crops need to be grown in rotation or as intercropping or as green manuring with main crops. This diversity is mainly beneficial to make efficient use of soil fertility and water, to break the life cycle of pests and to fulfil the diversified needs of small farmers. Some of the basic guidelines for crop rotation are:

- Legume crops should be followed by cereal crops.
- Crops of the same family – e.g. cotton and okra, potato and chilli or tomato – should not be planted in rotation.
- Deep-rooted crops should be planted after shallow-rooted crops.

A rotation can be of three to five years, followed by a fallow period for soil restoration, which may be two to four months in higher rainfall areas and four to eight months in low rainfall areas.

Agroforestry: The inclusion of trees in a crop field for conservation and sustainable production in a system is called agroforestry.

- Trees provide permanent cover for soil with their canopies and shed leaves which cover soil, while their strong root networks conserve soil and water, improving the micro-climate and in this way acting as permanent mulching.
- Several nitrogen-fixing trees transfer nitrogen from the atmosphere to the soil, the micorhiza associated with tree roots solubilises soil phosphorus, and tree roots promote microbial activities in soil.
- Trees provide shelter to several insectivorous birds, and these birds also drop nutrient-rich excreta on soil. Birds maintain diversity by the dispersal of seeds, and some birds are good pollinators.
- *Mixed farming*: When animals (cattle, goats, sheep, pigs, poultry, ducks, fish, honeybees etc) are integrated with crop production, and animals and crops have a complementary relationship, the system is called mixed farming. This is the advanced stage of organic farming to make a system self-reliant in both intensive and extensive farming (Roy and Sharma 2014).

Component arrangement and management:

- Mixed farming mainly depends on the agro climate of the location, the resources available to farmers and the individual farmer's managerial capacity. For example, in rice-growing areas rice-duck-fish farming is recommended from an agro climatic point of view, but adoption of this system by an individual farmer depends on his capacity.

- More skill is required in intensive irrigated farming, while in extensive rain-fed farming most of the farmers already have a mixed farming system. The need is to just increase their knowledge level by training and visits to model farms.
- Some animals, especially poultry (including ducks and turkeys), combine protection with production while others do only protection work – e.g. owls control rats, bats deal with nocturnal pests, sparrows protect against larvae etc. These creatures need either shelter or plant diversity.

13.6   Conclusion

Modified landscape management and alternative farming practices can contribute to biodiversity conservation in various ways. However, biodiversity in and of itself does not automatically translate into ecosystem services such as enhanced pollination or natural pest control. To optimise these benefits, we need to understand which biodiversity elements drive these ecosystem services. Based on this information, benefits to farm productivity can be generated through a rational design and management of agro-ecosystems and landscape structures. Such management strategies can range from an informed choice of non-crop vegetation – such as field margins, forests and hedgerows – and other non-crop elements – e.g. animals, fish and poultry – to conservation tillage, crop diversification and crop rotation.

References

Brenda, B., 2011: "Resilience in agriculture through crop diversification: Adaptive management for environmental change", in: *BioScience*, 61(3): 183–193.

Ploeg, J.D.; Laurent, C.; Blondeau, F.; Bonnafous, P., 2009: "Farm diversity, classification schemes and multifunctionality", in: *Journal of Environmental Management*, 90(2): 124–131.

Rockström, J.; Karlberg, L.; Wani, S.P., 2010: "Managing water in rainfed agriculture—The need for a paradigm shift", in: *Agricultural Water Management*, 97(4): 543–550.

Roy, M.M.; Sharma, A.K., 2014: "Farming systems research for economic and environmental security in hot arid regions of India", in: PDFSR, Modipuram (Eds.): *Research in Farming Systems* (New Delhi: Today and Tomorrow's Printers and Publishers): 201–228.

Sharma, A.K., 2001: *Handbook of Organic Farming* (Jodhpur: Agrobios India).

Chapter 14
Understanding Livelihood Diversification: A Case Study of Mushroom Farming in Bihar

Surya Bhushan, Piyush Kumar Singh, Sridhar Telidevara and Santosh Kumar

Abstract Mushroom cultivation has been identified as an important livelihood strategy that fits in very well with sustainable farming. This has several advantages, for example, it uses agricultural waste products, it works like a quick cash crop with no use of land, a high production per surface area can be obtained, and after picking, the spent substrate is still a good soil conditioner. The study also delineates the production economics of mushroom cultivation among the poorest of the poor.

A study was also conducted in three villages of Nalanda district in the state of Bihar to identify the determinants and constraints to livelihood diversification. The study has shown that educational level, asset position, access to credit, and rural infrastructure are some important driving forces towards livelihood diversification in the region. The resource-poor are particularly vulnerable and unable to diversify because of the entry barriers imposed by their weak asset base.

Keywords Sustainable livelihood farming · Livelihood diversification Mushroom cultivation

14.1 Introduction

Strategically, mushroom cultivation has emerged as one of the viable options for achieving sustainable agricultural systems in recent times, especially among landless or smallholder farming communities. The dual effects of income generation due

Surya Bhushan, Associate Professor, Development Management Institute (DMI), Patna, Bihar; Email: surya.bhushan@gmail.com

Piyush Kumar Singh, Assistant Professor, Indian Institute of Technology (IIT), Kharagpur, West Bengal; Email: piyushsingh.er@gmail.com

Sridhar Telidevara, Associate Professor, Associate Professor, Great Lakes Institute of Management, Gurgaon, Haryana; Email: sridhar.telidevara@gmail.com

Santosh Kumar, District Project Manager, Bihar Rural Livelihoods Programme Society (BRLPS), Nalanda, Bihar; Email: dpm_nalanda@brlp.in

© Springer Nature Switzerland AG 2019
A. K. J. R. Nayak (ed.), *Transition Strategies for Sustainable Community Systems*,
The Anthropocene: Politik—Economics—Society—Science 26,
https://doi.org/10.1007/978-3-030-00356-2_14

to cash crops and nutrition provision due to rich nutrient content mean it can help to reduce vulnerability to poverty and strengthen livelihoods security. It does not require access to land or any significant capital investment, as substrate can be prepared from any clean agricultural waste material; mushrooms can be produced in temporary clean shelters; they can be cultivated on a part-time basis, and require little maintenance (Marshall/Nair 2009). Bihar, one of the poorest and most populous states in India, is the twelfth largest in terms of geographical coverage (2.8% of the total land area) and third largest by population (8% of the total population of the country). With the lowest per capita income of any Indian state, the highest population density, and low nutritional security, it is struggling to get out of the low income equilibrium trap. The state is largely rural, with almost 90% of its inhabitants living in rural areas, and poor, with more than 30% of its total population living precariously. Furthermore, more than 90% of the farming population is marginal, with a holding of less than one hectare. As a consequence, when the horizontal expansion of agricultural production is difficult to attain, one of essential means of raising rural incomes and improving nutritional security is through livelihood diversification. The human development indicators of health and education are way below the national average of developing India. Against this backdrop, the government of Bihar, with the support of the World Bank, initiated *JEEViKA, the Bihar Rural Livelihoods Project Society* (BRLPS), with the objective of enhancing the social and economic empowerment of the rural poor, especially through microfinance among women. In this regard, mushroom cultivation through microfinance has emerged as a viable means of economic well-being for the farm and non-farm dependent population carrying a poor balance sheet.

The present study has been conducted on SHG members in three villages – Sarilchak, Surajpur, and Anantpur – of Nalanda district, Bihar. The overall objective of the study is to assess the viability of mushroom production through SHGs as an alternative means for the alleviation of rural poverty leading to inclusive growth. An attempt has been made to understand the determinants of livelihood diversification in the study area. The study was undertaken with the specific objective of understanding the factors that determine the level of diversification of rural households in general. What are the various constraints to livelihood diversification? The study is an attempt to answer these questions. The study is organised as follows: starting with a brief overview of the development narrative over recent decades, a case of microfinance has been discussed. The mechanism of mushroom cultivation in the region has been contextualised in the sustainable livelihood framework in the next section. The censored regression model has been used to understand the issue of diversification discussed above. The subsequent section, drawing upon the field study, tries to examine the case of livelihood diversification. The challenges, opportunities and policy implications are discussed in the concluding section.

14.2 Overview

The development narrative in the past half-century has followed a series of over-lapping transitions, ranging from community development in the 1950s to the emphasis on small-farm growth in the 1960s, extending to state-led rural development in the 1970s, furthering to market liberalisation in the 1980s, followed by the emergence of sustainable livelihoods through process, participation, empowerment and actor approaches as an integrating framework in the 1990s and 2000s, and mainstreaming rural development in poverty reduction strategy (Ellis/Biggs 2001). The contemporary development thinking recognises the benefits of livelihoods thinking and approaches, including, for example, its stress on the importance of people-centred change, a holistic approach, people's access to different assets, poor people's vulnerability, partnerships, sustainability, change, and the multi-faceted nature of livelihoods, all leading to the Millennium Development Goals (MDGs), subsequently universalised as Sustainable Development Goals (SDGs) in 2015.

Slightly modifying the Anna Karenina principle of Leo Tolstoy, "All rich households are alike; every poor household is poor in its own way", implies that the poverty and their livelihoods are heterogeneous and divergent in nature. The poor carry a weak balance sheet with no cash, no adequate collateral, no guaranteed income streams, and are unlikely to have access to sources of finance. Using social capital, defined as "the collective value of all social networks (*who people know*), and the inclinations that arise from these networks to do things for each other (*norms of reciprocity*)" (Sander 2015) as collateral, a number of initiatives on the part of state and non-state organisations have tried to strengthen the balance sheet of the poor by lending to groups rather than individuals. An important manifestation of social capital is the trust and norms of reciprocity people of a group or community have in each other, thereby minimising the problems of adverse selection on decisions on loans and the moral hazards of loan usage and repayment by peer monitoring, which any financial transaction faces. The poor organise themselves in groups and each participant accepts joint responsibility for the loan. Members of a group value their reputation within the group, as they will be chosen for valuable interaction or given discretionary power if they are deemed trustworthy or reliable. A well-known example is the Grameen Bank in Bangladesh, but similar institutions exist in several developing countries, including India.

14.3 Microfinance Modalities

Microfinance has worked as an intervention made necessary by the conditions of material deprivation of millions of people in the not-so-developed part of the world (Nair 2005). A comprehensive overview of institutions, incentive considerations and empirical data in microfinance can be found in Aghion/Morduch (2005). Further, the ideological construct that supporting poor women through microfinance is central to poverty reduction strategies across the developing world, has been due to public policy and business cases (Ackerly 1995). Women comprise the majority of the rural population because there are high levels of male outmigration in developing countries, including India. Thus, targeting female borrowers makes sense from a public policy standpoint. A typical rural female borrower uses the income to enhance the nutritional status of her family, educates her children and begins to participate in major family decisions (Garikipati 2008). The business case for focusing on female borrowers is therefore substantial, as they register higher repayment rates (UN 2009: 57).

The *JEEViKA* intervention was carried out through the formation and mobilisation of *self-help-groups* (SHGs) for extending credit support and the promotion of thrift to foster viable economic activities. An SHG, generally comprised of ten to twenty homogenous people, encourages its members to make voluntary personal economies on a regular basis in order to pool resources to make small interest-bearing loans to their members. Once an SHG has demonstrated the commitment of its members to contributing to regular savings, and the capacity to keep records of these finances, it becomes eligible for access to an initial capitalisation fund of ₹50,000 to lend to its members. Additional interventions, such as training in intensive farming and animal husbandry, are also planned for targeted populations. The process helps them to absorb the essentials of financial intermediation, including prioritisation of needs, setting terms and conditions, and book-keeping. This gradually builds financial discipline in all of them, in addition to equipping them to handle resources of a size that is much beyond the individual capacities of any of them.

The broader goal of the microfinance intervention is to create an ambience of faith aimed at creating institutions which the poor are sufficiently empowered to run on the principles of financial prudence and sustainability.

Further, a broadening of income and livelihood strategies away from purely crop and livestock production towards both farm and non-farm activities to generate additional income is considered desirable (Hussein/Nelson 1998). Putting this sort of diversification in the 'livelihoods' has broadened the concept to include "the process by which rural families construct a diverse portfolio of activities and social support capabilities in their struggle for survival and in order to improve their standard of living" (Ellis 1998). Overall, successful implementation of the *JEEViKA* programme in mushroom cultivation among the poor in the villages of Nalanda, Bihar, offers a compelling case to explore further. The study has been segmented as follows: the production mechanism of mushroom cultivation has been described with the lifecycle. The key business processes involved are discussed thereafter. The *sustainable livelihoods approach* (SLA) framework is used to identify the entry points of livelihood intervention. The opportunities and challenges are identified in the concluding part of the study.

14.4   The Mushroom Cultivation Mechanism

Mushrooms belong to the kingdom of Fungi. Fungi, unlike the leaves of green plants, which contain chlorophyll to absorb light energy for photosynthesis (the process by which plants convert carbon dioxide and water into organic chemicals), rely on other plant material (the substrate, the material in which the mycelium grows) for their food, and absorb nutrients from the organic material in which they live.[1] The living body of the fungus is mycelium, made out of a tiny web of threads (or filaments) called hyphae. These sexually compatible hyphae fuse to form spores under specific conditions. Released from the gills, spores germinate and develop to form hyphae.

Hyphae are the main mode of vegetative growth in fungi, which are collectively referred to as mycelium. These feed, grow and ultimately produce mushrooms, and as a mushroom produces several million spores in its life, this life cycle is repeated each time the spores germinate to form the mycelium.

The three basic sources of nutrients for mushrooms are:

- *Saprobic*: grows on dead organic matter. Saprobic edible fungi can be wild-harvested, but are most widely valued as a source of food and medicine in their cultivated forms. They need a constant supply of suitable organic matter to sustain production and, in the wild, this can be a limiting factor in production.
- *Symbiotic*: grows in association with other organisms. The majority of wild edible fungi species (e.g. chanterelles – Cantharellus and Amanita species) are symbiotic and commonly form mycorrhizas with trees. The fungus helps the tree gather water from a wider catchment and delivers nutrients from the soil that the tree cannot access, and the tree provides the fungus with essential carbohydrates.
- *Pathogenic* or parasitic: plant pathogenic fungi cause diseases in plants and a small number of these microfungi are eaten in the form of infected host material.

Further, mushroom species can be cultivated in two ways:

- *Composted substrates*: wheat and rice straw, corncobs, hay, water hyacinth, composted manure, and various other agricultural by-products including coffee husks and banana leaves.
- *Woody substrates*: logs or sawdust.

Thus the key steps in mushroom production – a cycle that takes between one to three months from start to finish depending on species– start with some mushroom spores, which grow into mycelium and expand into a mass sufficient in volume and stored-up energy to support the final phase of the mushroom reproductive cycle, which is the formation of fruiting bodies or mushrooms.[2] The BRLPS intervention

---

[1]Mycelium is the network of hyphae that form the vegetative body of the fungus. Mushrooms are the fruiting bodies of the mycelium.

[2]The mycelium will form mushrooms in its reproductive stage. This is called fruiting, as the mushrooms are actually the fruiting bodies of the mycelium.

follows the cheapest cultivation system: using composted substrate, mushrooms are grown in plastic bags (which can be sterilised and re-used with new substrate) containing substrate or compost, in a simple building to provide controlled growing conditions.

14.4.1 Step I: Making Mushroom Compost

The compost preparation usually occurs outdoors, although an enclosed building or a structure with a roof over it may be used. A concrete hemisphere slab is required for composting. 40 ml of formalin added to one litre of water are used for pasteurisation and sterilisation.[3] After this, 45 l of fresh water are added to a mixture consisting of 125 ml formalin, 4 g of Bevestin, 100 g of processed lime (gypsum), and 10 kg of straw. Once the pile is wetted and formed, aerobic fermentation (composting) commences as a result of the growth and reproduction of the microorganisms which occur naturally in the bulk ingredients. Heat, ammonia, and carbon dioxide are released as by-products during this process. After 24 hours, this straw mixture is dried completely in the sunlight.

14.4.2 Step II: Production

The next step is to put this dried straw into a 40 micron polythene bag of 1.5 kg capacity with 100 g spawn (Latin expandere = to spread out, also called Mushroom seed) spread through system root intensification (SRI) methods at a distance of 2 in.. Once it is done tightly, it is pricked to take air.

14.4.3 Step III: Harvesting

The bags are kept in a dark quarantined room for 20 days. It is important to maintain optimal temperature, moisture, hygiene and other conditions for mycelium growth and fruiting, which is the most challenging step. It is essential to maintain hygienic conditions over the general cropping area, in order to protect the crop from contamination. After 20–23 days, fruits can be cropped in three batches. The first crop can be harvested on the 25th day, the second crop on the 32nd day and the third on the 45th day.

---

[3]A 30% solution of formaldehyde is used to sterilise areas. The gases kill living micro-organisms and spores.

14.4.4 Sustainable Livelihood Context

Drawing upon the idea of "Sustainable Livelihoods" with the intention of enhancing the efficiency of development co-operation advocated by Robert Chambers in the mid-1980s, the British Department for International Development (DFID) developed the Sustainable Livelihoods Approach (SLA), which visualises five types of household assets – natural, social, financial, physical and human capital – as the main factors of influence, to understand the livelihoods of the poor.

Key assets or resources associated with mushroom cultivation are described below.

Natural assets: Due to the minimal role of land and climate in mushroom cultivation, the livelihood enterprise is suitable for farmers with limited land, as well as the landless. Access to sufficient, suitable and locally-sourced substrate and spores are key determinants of successful and sustainable mushroom cultivation. The agricultural by-products, used as substrate, are available easily and cheaply. Once inoculated with spawn, a sterilised composted substrate can be used for three harvests and then recycled by incorporating it as an organic mulch or fertilizer in other horticultural or agricultural systems, which can improve soil structure, or it can be used as a nutritious fodder for the backyard poultry commonly kept in the region.[4]

Social assets: The formation of social capital through JEEViKA, including networks and support from families, friends, organisations and membership of SHG groups, helps people meet their livelihood objectives. Social capital strengthens the individual by helping them and their communities access information and resources, including technical information, basic training, sources of mushroom spores, and marketing outlets to sell their crop. As a result of the high perishability of mushrooms, it can be of great benefit for small-scale cultivators selling their crop to form organisations with other growers and share transport costs, market contacts, etc. In addition, working in collaboration with other growers may enable cultivators to establish local production, processing, drying or packaging facilities to increase harvest output or product shelf-life. Mushroom cultivation represents a very suitable and empowering income-generating option for women in particular, because it can be combined with traditional domestic duties and can be undertaken at home. Several programmes related to rural mushroom production have given women the opportunity to gain financial independence, farming skills and higher self-esteem.

Human assets: Human assets relate to the skills, knowledge, ability to work and level of health that people need to pursue different livelihood strategies and to achieve their objectives. Being non-labour-intensive, mushroom cultivation can be undertaken as an additional livelihood activity which fits around other household or

---

[4]Inoculation helps in transferring an organism into a specific substrate.

productive tasks. Qualities identified as being useful for mushroom cultivators include the ability to carry out operations on time, attention to detail, vigilance (to prevent and mitigate pest invasions), and, for marketing purposes, excellent skills in public relations.

Physical assets: Many of the physical assets required to undertake mushroom cultivation are not exclusive items, but rather assets which help meet livelihood needs in general, including the transport and communication infrastructure, clean water, a source of energy, and buildings for shelter and storage. Mushrooms are best cultivated indoors in a dark, cool, sterilised and enclosed building. This enables the most suitable growing conditions for mushrooms to be maintained, in terms of temperature, humidity, uniform ventilation and substrate moisture levels. Unwanted contaminants, moulds and sunlight can also be kept away from the crop. Any small room with ventilation and a cement floor can be used. The interior should be arranged so that it is easy to clean at the end of each cropping cycle. The mushroom house should be well insulated (by using, for example, fibreglass wool or expanded polystyrene) to maintain a steady temperature, and concrete or clay tiles are preferable to corrugated metal for roofing.

Financial assets: Mushroom cultivation is attractive to the resource-poor for two reasons. Firstly, because mushroom cultivation can be done on any scale, the initial financial outlay to establish a basic cultivation system need not be very great, and substrate materials are often free. An example from a village illustrates the point: a mushroom house with an area of 100 sq. ft. is large enough to hold 100 mushroom bags with a yield of 3–4 kg per day, yet could cost less than ₹200, utilising the materials available locally. Secondly, compared to many agricultural and horti-cultural crops, mushroom production systems have a short turnaround; a har-vestable crop can be produced and sold within two to four months, which is very helpful for small-scale producers.

14.4.5 Illustrative Production Economics of Mushroom Cultivation in the Region

Mushroom cultivation has provided an important stepping stone in improving livelihood security in the region. The poor cultivators have added value to the product innovation, such as dried mushrooms, mushroom pickles, mushroom jelly, mushroom powder, and so on. They have also added a few more recipes to their kitchen, such as Mushroom Halwa and Mushroom Pooaa. Table 14.3 shows the average production economics of the mushroom cultivation in the region (Tables 14.1 and 14.2).

Table 14.1 Production economics

Spawn production		Mushroom production	
Particulars	Description	Particulars	Description
Capacity	250 kg/month	Capacity	100 bags/100 sq ft
Rate	Rs. 80/kg	Yield	3 kg/day
Income	Rs. 20,000/month	Cost of cultivation	Rs. 100
Cost of inputs	Rs. 10,000/month	Income	Rs. 300/day
Net profit	Rs. 10,000/month	Net profit	Rs. 200

Source The authors

Table 14.2 Mushroom product innovation

Products	Amount (kg/year)	Rate (Rs.)	Income (Rs.)	Cost of inputs (Rs.)	Net profit (Rs.)
Dried mushroom	20	1,000/kg	20,000	12,000	8,000
Mushroom pickle	50	300/kg	15,000	10,000	5,000
Mushroom jelly	50	200/kg	10,000	6,000	4,000
Mushroom powder	10	1,000/kg	10,000	6,000	4,000
Additional income through added value			55,000	34,000	21,000

Source The authors

14.4.6 Constraints on Mushroom Cultivation

Mushroom cultivation is an important livelihood strategy for the rural households in the region. However, several constraints on mushroom cultivation have been identified, especially socio-economic, technological, institutional and policy constraints. The major constraints were: poor asset base, lack of credit facilities, lack of awareness and training facilities, fear of taking a risk, lack of rural infrastructure, lack of marketing opportunities for the product, and lack of opportunities in non-farm sector diversification (Fig. 14.1).

Because the units are small, mushroom cultivation by the rural poor will not generate a huge income. However, it can make a valuable contribution to sustainable livelihoods for them, because it is highly compatible with other livelihood activities and requires minimal physical and financial inputs and resources.

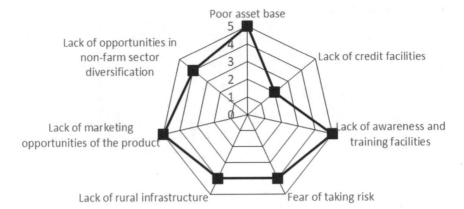

Fig. 14.1 Constraints to mushroom cultivation. *Source* Author estimation based on focus group discussions with cultivators in the three villages participating in the study

14.5   Study Sites, Data and Methodology

The study was based on collaboration with *JEEViKA* in Nalanda district, Bihar. The three villages – Anantpur (Chandi Block), Surajpur and Sarilchak (both of Silao Block) – were purposely chosen for twin reasons. Firstly, these villages had a strong SHGs presence, thanks to the Jeevika intervention and support, and secondly, the villages have adopted mushroom cultivation as one of their dominant livelihood strategies.

14.5.1   Definitions

Livelihood diversification: Livelihood diversification, different from income diversification, can be defined as "the process by which rural families construct a diverse portfolio of activities and social support capabilities in their struggle for survival and in order to improve their standards of living" (Ellis/Biggs 2001: 5). The diversification can take place through both agricultural diversification – i.e., production of multiple crops or high-value crops – and non-agricultural livelihood diversification – i.e., undertaking small enterprises, or choosing non-agricultural sources of livelihood, such as casual labour or migration.

Livelihood diversification index: There are various indicators and indices to measure livelihood diversification, such as the number of income sources and their share, Simpson's Diversity Index, the Herfindahl Index, the Ogive Index, the Generalised Entropy Index, the Modified Entropy Index and the Composite Entropy Index (Shiyani/Pandya 1998). In this study the Simpson Diversity Index is used because of its computational simplicity, robustness and wider applicability. The

Simpson index (SI) is: $SI = 1 - \sum_{i=1}^{N} P_i^2$ where N is the total number of income sources and P_i represents the income proportion of the i-th income source. Its value lies between 0 and 1. The value of the index is zero when there is complete specialisation and approaches one as the level of diversification increases.

14.6 Tobit Analysis

As the constructed dependent variable takes a value of between 0 and 1 or, in other words, censored, the *ordinary least squares* (OLS) estimate may be biased and inconsistent (Wooldridge 2016; Chap. 17.2). To identify the major drivers of livelihood diversification, we use a latent dependent variable model, also called the Tobit censored regression model. The structure of the model believes that there is an unobservable variable and the values of the sample are testifying at a certain threshold instead of at the actual values. The vector of the independent variables determines the latent variable, and the normally distributed error term is there to capture the impact of this association.

$$Y_i = \alpha + \beta' X_i + \epsilon_t$$

and the observable variable complies with

$$Y = Y^*, \ if \ Y^* > A$$

$$Y = A, \ if \ Y^* \leq A$$

for certain threshold A.

Where Y_i is a latent variable implying the livelihood diversity of i-th observation, vector of estimated coefficients and $\beta' X_i$ independent variables and ϵ_i. is an error term which is assumed to be $\epsilon_i \sim IID(0, \sigma^2)$. The latent variable Y^* would have minimum value if the diversity is lowest or else maximum if diversity is increasing.

14.6.1 Results and Discussion

A scatter plot between diversity index and per capita income shows a positive association between the two and validates Ellis's and Biggs (2001) proposition (Fig. 14.2).

The main livelihood groups, their level of livelihood diversification and the contribution of different sources of income in the study area are given in Table 14.3.

Table 14.3 shows that the level of diversification is highest for salaried groups as well as farm groups overall, and it is less diversified for the labourer groups. This

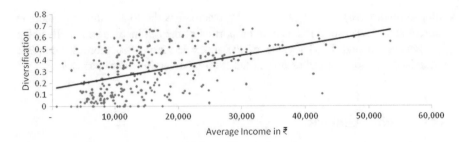

Fig. 14.2 Average Income vis-à-vis Extent of Diversification. *Source* Author estimation

Table 14.3 Level of livelihood diversification and per capita income

Village	Indicator	Farm	Non-farm	Casual labourer	Petty business	Salaried
Anantpur	Diversity index	0.35	0.13	0.24	–	0.48
Sarilchak		0.36	0.44	0.24	0.37	0.39
Surajpur		0.29	0.27	0.25	0.40	0.31
Overall		0.34	0.32	0.25	0.39	0.36
Anantpur	Per capita income	20,859	21,000	11,706	–	21,942
Sarilchak		19,645	14,241	16,400	18,545	15,170
Surajpur		29,071	18,455	13,919	25,277	19,395
Overall		22,120	17,203	14,782	21,490	18,107

Source Author estimation

may be due to a relatively stronger asset base among salaried and farm groups. Resource-poor households, on the other hand, lack assets, skill and education, which hinders them from engaging in remunerative activities, which in turn forces them to diversify in low-return activities for survival. This, however, varies across the villages.

14.6.2 Tobit Results

The results of Tobit regression estimates are presented in Table 14.4. The likelihood ratio chi-square of 24.19 (df = 7) with a p-value of 0.001 tells us that our model as a whole fits significantly better than an empty model (i.e., a model with no predictors). All the estimated coefficients, except per capita land availability, have the expected signs and are statistically significant.

We hypothesised that the livelihood diversification index is a function of a set of factors that include the following variables in the model: mobility medium (vehicle in the house as proxy for mobility), household, education, per capita income, dependency ratio, per capita land, and borrowing. As the region lacks a basic mobility infrastructure, a vehicle, including a bicycle, in the house is considered to

Table 14.4 Determinants of livelihood diversification

Variables	Coefficient	Std. Err.	t-stat	P > t
Intercept	0.083413	0.054395	1.53	0.126
Mobility	0.146159	0.074417	1.96	0.05
Household	0.036757	0.026346	1.40	0.164
Education	0.014358	0.010344	1.39	0.166
Dependency ratio	0.012512	0.011185	1.12	0.264
Per capita income	1.68E−06	7.51E−07	2.23	0.026
Borrowing	0.013529	0.005112	2.65	0.009
Per capita land	−0.00128	0.001731	−0.74	0.46
LR chi^2(5)	24.19			
Prob > chi^2	0.001			

Source The authors' calculations

be one of the most important components for livelihood enhancement, as it increases access to a wider network. The education level is an important determinant for livelihood diversification as it creates an entry point to the non-farm sector. Similarly, the availability of household, per capita income and per capita land can constitute a household asset base. This offers a store of wealth as well as providing opportunities to invest in alternative livelihoods. The lack of these assets may prove a barrier to diversification for resource-poor households. However, the negative and statistically insignificant association of per capita land and livelihood diversification shows that a decrease in its value creates overpressure on land, which in turn results in very low or zero marginal productivities (i.e. disguised unemployment). As a consequence, this surplus labour is finding jobs in the non-farm sector. The dependency ratio, or percentage of non-working population, has a positive influence on livelihood as this improves the ability to meet household needs. The positive association of borrowing with livelihood diversification shows that since the resource-base is very poor for most of the households, providing credit to households will improve their livelihood.

14.6.3   The Way Forward

In the case of mushroom cultivation, since the basic skills required to grow medicinal mushrooms are the same as those for growing edible mushrooms, growers can diversify from growing edible mushrooms to producing mushrooms with medicinal values. Another diversification option may be to integrate mushroom cultivation with other livelihood activities practised in the region, for example, the spent mushroom substrate acts as an organic mulch in growing other horticultural crops, e.g. vegetables, and can supply protein for backyard poultry farming. This may result in not only diversification for securing additional income but also in recycling the organic waste created from mushroom cultivation. The recycling process makes this sustainable farming. Furthermore, the mushroom farmers have mainstreamed mushroom produce in their food consumption basket with wonderful indigenous recipes, like Mushroom Pickles, Child Mushroom food, and so on.

However, mushroom cultivators need organisational strength to manage post-harvest storage and obtain the best prices. Economies of scope and scale improve their overall strategy. A way forward is setting up member-based Producer Organisations (POs), which can leverage bargaining power to access financial and non-financial inputs and services and appropriate technologies, reduce transaction costs, tap high-value markets, and help the members to move further up the value chain, entering into post-harvest management. Direct retailing, value-added services, storage and processing, and engaging in contract production of primary agricultural produce are other possibilities.

14.7 Summary and Policy Implication

14.7.1 Opportunities

Since it produces tradeable goods of high nutritional content, mushroom cultivation provides quick cash and so strengthens livelihood security for the poor. A good understanding of mushroom cultivation, whether based on local knowledge or acquired through external support, allows cultivators to provide consistent and predictable quantities and qualities of mushrooms, thereby attracting buyers more easily.

14.7.2 Challenges

One of the most important aspects of growing mushrooms by subsistence farmers for commercial purposes is the ability to maintain a continuous supply for chosen market outlets. This is labour- and management-intensive, given the fact that mushroom production systems, to some extent, are vulnerable to sporadic yields, invasions of 'weed' fungi, insect pests, and unreliable market prices for traded goods. The formation of co-operative or community groups can help to streamline production costs, harvesting and marketing, and reduce vulnerabilities and risks. Public orientation can also drive demand. Marketing information, business and entrepreneurial skills, and training in cultivation, processing and productivity challenges are also essential for success.

14.7.3 Policy Implications

From the public policy point of view, the provision of public goods, such as roads, electricity, telecommunications and rural markets, can be regarded as an ecosystem to enable small-scale producers to improve their livelihoods. Public investments which have an impact on people's capabilities to carry out livelihood activities are therefore highly desirable. As their capabilities increase, so does efficiency, while costs, risks and vulnerability reduce.

References

Ackerly, B.A., 1995: "Testing the tools of development: credit programmes, loan involvement and women's empowerment", in: *IDS Bulletin*, 26(3): 56–68.
Aghion, B.A.de; Morduch, J., 2005: *The Economics of Microfinance* (Cambridge, MA: The MIT Press).
DFID, 1999: *Sustainable Livelihoods Guidance Sheets* (London: Department for International Development).
Ellis, F.; Biggs, S., 2001: "Evolving themes in rural development 1950s–2000s", in: *Development Policy Review*, 19(4): 437–448.
Ellis, F., 1998: "Household strategies and rural livelihood diversification", in: *The Journal of Development Studies*, 35(1): 1–38.
FAO; World Bank, 2001: *Farming Systems and Poverty—Improving Farmer's Livelihoods in a Changing World* (Rome: Washington D.C.).
Garikipati, S., 2008: "The impact of lending to women on household vulnerability and women's empowerment: Evidence from India", in: *World Development*, 36(12): 2,620–2,642.
Hussein, K.; Nelson, J., 1998: *Sustainable Livelihoods and Livelihood Diversification*, IDS Working Paper 69 (Sussex: Institute of Development Studies).
Marshall, Elaine; N.G. (Tan) Nair, 2009: *Rural Infrastructure and Agro-Industries Division*, (Rome: Food and Agriculture Organization (FAO) of the United Nations).
Nair, T.S., 2005: "The transforming world of Indian microfinance", in: *Economic and Political Weekly*, 23 April: 1,695–1,698.
Sander, T., 2015: "About Social Capital", Saguaro Seminar: Civic Engagement in America, John F. Kennedy School of Government at Harvard University.
Shiyani, R.L.; Pandya, H.R., 1998: "Diversification of agriculture in Gujarat: A spatio-temporal analysis", in: *Indian Journal of Agricultural Economics*, 53: 627–639.
UN, 2009: *Woman's Control over Economic Resources and Access to Financial Resources, Including Microfinance* (Washington DC: World Survey on the Role of Women in Development).
Wooldridge, J., 2016: *Introductory Econometrics* (Boston: Cengage Learning).

Chapter 15
Forest Accounting and Sustainability

Parashram J. Patil

Abstract There is a nexus between forest accounting and sustainable development. Forests are a valuable resource as they constitute an important component of the terrestrial environmental system and larger resource base. They provide different basic inputs to the global economic cum ecological system in a multi-dimensional way. They provide timber, fuel, pulpwood, fodder, fibre grass and non-wood forest products, and support industrial and commercial activities. The following are the specific objectives of the present research.

- To identify the reasons for ecological ecosystem loss.
- To explore the relationship between forest accounting and the rationality of the ecological ecosystem in the context of consumption, production, energy, biodiversity, economics, and ecological balance and sustainability.
- To identify and measure the ecological ecosystem loss.
- To develop theoretical modelling of ecological ecosystem sustainability.

The present study is an explorative study on ecological rationality and sustainability through developing a forest accounting system. Essential data has been collected and analysed to find out the present nexus.

Keywords Rationality · Ecology · Ecosystem · Forest accounting
India · Sustainability

15.1 Introduction

Natural resources are the real wealth of a nation and every citizen of the country has an equal right to them. They include agriculture and land, water, minerals and petroleum, fisheries and forests. Of all the resources, the forest is one of the most vital, being one of the important components of the terrestrial environmental system and larger resource base. Forests account for the second largest land use after

Parashram J. Patil, Researcher, Institute for Natural Resources, Maharashtra India; Email: patilparashram9@gmail.com

© Springer Nature Switzerland AG 2019
A. K. J. R. Nayak (ed.), *Transition Strategies for Sustainable Community Systems,*
The Anthropocene: Politik—Economics—Society—Science 26,
https://doi.org/10.1007/978-3-030-00356-2_15

agriculture and provide different basic inputs to the global economic cum ecological system in a multi-dimensional way. They provide timber, fuel, pulpwood, fodder, fibre grass and non-wood forest products, and support industrial and commercial activities. They also maintain the ecological balance and life-support systems which are essential for food production and health as well as the overall development of humankind. Forests control the wealth of adjoining land use systems. They also improve the wealth of urban areas. However, there is no proper accounting and valuation system available to monitor this very important resource. This could lead to a failure to notice warning signs and ensuing loss of biodiversity, with a consequent risk of disaster.

Natural Resource Accounting is closely related to Environment-Economic Accounting. Natural Resource Accounting (NRA) means taking stock of natural resources and noting changes in them because of natural processes or human use. The rapid industrialisation and progressive economic growth in recent years have led to phenomenal environment degradation and depletion of natural resources in India. The literature reveals that Natural Resource Accounting has been undertaken for many resources but that it is now essential to carry out more accounting work on forest areas.

In India, forests have played a significant role in rural industry as well as improving major environmental resources. India is one of the top ten most forest-developed countries in the world, along with the Russian Federation, Brazil, Canada, the United States of America (USA), China, the Democratic Republic of the Congo, Australia, Indonesia and Sudan. Along with India, these top forest-rich countries account for 67% of the total forest area in the world (India State of Forest Report 2011). In India, forest cover improved at 0.22% annually over the decade 1990–2000 and the forest-cover area grew at the rate of 0.46% per year during the decade 2000–2010 (Global Forest Resource Assessment 2010). According to the Forest Survey of India 2013, the forest cover increased to 69.8 million hectares by 2012 per satellite measurement. This performance shows increases of 5.871 sq. km of forest cover in the last two years (State of Forest Report 2013).

15.1.1 Literature Review

Reviewing the literature on forest accounting makes it easier to understand inter-linkages between the various components of forest accounting. Gundimeda et al. (2007) set out and apply a System of Environmental Economic Accounting based methodology to show the true value of forest resources in India's national as well as state accounts. Their study is focused on four components of value creation in forests: timber production, carbon storage, fuelwood usage and the harvesting of non-timber forest products. The study finds that there is a need to integrate national resources accounting into the national accounting framework. This is important to generate appropriate signals for sustainable forest management. The authors conclude that existing measures of national income in India underestimate the contribution of forest income. The incomes of North-Eastern states, in particular, are highly underestimated by these traditional (GDP/GSDP) measures. The study has shown that if the limitation of the current data on production and prices are addressed, the income from forests will be much more than today.

Ykhanbai (2009) has measured and assessed forest degradation in Mongolia. The aim of his study was to collect information and review forest degradation accounting. For the study, he used different methods such as the degradation (depreciation) method, the total rent approach and the user cost method. He also studies physical and monetary accounting of forest resources degradation, but he does not consider accounting for the depletion of forest environmental services. The study shows that forest degradation has increased year by year over the study period. The reduction of forest degradation is due to expansion in economic activity and increased global climate changes. He suggests that there should be an improvement in forest resource management policies for the future sustainable development. He concludes that to improve the situation of national resource degradation, green accounting should be adopted and institutionalised in the country.

Harris and Fraser (2002) critically examine the theory and practice of natural resource accounting. The prime aim of their study is to provide an extensive review of the theoretical and applied literature on natural resource accounting. They also explain the economic theory that underpins natural resource accounting, counselling welfare and sustainability of the policy goals. In the study, they present various different concepts of national income. They have found that there is a fundamental difference in economic and national accounting methodology. They conclude that the data shows that insufficient attention has been paid by economists to the revisions to the SNA. Furthermore, they suggest that a growth theory model should be used to solve this particular technical problem.

Parikh and Ghosh (1991) discuss natural resource accounting for the soil in order to estimate the cost of soil degradation in India. They have analysed the soil resource as an empirical estimation of the cost of soil degradation by using the soil quality index for the states in India. According to the researchers, soil productivity is a function of measurable soil properties/assets. They conclude that soil salinisation has contributed to soil degradation and the reduction in soil productivity.

15.1.2 Problems to be Studied

Forests play a significant role in the balance of natural resources. The global concern about forest degradation and depletion is related to two main problems: destruction of the carbon sinks affecting the global climate, and extinction of species affecting the biodiversity. In this context, it is relevant to study forest accounting. These issues raise some problematic statements in the researcher's mind, such as:

- No proper accounting for forest resources in the *Systems of National Accounts* (SNA).
- Management cannot understand the actual forest assets in most areas.
- Problems of awareness and knowledge about the forest service.
- Lower flow of goods and services in forest products.
- Lower income from goods and services produced by forest areas.
- Lack of availability of fodder and reduction in productivity of livestock population.
- Lack of awareness about natural resource economics in terms of the forest.
- Lack of proper valuation methodology of forest resources.
- Lack of investment in the forest sector.
- Lack of awareness about the business opportunity in the forest sector.
- The issue of remunerative pricing.
- Value additions in forest produce – bamboo, jute, wool, etc.
- Agroforestry interfaces.
- Trade-offs between agriculture and environment.
- Biodiversity valuation.
- Nutrition issues in forest products.
- Measures of biodiversity loss and risk of disaster.

There is no proper or scientific way of accounting for forest areas. This creates many problems in the management of forest resources and hence does not show the impact of forests on biodiversity and the economy.

15.1.3   Objectives of the Study

Current research focuses on ecological sustainability. The specific objectives of the present research are:

- To explore the relationship between forest accounting and the rationality of the ecological ecosystem in the context of consumption, production, energy, biodiversity, economics, and ecological balance and sustainability.
- To identify and measure the ecological ecosystem loss.
- To identify the reasons for ecological ecosystem loss.
- To develop theoretical modelling of ecological ecosystem sustainability.

The specific hypothesis of this study is that forest accounting could make an impact on sustainability.

15.1.4   Research Methodology

This chapter is a descriptive study and depends primarily on secondary sources of data.

- Published sources: The researcher has collected the data from sources such as the Directorate of Economics and Statistics (DES), the Forest Department of the Government of Maharashtra, the Administrative Report of the Forest Department, the Forest Survey of India, the National Sample Survey of Organisation (NSSO), the Central Statistical Organisation (CSO) and also books and research papers published in journals and on websites.
- Unpublished sources: Ph.D. thesis, M.Phil. dissertations and other unpublished sources.

15.2   Forest Accounting

Forest accounting it is multi-disciplinary area including mathematics, physics, life sciences, chemistry, statistics, accounting and finance, and economics. Therefore it is necessary to examine different aspects of it. There are various parameters on which forest accounting systems could be developed. Exploration of all these parameters would give a clear picture of forest accounting in a better way. It is a challenge to develop forest accounting, since it involves a complex ecosystem and its invisible services to society. Forest accounting entails counting nature's services, especially forests and their economic value. It is essential to develop mechanisms that include the value of all economic contributions made by forests and linked to the economy. The parameters that need to be considered while undertaking forest accounting are as follows:

- Actual/Economic Accounts:

  - Physical Account
  - Monetary Account
  - Flow Accounts

- Financial Performance:

  - Income
  - Expenditure

- Ecological Classifications:

  - Legal Classification of Forests
  - Forest Types
  - Species
  - Animal/Fauna
  - Forest Products

- Valuation Methods:

  - Historical Cost Method
  - Market Price Method
  - Net Present Value Method
  - Discounted Cash Flow Method
  - Scholastic Discounted Cash Flow Method
  - Real Option Pricing Methods
  - Sensitivity Analysis

- Forest Economics:

  - Forest Resources
  - Goods and Services
  - Employment
  - Business

These parameters form the building blocks for developing a forest accounting system. Therefore careful analysis of them to gain a comprehensive understanding of the situation is an essential part of the process of preparing forest resources accounts.

15.3 Actual and Economic Accounts

The following are the economic parameters of forest accounting.

- *Physical Accounting*: Physical accounting refers to "the natural resource and environmental accounting of stocks and changes in stocks in physical (non-monetary) units" – for example, weight, area or number. Qualitative measures – expressed in terms of quality classes, types of uses, or eco-system characteristics – may supplement quantitative measures (Statistic New Zealand, 2002). Forest resources stock is changing every year.
- *Monetary Accounting*: In monetary accounts "the entries correspond to the physical accounts but contain an additional entry for revaluation, which records the change in asset value due to changes in prices between the beginning and end of the period, (Statistic New Zealand, 2002).
- *Forest Flow Accounts*: Forest flow accounts include supply and use tables detailing forest products by sector (wood and non-wood, marketed and non-marketed). They are linked to the *input/output* (I/O) and also include measures of forest ecosystem services and environmental degradation associated with forest use (Statistic New Zealand, 2002).

15.3.1 Financial Performance

When measuring financial performance, the following two parameters could be used:

- Income: Forest departments obtain revenue from the forest in different ways, such as renting land under temporary cultivation, fees for duplicate permits, sale proceeds from licences to catch hawks, and sale proceeds from condemned tents, furniture and other stores, livestock, tools and plants. When evaluating the financial performance of the forest department, aggregation of all income derived from the forest needs to be considered.
- Expenditure: Forest departments inevitably incur various costs when undertaking essential forest maintenance, such as plantation, forest protection, harvesting, organisation, and the improvement and extension of forests, livestock, stores, tools and plants etc. The salaries of forest staff also need to be taken into account. The aggregation of these all expenses needs to be deducted from the revenue of the forest department.

15.3.2 Ecological Classifications

For forest accounting, ecological classification is a must. These are some parameters which are used for it:

- Legal Classification of Forests: It is important to understand what constitutes 'forest' in a legal sense in order to distinguish forest from other land uses (D. Venkateswarlu). In India there are two relevant acts: the Indian Forest Act 1927 and the Forest Conservation Act 1980. It is necessary to take into consideration everything which comes under the purview of 'forest', otherwise an incorrect picture would be obtained. The identification of forest resources must be under the ambit of the law. This is essential for carrying out forest accounting in fair manner.
- Forest Types: Forest type is a group of forest ecosystems of generally similar composition that can be readily differentiated from other such groups by their tree and under-canopy species composition, productivity and/or crown closure (D. Venkateswarlu). Forest valuation depends on the type of forest, especially tree species, their utility, availability and demand etc. These issues are a matter for forest accounting because they impact on economic transactions.
- Species: Forests do not just consist of trees. There are various other things present. Hence forest accounting has to take care of all these visible and invisible matters. Forests are home to 80% of the world's terrestrial biodiversity. These ecosystems are complex webs of organisms that include plants, animals, fungi and bacteria. Forests take many forms, depending on their latitude, local soil, rainfall and prevailing temperatures. Coniferous forests are dominated by

cone-bearing trees, like pines and firs that can thrive in the northern latitudes where these forests are often found. Many temperate forests house both coniferous and broad-leafed trees, such as oaks and elms, which can turn beautiful shades of orange, yellow and red in autumn. The economic invisibility of nature and its impact on society poses a challenge to forest accounting because the valuation of forests varies, depending on the species present.

- Fauna and Flora: India has a rich diversity of flora and fauna. Flora refers to plant species and fauna refers to animal species. The term biota includes both plants and the domesticated and wild species of animals. There are over 45,000 plant species in India and 81,251 animal species. This represents about 7% of the world's flora and 6.5% of the world's fauna. Plants are the main source of food, fodder and other useful things, such as fuel (firewood), fibre, timber, medicine, gums and tannin. The Indian fauna includes a variety of animal life, such as mammals, birds, reptiles, fishes and insects – i.e. about 800 species of mammals, 2,000 species of birds, 420 species of reptiles, 2,000 species of fish, 50,000 species of insects, and 4,000 species of molluscs (Agarwal 1998). This biodiversity exists because of the forests. It makes its own tremendous socio-economic contribution to society. Therefore taking this parameter into account is very important when undertaking forest accounting.
- Forest Products: A forest product is any material derived from forestry for direct consumption or commercial use, such as lumber, paper, or forage for livestock. It is used for many purposes, such as fuel, or the finished structural materials used in the construction of buildings, or as a raw material in the form of the wood pulp used in the production of paper. All other non-wood products derived from forest resources are collectively described as non-timber forest products. (Thus the forest is the home of various products which have great economic value. Therefore, the potential of a forest to produce such products and the availability of and demand for such products – in other words, the output from the forest in the form of products of use to society – would be among the essential criteria for determining the economic value of the forest. Forest accounting considers all forest products and their economic value.

15.4 Valuation Methods

There are various forest valuation methods, such as:

- *Historical Cost Method*: The value of a forest is determined by adding up all the accrued investment, management, and operating costs. The basic approach consists of totalling all the costs that have been incurred since the acquisition of the forest (Wagniere 2011). In this forest valuation technique the historical cost is taken as the baseline for forest resources valuation.
- *Market Price Method*: The value of a forest is determined by adding up all the accrued investment, management, and operating costs on the basis of the costs

that would accrue if the forest had to be established again under current market conditions – i.e. current market price or replacement cost. Hence the value of a forest depends on the current observable market prices (Wagniere 2011). In this forest valuation technique the current market price of forest resources is taken into consideration for valuation.

- *Discounted Cash Flow Method*: A *Discounted Cash Flow* (DCF) is a valuation method used to estimate the attractiveness of an investment opportunity. DCF analysis uses future free cash flow projections and discounts them to arrive at the present value estimate, which is then used to evaluate the potential for invest-ment. If the value arrived at through DCF analysis is higher than the current cost of the investment, the opportunity may be a good one (Investopedia). The DCF calculation of a forest starts with an inventory of the current forest stand, fol-lowed by identification of the marketable forest products that will provide future cash flows. After that follows a projection of the future yields of the forest with regard to the identified products and the forecast of the net cash flows based on cost estimations and revenue estimations. The final step is discounting the net cash flows in the present, using an appropriate discount rate (Wagniere 2011). Although finding the appropriate discount rate can prove a challenge, DCF tends to provide more reliable results than the previously described techniques and at present is the method of forest valuation most widely used by experts.
- *Real Options Pricing Methods*: Also termed Real Options Valuation (ROV) or Real Options Analysis (ROA), these methods apply option valuation techniques to capital budgeting decisions. Real options confer the right, but not the obli-gation, to undertake certain business initiatives, such as deferring, abandoning, expanding, staging, or contracting a capital investment project (Wagniere 2011). When these models are applied to the domain of forestry, the forest owner's option constellation is provided by a forest property.
- *Sensitivity Analysis*: A sensitivity analysis is a technique used to determine how different values of an independent variable impact a particular dependent vari-able under a given set of assumptions. This technique is also useful in forest valuation because the parameters of forest valuation – such as the price of forest products, the discount rate, the length of the rotation, growth and yield assumptions, and the presumed management regime of the forest – are extre-mely sensitive (Wagniere 2011). In this valuation technique sensitive factors of forest resources are taken into consideration for valuation.

15.5 Forest Economics

This section contains more detailed information about the parameters which affect forest economics.

- *Forest Resources*: This means the various types of vegetation normally growing on forestland, the associated harvested products and the associated residue, including but not limited to brush, grass, logs, saplings, seedlings, trees and slashing. Forests are rich in various valuable resources. The economics of forest resources are very strong. Forest accounting must include all the forest resources and the related cost benefit analysis. This will make it easier to understand the contribution made by forest resources to the mainstream economy.

- *Goods and Services*: Forests produce significant products, such as paper, plywood, sawnwood, timber, poles, pulp, matchwood, firewood, sal seeds, tendu leaves, gums and resins, cane and rattan, bamboo, grass and fodder, drugs, spices and condiments, herbs, cosmetics and tannins. Additionally, forests produce various invisible ecosystem services, such as climate regulation, water regulation, pollution control, biological control, pollination, hazard control, biodiversity and soil erosion. All these goods and services have tremendous economic value and make a significant contribution to the economy. In forest accounting goods and services constitute a major component to be considered.
- *Business*: Forestry is a significant element of rural industry. It has tremendous potential to generate business, such paper mills, animal fodder, food processing, medicine, sawmills, furniture and crafts. Thus the business potential of forests is a significant factor in forest accounting.
- *Employment*: Forests generate huge employment for poor people. Millions and millions of people and animals are directly dependent on forests for their livelihoods. Therefore employment generation from forests is also one of the important aspects of forest accounting.

The above-mentioned forest accounting parameters provide excellent grounds for undertaking forest accounting and would definitely be helpful when developing a good forest accounting system. However, these parameters alone are not sufficient; there is a need to take additional factors into account.

Challenges of Forest Accounting: While forest accounting is very important, it is a difficult task. These are the challenges which are emerging in forest accounting:

Identification of Biological Assets: Identification of biological assets in forest accounting is a real challenge. Biological assets produce invisible economic ecosystem services which support life systems. They are hidden and versatile (like micro-organisms) and hence very difficult to identify.

Valuation of Biological Assets: After biological assets have been identified, they need to be valued. However, no systematic or perfect method exists for making a valuation of such biological assets.

Measurement of Bio-diversity Loss: An important component of forest accounting is the identification and measurement of biodiversity loss. Again there is no fixed standard available which measures biodiversity loss in monetary terms.

Compensatory Value: Again identifying the compensatory value of ecosystem services is difficult. There is no perfect methodology available for calculating the compensatory value for nature services.

Availability of Data: There is a lot of complexity and greyness in data related to forest resources and forest ecosystems. Unavailability of data leads to limitations in making an accurate analysis and interpretation.

Internalisation of Externalities: Forest resources and forest ecosystems are versatile phenomena on which external and internal aspects make an impact. In forest accounting, internalisation of externalities is a problem since there is no standard methodology available for assessing the impact of externalities in monetary terms.

Generally the above-mentioned challenges are dealt with by researchers in natural resources accounting. Resolving them is difficult because it involves various invisible services which are difficult to quantify or assign values to. However all these services have economic value.

15.6 Forest Accounting and Biodiversity

Forest biological diversity is a wide term that refers to all life forms found within forested areas and the ecological roles they perform. As such, forest biological diversity encompasses not just trees, but the multitude of plants, animals and micro-organisms that inhabit forest areas and their associated genetic diversity. Forest biological diversity can be considered at different levels, including the ecosystem, landscapes, species, populations and genetics. Complex interactions can occur within and amongst these levels. In biologically diverse forests, this complexity allows organisms to adapt to continually changing environmental conditions and to maintain *ecosystem functions*. The significant contribution of forest accounting helps to maintain biodiversity. There is a strong nexus between forest accounting and biodiversity. Forest accounting explores different services provided by forests to biodiversity and measures their impact on it.

Climate: Forests make a significant contribution to climate regulation and are essential for biodiversity. Forests are the only major ecosystem where the amount of carbon stored in the biomass of the plants exceeds that in the soil; deforestation therefore also affects climate regulation (Elmqvist et al. 2011). These regulatory services are provided by forests free of charge. This needs to be taken into consideration in forest accounting. Forest accounting would be of significant help in monitoring the financial cost of climate change, thereby fostering action to reduce the impact of climate change on the economy. Identification of the services to climate regulation provided by forests, and quantification of their monetary value, would be concrete steps in understanding and preserving biodiversity in the long run.

Water: Forest services play a major role in maintaining the water cycle. Without this, water regulation would not be possible. Forest and wetlands with intact groundcover and root systems are considered very effective at regulating water flow and improving water quality (Elmqvist et al. 2011). Water is an essential component of biodiversity preservation. Again, forest services for water regulation are invisible and not counted in monetary value. With forest accounting, these services could be quantified and their economic value identified so that the nexus between forest, water regulation and biodiversity could be recognised.

Biological Control: The loss of biodiversity can be attributed to deforestation, which affects biological processes. Forests protect biodiversity by providing a habitat for diverse species of flora and fauna whose own biological processes depend directly and indirectly on forests.

Erosion Prevention and Hazard Control: Vegetation cover is the key factor for preventing soil erosion (Elmqvist et al. 2011). Forests protect against landslides by modifying the soil moisture regime (Sidle et al. 2006). Thus forest services directly contribute to erosion prevention and hazard control, which are essential for biodiversity and sustainability. Accounting of forest ecosystem services recognises the value provided by forests to society.

Pollution Control: Forests play multiple roles when it comes to local air pollution. Trees in general help to reduce air pollution, including absorbing the greenhouse gas carbon dioxide, but some species contribute to local smog by emitting volatile organic compounds (VOCs). Selecting the planting location of individual trees and species makes a difference to the overall pollution balance (Lenart 2015). Forests are essential for reducing pollution in urban as well as rural areas. These forest services which absorb carbon dioxide directly contribute to biodiversity and sustainability.

Pollination: One of the most significant forest services is pollination. Pollination promotes biodiversity, and forests are responsible for assisting over 80% of the world's flowering plants. Pollination provides food to humans and animals, without which it would not be possible to meet food demand. It is an essential component of ecological survival. Without pollinators, the human race and all of earth's terrestrial ecosystems would not survive (USDA). Hence quantification of pollination services in monetary value is an essential aspect of forest accounting.

Biodiversity Loss: Unless biodiversity loss is identified and measured, adequate preventative action cannot be taken. Forest accounting using different parameters measures the loss of biodiversity. Identification and quantification of actual biodiversity loss would assist concrete policy-making to reduce biodiversity loss. Therefore, forest accounting is a useful tool in the quest for biodiversity preservation.

Trade-Off Between Agriculture and Environment: The impact of an undesirable trade-off could be assessed through forest accounting. 35% of the Earth's land surface is used for growing crops or rearing livestock (Lenart 2015). Due to industrialisation and agriculture, trade-offs between agriculture and the environment have become a complex issue which is impacting badly on biodiversity. Forest accounting would help to assess the trade-off situation.

Risk of Disaster: Forests play a significant role in reducing the risk of natural disasters, including floods, droughts, landslides and other extreme events. At global level, forests mitigate climate change through carbon sequestration, contribute to the balance of oxygen, carbon dioxide and humidity in the air, and protect watersheds, which supply 75% of freshwater worldwide. Natural disasters cause huge biodiversity loss and damage to the ecosystem that has negative implications for the economy. Forest accounting would help to identify the risk of disaster.

Sustainable Development: The practices of sustainable development are biodiversity-friendly. Forest ecosystem services support sustainable development by regenerating the ecosystem. The forest itself shows various ways of practising sustainable development. Forest accounting would also help in sustainable development.

Thus forest accounting offers significant input in various ways, such as protecting biodiversity, sustainability, ecosystems, livelihoods, economic development, industry and trade; employment generation; reduction of biodiversity loss and risk of disaster; true measurement of economic development; and generating accurate data for policy-makers etc. In the present context bio-economics is increasingly important in national policy-making. Green economies are the way ahead, and forest accounting can play a major role in helping to find the right path to transform existing economic models.

15.7 Conclusion

In India the forest sector is the second largest use of land after the agricultural sector. Forests help to maintain the ecological balance of the environment. Forest accounting is crucial for the proper maintenance of natural resources and the monetary accounting of forest resources. It is important for the management of forest areas and for understanding the availability of natural assets and the income earned from these assets. It demonstrates the importance of green economies and different methods of calculating the value of natural resources.

This study is also helpful to societies and industries dependent on forest products. They can understand and be aware of the forest resources they use for their own purposes. The forest is beneficial to agricultural systems and plays an important role as foster-mother in promoting the agro-industrial economy. Forests make a significant contribution to the Indian economy and to the state domestic production.

The specific purposes of forest accounting are: (1) To reduce loss of biodiversity. (2) To mitigate inflated economic production figures. (3) To enable value chain and supply chain accounting starting with net forest produce. (4) To enable the Gross National Happiness (GNH) calculation that is dependent on forest living and environmental standards. (5) To enable balanced economic growth keeping future economic concerns. (6) To enable balance in regional economic diversity. (7) To safeguard biodiversity (both plant and animal). (8) To assess the trade-offs between agriculture and environment preservation exercises. (9) To assess the nature of food safety networks based on area-specific nutrition availability and generate economic measures for balanced nutrition in all regions. (10) To promote rational international economic and diplomacy dialogues based on hard data. (11) To measure economic sustainability.

Hence it the necessity of making an in-depth theoretical analysis of forest accounting. It is an instrument to understand biodiversity, ecosystem services and the risks associated with the loss of them, so that potential disasters can be prevented. Investigating sustainability by undertaking proper forest accounting will definitely make difference.

References

Agarwal, K.C, 1998: *Environmental Biology* (Bikaner: Agro Botanical Publishers).
Elmqvist, T.; Tuvendal, M.; Krishnaswamy, J.; Hylander, K., 2011: *Managing of Trade Off in Ecosystem Services* (Nairobi: The United Nations Environment Programme).
Food and Agriculture Organization of the United Nations, 2011: Global Forest Resources Assessment, 2010: FAO Forestry Paper 163 (Rome: FAO): 12–13.
Gundimeda, H.; Sukhdev, P.; Sinha, R.K; Sanyal, S., 2007: "Natural resource accounting for indian states—Illustrating the case of forest resources", *Ecological Economics*, 61(4): 635–649.
Harris, M.; Fraser, I., 2002: "Natural resource accounting in theory and practice: A critical assessment", in: *The Australian Journal of Agricultural and Resource Economics*, 46(2): 139–192.
Haripriya, G.S., 2000: "Integrating forest resources into the system of national accounts in Maharashtra", *Environment and Development Economics* 5: 143–156 (Cambridge: Cambridge University Press).
Investopedia; at: https://www.investopedia.com.
Jordon, S.J.; Hayes, S.E.; Yoskowitz, D.; Smith, L.M.; Summers, J.K.; Russell, M.; William, B.H., 2010: "Accounting for natural resources and environmental sustainability: Linking ecosystem services to human well-being", *Environmental Science and Technology* (Washington D.C.: American Chemical Society).
Lenart, M., 2015: "Urban forest and pollution University of Arizona"; at: http://articles.extension.org/pages/58387/urban-forests-and-pollution (7 April 2016).
Lenart, M.; Jones, C., 2014: "Perceptions on climate change correlate with willingness to undertake some forestry adaptation and mitigation practices", in: *Journal of Forestry*, 112(6), 553–563.
Meyerson, L. et al., 2005: "Aggregate measures of ecosystem services: Can we take the pulse of nature?", in: *Frontiers in Ecology and the Environment*, 3(1): 56.
Mabugu, R.; Chitiga, M., 2002: "Accounting for Forest Resources in Zimbabwe", Centre for Environmental Economics and Policy in Africa (CEEPA) Discussion Paper Series, No. 7: 1–53 (Pretoria: University of Pretoria).
Mkanta, W.N.; Chimtembo, M.M.B., 2002: "Towards Natural Resource Accounting in Tanzania: A Study on the Contribution of Natural Forests to National Income", CEEPA Discussion Paper Series, No. 2: 1–53.
Parikh, K.S.; Ghosh, U., 1991: "Natural Resource Accounting From Soils: Towards an Empirical Estimation of Costs of Soil Degradation for India", Discussion Paper No. 48 (Bombay: Indira Gandhi Institute of Development Research [IGIDR]).
Sidle, R.C.; Ziegler, A.D.: Negishi, J.N.; Nik, A.R.; Siew R.; Turkelboom, F., 2006: "Erosion processes in steep terrain—Truths, myths, and uncertainties related to forest management in Southeast Asia", in: *Forest Ecology and Management*, 224(1–2): 199–225 (United States Department of Agriculture).
Wagniere, S., 2011: *Forest Valuation: A Knowledge-Based View* (St. Gallen: University of St. Gallen).
Ykhanbai, H., 2009: "Forest Resources Degradation Accounting in Magnolia", Forest Resources Assessment Working Paper 176 (Rome: FAO).

Chapter 16
Moringa Production and Consumption: An Alternative Perspective for Government Policy-Making

Goutam Saha and Mou Sen

Abstract Drumsticks (Moringa) have a tremendous potential to solve many issues of the world related to nutrition and general diseases with the support of proper institutions (The authors acknowledge all the respondents, primarily from the marginalised segment of West Bengal, for their valuable time and insights. The authors also acknowledge Professor Lipsa Mohapatra's efforts regarding data assimilation.). We scanned the literature to understand different benefits of Moringa with a major focus on its nutritional benefits. We analysed the situation of under-nutrition in India, strengths and weaknesses of different Moringa-producing organisations across countries, governments' policies on Moringa production and consumption and interviews with – middle-class/poor rural/semi-urban Indian women through questionnaires. We studied Moringa production and consumption to address the issue of undernutrition and interviewed different government officials who worked in this area. The literature review and empirical evidence collected helped the authors to argue for a holistic, sustainable community-driven organi-sation that will be deep-rooted with local market networks. This will help in effective Moringa production and consumption in India to address the malnutrition challenges of our country.

Keywords Moringa · Malnutrition · Sustainable community development Local economy · Nutritional needs

16.1 Introduction

The idea and practice of industrial agriculture have limitations with regard to solving the increasing food, water and environmental crisis. Hence, sustainable agriculture has arisen to solve these crises. Sustainable agriculture aims to provide environmental stewardship and farm profitability in a more efficient way.

Goutam Saha, Associate Professor, NIFT, Bhubaneswar, India; Email: goutam.saha@gmail.com

Mou Sen, Joint Director, Department of MSME&T, Government of West Bengal, India; Email: mailmousen@gmail.com

© Springer Nature Switzerland AG 2019
A. K. J. R. Nayak (ed.), *Transition Strategies for Sustainable Community Systems*,
The Anthropocene: Politik—Economics—Society—Science 26,
https://doi.org/10.1007/978-3-030-00356-2_16

Sustainable agriculture also develops a prosperous farming community. To attain these objectives, Moringa is an excellent plant to study. Moringa has exceptional nutritional value, as it is a very rich source of protein, vitamins and minerals. The shelled Moringa oleifera contains 36.7% proteins, 34.6% lipids, 5% carbohydrates, and all essential amino acids. It has great medicinal values as well. In countries like India, where undernutrition, poor health and unemployment are very significant issues, producing and consuming Moringa leaves and seeds through sustainable farming may lead to the development of a sustainable community system.

The most popular species of the Moringaceae family is *Moringa oleifera*. This tree is native to India. It has primarily spread to different parts of the world from India and it has high ecological plasticity. Because of its high ecological plasticity, it can be grown in different sorts of soil and has the capacity to adapt to different temperature, humidity and rainfall conditions. For farming, Moringa has a very low input requirement. Although the Moringa tree is very adaptable, it grows particularly fast in tropical regions. It bears fruit in less than a year and grows ten metres high in two years. Due to its fast growth rate, Moringa is harvested multiple times in an economic way without too much external input. All parts of the Moringa tree are useful. Apart from its nutritional and medicinal values, Moringa is also used as a fertilizer. Due to its high carbon dioxide absorbency factor – about fifty times that of the cedar tree – it is very good for the environment and consequently highly effective in urban forestry, farming and sustainable community development. Moringa grows five metres a year, producing nourishing leaves and seeds in an efficient and productive way. After oil extraction, the powdered seeds can be used as a coagulant to purify water. The sediment which remains after the purification process is an extremely suitable manure for organic agriculture. All these uses raise farm income. Therefore, we find Moringa an excellent plant to study for building sustainable agriculture and sustainable community systems in rural and urban pockets of India and many other countries where Moringa grows extensively and effortlessly.

16.2   Moringa: The Great Nutrient

Moringa oleifera Lam (popularly known as the drumstick tree) has exceptional nutritional and therapeutic properties. Globally it is considered to be one of the foods that can potentially eradicate malnutrition and contribute significantly to preventative health care. So it may play a crucial role in the welfare of the citizens of Third World countries. *Moringa oleifera* contains 36.7% proteins, 34.6% lipid and 5% carbohydrate. The unshelled Moringa Oleifera contains 27.1% proteins, 21.1% lipids, and 5% carbohydrate. Many researchers have documented this (Dixit et al. 2016). It can have a greater importance in the diet as it contains all amino acids and antioxidants (ascorbic acids, flavonoids, phenols and carotenoids), high levels of valuable vitamins (like vitamins A, B1, B2, B3, C, E and K), and important minerals like calcium, iron, potassium, copper, magnesium and zinc.

Apart from leaves, Moringa flowers are rich in calcium and potassium. "For a child aged 1–3 years, a 100 gm serving of fresh Moringa leaves would provide all his daily requirements of calcium, about 75% of his iron and half his protein needs, as well as important supplies of potassium, Vitamin B complex, copper and essential amino acids. As little as 20 gm of fresh leaves would provide a child with all the vitamins A and C he needs" (Gopalakrishnan et al. 2016).

Additionally, 100 gm of fresh Moringa leaf contains four times the vitamin A of carrots, seven times the vitamin C of oranges, four times the calcium of milk, three times the potassium of bananas, three-quarters of the iron of spinach, and twice the protein of yogurt. When leaves are dried, some of the vitamin levels decrease. However, drying significantly increases the amount of most other nutrients present in the leaves. For example, 100 gm of dried Moringa leaf has ten times the vitamin A of carrots, half the vitamin C of oranges, seventeen times the calcium of milk, fifteen times the potassium of bananas, 25 times the iron of spinach, and nine times the protein of yogurt. Cooked Moringa leaves lose 32% of the vitamin C content. However, the availability of beta-carotene is enhanced when Moringa leaves are cooked. The cooked Moringa leaves also improve the stimulation of iron (Moringa News Association of Ghana). Moringa fruit is also rich in protein, essential amino acids, and multivitamins.

Dried Moringa leaf powder (ground dried leaf) has emerged as a nutritional supplement in the Western market and is positioned as a "green superfood" which is available in health stores and online. It has thus become a premier product for high-end consumers.

16.2.1 Medicinal Properties of Moringa

There are some other nutritional facts that have made Moringa a superfood and a very important food for preventative health care. Various chemical tests conducted in the last few years in different countries have proved that Moringa has the properties of an antioxidant, which reduces the process of ageing. It works well in managing respiratory, cardiovascular, gastrointestinal, and endocrine and central nervous system related diseases. It helps in building the immune system and has medicinal uses as an antibacterial, antipyretic, antispasmodic agent. It is useful for improving liver and renal function, helps to dissolve kidney stones, and prevents hepatotoxicity. It has an antifungal effect and works well against skin-related diseases. It brings down inflammation, manages rheumatoid arthritis and has a protective effect against ovarian cancer. In many regions of Africa, Moringa is used to treat patients with diabetes, hypertension and immunodeficiency (Ramos 2015).

Let us discuss its different medicinal properties in the following five important categories:

- *Antimicrobial Activity*: The antimicrobial activities of *Moringa oleifera* leaves, roots, and barks have proved effective to human beings. The antibacterial effect of an aqueous and ethanolic extract of *Moringa oleifera*was also proved in the laboratory set-up. Many of the currently used antibacterial products are associated with negative effects, such as toxicity, hypersensitivity, immunosuppression and drug resistance. These disadvantages necessitate the search for alternative remedies to treat bacterial diseases. Moringa can be a very useful plant in this regard.

- *Anti-inflammatory Activity*: In most cases, inflammation is the body's response to another external process, rather than a disease or illness in its own right. This response is an anti-immune response and is a necessary part of the healing process. However, it creates serious problems when inflammation becomes chronic or extreme in its duration and extent. Moringa helps to control such types of inflammations.
- *Anticancer Activity*: *Moringa oleifera* has many characteristics that make it a superior complement to cancer treatment or prevention plans. It is loaded with nutrients like protein and different types of vitamins and minerals that promote a healthy body, which is an effective tool for fighting cancer. It is known to have anti-viral, antioxidant, anti-allergenic and pain-relief properties, and it can be used to fight a variety of infections. These properties make Moringa very important for cancer treatment.
- *Anti-cholesterol activity*: *Low-Density Lipoprotein* (LDL) is known as bad cholesterol. It causes lipid deposits in blood vessels and contributes to heart disease, stroke, and other cardiovascular diseases. Moringa leaf extract contains powerful diuretic medicine that can reduce the level of bad cholesterol in the blood.
- *Anti-diabetes activity*: Diabetes is a disease that is related to the secretion of the hormone insulin. In some cases, the pancreas does not produce any insulin at all. In other cases, the body does not react in the right way to insulin. Ultimately, the pancreas produces an insufficient volume of insulin. In both the cases, Moringa is used to treat and manage the symptoms of diabetes for years (Dixit et al. 2016).

However, in this paper, we have primarily focused on Moringa as a nutrient for addressing undernutrition, as it is the single biggest health problem of the world, and affects both human development and economic development in India very significantly (Antony/Laxmaiah 2008).

With a plethora of medicinal properties and nutritional element, Moringa has the power to overcome the problem of malnutrition and contribute significantly to health care in a very cost-effective way. In 1997–98, Alternative Action for African Development and Church Worldwide Service proved the ability of Moringa leaf powder to prevent and cure undernutrition in pregnant and breastfeeding women and their children in Southern Senegal (Dhakar et al. 2011). Unfortunately, many people in the areas of the Earth where Moringa grows are significantly affected by malnutrition. India, the original land of Moringa, where it grows abundantly with very little human input, is also the home of the world's largest number of children with acute undernutrition (Deaton/Drèze 2009).

16.2.2 Undernutrition in India

Undernutrition has emerged as a major challenge in modern India, though the country has grown economically fast in the last couple of decades. Undernutrition levels in India remain higher than the most countries of Sub-Saharan Africa, though these countries are poorer than India and have grown very slowly in comparison to India. In the list of countries with the highest level of child undernutrition, India's rank was third. Even countries like Yemen, Sudan, Burundi and Afghanistan have been better than India at combating child undernutrition (Deaton/Drèze 2009). Nutritionists have been raising their voices about a range of nutrient deficiencies, including essential minerals and vitamins, amino acids and fats. Gopaldas (2006) argued that all members of the low-income group (and even middle-income group) Indian families are likely to be deficient in vitamins and minerals. In the age group of four to six years, the ratio of average intake to "recommended daily allowance" is only 16% for vitamin A, 35% for iron and 45% for calcium. Today, anthropometric indicators for both adults and children are the worst in the world. In this context, we studied government policies to address undernutrition in India.

The Sustainable Development Goals 2030 gave global leaders an opportunity to develop strategies to end undernutrition. It is important to know that investing in nutrition has both a development and a human rights perspective. It improves the human development indicator on the one hand and, on the other hand, protects marginalised and vulnerable people and promotes health, education, employment and dignity of life.

Nutrition has been part of the national policy since the Eighth Five-Year Plan. In 1993 the National Nutrition Policy advised a comprehensive intersectoral strategy for combating the problems of malnutrition and towards improving the nutritional status of children, adolescents, girls and women. However, the National Family Health Survey (NHFS-3 2005–06) data said that there is a long way to go, as children in India continue to suffer from a different kind of malnutrition. About 43% of five-year-old children were underweight, 48% too short for their age, and 20% too thin for their weight. However, the proportion of underweight children below three years old decreased from 34% in 1998–99 to 31% in 2005–06. The decrease was marginal in rural areas, from 45% in 1998–99 to 44% in 2005–06 (International Institute for Population Sciences 2007).

The figures are higher in rural areas than urban areas. The NFHS-3 data also emphasises two significant causes:

- The inter-relationship between the nutritional status of children and their mothers.
- Direct correlation between the decrease in undernutrition and the wealth index of a household.

The Eleventh Five-Year Plan accorded high priority to addressing both child and maternal nutrition with a multisectoral approach. In 2012 the Ministry of Women and Child Development focused on adolescent anaemia control. Undoubtedly, there has been progress; however, the rate of progress has been slow. Even the then Prime Minister, Dr Manmohan Singh, lamented this issue (Nisbett 2017). Taking into consideration the gaps of implementation in the earlier phases, the Twelfth Five-Year Plan focused on nutritional security for all. The Twelfth Five-Year Plan emphasised household food security, the strengthening of institutional arrangements, and inter-departmental convergence to address undernutrition through a life-cycle approach.

According to NFHS-4 (2015–16) data, there have been improvements in the areas of children being stunted, wasted and underweight, but it still remains an uphill and a long journey (Sengupta et al. 2016).

Overcoming this massive nutrition deficiency and condition of the bottom of the pyramid, we need to prioritise this issue and think creatively. In this backdrop, Moringa production and consumption may provide a sustainable solution.

16.2.3 Objectives of This Chapter

This chapter aims to find the organisations across the world that are considered to be Moringa producers/sellers in the global and local market and analyse their focus with regard to addressing the undernutrition issues of the local community.

This paper also analyses the policies of different governments regarding the promotion of Moringa production and consumption to combat undernutrition.

This paper additionally aims to discover the awareness level of the rural and urban poor of West Bengal regarding Moringa as a great nutrient and analyse their willingness as a segment to produce and consume Moringa.

This paper further aims to collect government officials' opinions regarding the current practices of governmental and NGO initiatives on Moringa production and consumption to combat malnutrition.

This paper scanned the evidence across the globe and provides a few policy guidelines to both National and State Governments of India regarding Moringa production and consumption for addressing malnutrition (Table 16.1).

16.2.4 Indian Moringa Producing Organisations

We found that Indian Moringa-producing organisations are primarily export-driven. However, some of them have involved rural communities in their production process. They are also efficient, as 80% of global Moringa exports come from India. However, these organisations hardly fight against undernutrition in local communities.

Table 16.1 A few Moringa powder producing organisations

Name of the organisation	Description
Bacca Villa Organic Co.	Bacca Villa Organic Co. started at Siem-Reap, Cambodia in 2008, is a Dutch non-profit organisation which helps the Cambodian population by means of various projects. The projects aim to make people self-dependent. Since 2010, they have focused on making *Moringa oleifera* tea, powder, tablets, and oil soaps with certification. They cater for both the Cambodian domestic market and the global market (Ramos 2015)
Small Holder Farmers Alliance (SHFA)	60% of the farming land of Haiti is occupied by small-scale farmers and they contribute 40% of the country's GDP. After major deforestation, Haiti focused on reforestation. With the Government's support, SHFA focuses on the Moringa tree to address malnutrition issues and reforestation. After successful commercial production and testing the nearby market, it has begun exporting Moringa powder and oil (Smallholder Farmers Alliance 2015)
Dago Network Youth Group	This is a Kenya-based business organisation that produces *Moringa oleifera* nutritional supplements like Moringa powder and porridge to address malnutrition in the schoolchildren of Kenya in a Government-aided feeding programme for schools (Odeyo 2014)
Moringa Connect	Moringa Connect is a social enterprise of Ghana that has trained, supported and financed 1,500 farmers in Moringa planting. Moringa Connect has developed Moringa base oils, food which they sell globally through an organised network (http://moringaconnect.com).
Kulikuli	The Oakland, U.S. based social enterprise farm Kuli Kuli makes a special nutrition bar made from *Moringa oleifera*. Moringa grows in predominantly subtropical areas of West Africa where malnutrition is most prevalent. Kuli Kuli has addressed this anomaly. Made from raw Moringa, Kuli Kuli bars are gluten-free, low in calories and contain a high level of fibre, protein, and vitamins. Kuli Kuli's founder and CEO Lisa Curtis stated about their Nigerian operations "We came up with the idea for Kuli Kuli to support women co-operatives to grow more Moringa to nourish their communities and sustain their efforts by selling a small portion of their harvests in the United States in the form of delicious nutritional bars." www.kulikulifoods.com
Sakala	As well as exporting Moringa leaves as a superfood, Sakala, a Haitian NGO based in Cité Soleil, explores the commercialisation of Moringa leaf powder in a lower income group of people living in Cité Soleil. The organisation deploys local youths to sell one serving package of Moringa leaf powder at USD 0.08, a price low enough to be affordable for the local community (Smallholder Farmers Alliance 2015)

Source The authors

16.2.5   Different Government Policies for Moringa Production and Consumption

Ghana: As malnutrition is a big challenge in Ghana, the Government found the need for food fortification and usage of food supplements. They focused on Moringa for this purpose.

Zambia: The Zambian Government focused on Moringa production and consumption aims to reduce malnutrition and to develop their economy from other than the dependence on mining sector (Stolarz 2015).

India: The Indian Government is going to establish nutria-farms in the districts most affected by undernutrition especially among children. The Government has also identified a few bio-diversified food crops enriched with critical micronutrients to combat malnutrition. Moringa is one of them. At state level, Tamil Nadu Government is already encouraging cultivation by granting 1.5 crores for Moringa production, which will primarily cater for export and high-end Indian consumers (Shivarajah 2015). In some parts of Andhra Pradesh and Kerala Moringa production is also gaining popularity.

Cuba: Cuba is investing heavily in Moringa and started mass production of it with the Government's intervention. They have established a China-Cuba Moringa Olivera Science and Technology Centre Co-operation with research facilities in Yunnan, China and Havana, Cuba. The Cuban Biotechnology Institute has been extensively researching the anti-cancer benefits of Moringa. Cuba has created a condition whereby large-scale Moringa production is possible. However, Cuba produces Moringa primarily for its own people, not for export (Toensmeier 2012). In this context, the Cuban approach is: "The hungriest people grow food for a living. Investing in agriculture is the most effective method of reducing poverty. Raising awareness of Moringa's nutritional value will create a rise in demand and production and consequently consumption will occur in the communities where it is grown. This creation of a market for Moringa will create economic opportunities in the places it grows – the poorest and most malnourished parts of the world" (Blog 2015).

16.2.6   Opinions of Lower-Middle-Class Poor Women About Moringa Production and Consumption to Address Undernutrition

We have collected opinions regarding Moringa production and consumption from fifty randomly selected women of rural/semi-urban areas of West Bengal. They work primarily as maids/cooks in different houses and their incomes lie between Rs. 4,000 and 12,000 per month. Their family size is around two to six and their age group is 30–55. Their opinions are as follows:

16.2.6.1 Observations

- 84% of respondents are very concerned about the nutritional welfare of their family members.
- 50% of respondents think nutritious foods are very costly and unaffordable.
- 52% of respondents think that they cannot produce nutritious food in their kitchen garden or their backyard or community garden.
- 84% of respondents do not have complete information that Moringa leaves contain a huge amount of proteins, fat, minerals, vitamins and amino acids which can meet much of their nutritional requirements.
- 84% of respondents do not have complete information that Moringa leaves, flowers, and seeds have the potential to manage hypertension and cardiovascular, respiratory, renal, gastrointestinal and nerve-related diseases.
- 50% of respondents like to include Moringa leaves regularly in their diet.
- 50% of respondents think there will be much smaller supply of Moringa leaves if they start taking them regularly.
- 66% of respondents think that producing Moringa leaves at home by planting a Moringa tree in their courtyard, rooftop or kitchen garden will enable them to meet the family demand for Moringa leaves.
- 80% of respondents have a small yard to plant a Moringa tree.
- 80% of respondents know that they can grow this plant in their kitchen garden or on their rooftop.
- 80% of respondents are interested in producing Moringa at home in a kitchen garden/rooftop/community garden in their locality.

16.2.7 Views of Government Officials

We have interviewed a few government officials regarding Moringa production and consumption. Most of the Government officials agreed to encourage citizens to produce and consume Moringa seeds, flowers and leaves to obtain their nutritional requirements. The government officials of West Bengal have agreed on the possibility of involving self-help groups to produce Moringa leaf powder commercially and sell it to the local market to address nutrition issues in the local community. However, some of them have recommended production of a hybrid variety of Moringa in addition to the local variety as per the agro-climatic conditions of that area. And a few of them have identified a need for the Government to plant a sizable number of Moringa trees before self-help groups venture into commercial production of Moringa leaves. Government officials have provisionally agreed to encourage businesses to brand and market Moringa. However, they have also emphasised the importance of developing the bargaining capacity of producers through orientation and training.

There is also immense potential for the production of Moringa in community gardens, roadsides, backyards and kitchen gardens. But forward linkages need to be developed to make sure it will sell in nearby cities, towns and villages. Also, a massive awareness campaign is required to develop consumers in the local area.

16.2.8 Inferences

We have considered the respondents from lower-middle-class and below the lower-middle-class backgrounds. In those segments of the Indian population, there is a considerable threat of malnutrition, and 84% of our respondents are very concerned about the nutritional welfare of their family members.

Our dataset reveals that 83.7% of the respondents are not fully aware that Moringa leaves, which are readily available in their locality, have the power to combat malnutrition and prevent and cure many diseases. So there is a requirement to disseminate information about the benefits of Moringa leaves, as occurred in earlier successful promotional programmes like the polio eradication campaign in India.

At the same time, our respondents are interested in producing Moringa leaves in their kitchen gardens or country yards (the mostly widely available areas of the requisite size for Moringa tree cultivation). Hence, mass production of Moringa leaves for the common people should be encouraged to combat malnutrition issues. We found only 50% of respondents are interested in adding Moringa leaves to their daily menu, even after providing them with adequate knowledge about the benefits of Moringa leaves, mainly because of their tasteless character. So, promoting the different types of delicious dishes that can be made from Moringa leaves while keeping the nutritional value intact may be necessary and may open another arena of food processing research.

16.3 Concluding Remarks and Scope

India supplies 80% of the world's Moringa leaf powder requirement. As Moringa leaf powder and other varieties are being popularised as a superfood in the US and European market, Indian exports of Moringa are growing exponentially. India, the highest producer of Moringa, could provide a significant focus on Moringa production and consumption by rural and urban masses to combat undernutrition. At midday meal and ICDS centres, a successful governmental step to bring poverty-ridden kids into the education system, Moringa powder and soup could be introduced as an effective practice to address the undernutrition issue. We also need to create social organisations like Sakala to produce Moringa bars and powder primarily for the local community by self-help groups of farmers. However, a small part of their production could be exported for the sustenance of the business. In

India, we have found a number of export organisations who are doing very well at Moringa powder exporting. These organisations may grow more and can create huge job opportunities if they focus on the rural and urban local middle class and the bottom of the pyramid market in India. But we did not find any organisation which aims to overcome the undernutrition issues of poor Indians by producing Moringa powder; and, from a Government policy perspective, we have found hardly any substantial policy to combat malnutrition through Moringa powder production and consumption.

However, future work on Moringa production has a wide application in India. It can be used as a weapon to combat undernutrition among children in rural and urban communities.

We need holistic, community-determined organisations with deep local market networks for effective Moringa production and consumption to address the malnutrition challenges of India. There is huge scope for research into the feasibility and success criteria of these community-driven organisations, which use local resources and local knowledge to provide low-cost value-added solutions to local problems. From the policy-makers' perspective, education and research infrastructure need to be built locally to provide knowledge support to these sustainable community-driven organisations.

References

Antony, G.M.; Laxmaiah, A., 2008: "Human Development, Poverty, Health and Nutrition Situation In India", in: *The Indian Journal of Medical Research*, 128 (August): 198–205.

Blog, K., 2015: "What Kuli Kuli and Fidel Castro Have in Common", at: https://blog.kulikulifoods.com/2015/04/28/fidel-castro-moringa/ (12 May 2016).

Deaton, A.; Drèze, J., 2009: "Nutrition in India: Facts and Interpretations", in: *Economic and Political Weekly*, 44 (February): 42–65.

Dhakar, R.; Pooniya, B.K.; Gupta, M.; Maurya, S.D.; Bairwa, N.; Sanwarmal, 2011: "Moringa: The Herbal Gold to Combat Malnutrition", in: *Chronicles of Young Scientists*, 2 (3): 119, at: https://doi.org/10.4103/2229-5186.90887 (5 June 2016).

Dixit, S.; Tripathi, A.; Kumar, P., 2016: "Medicinal Properties of *Moringa oleifera*: A Review", in: *International Journal of Education and Science Research Review*, 3 (April): 2,348–6,457, at: www.ijesrr.org (14 December 2016).

Gopalakrishnan, L.; Doriya, K.; Kumar, D., 2016: "*Moringa oleifera*: A Review on Nutritive Importance and Its Medicinal Application", in: *Food Science and Human Wellness* (Beijing Academy of Food Sciences), 5 (2): 49–56, at: https://doi.org/10.1016/j.fshw.2016.04.001 (13 December 2017).

International Institute for Population Sciences, 2007: "National Family Health Survey (NFHS-3), 2005–06: India: Volume I, in: *International Journal of Health Care Quality Assurance*, Vol. 1", at: https://doi.org/10.1108/ijhcqa.2005.06218gab.007 (14 November 2015).

Kuli Kuli Foods, n.d.: "Kuli Kuli Foods—A Moringa Superfood Company", at: https://www.kulikulifoods.com/ (3 October 2016).

NN, 2005: "Improving Lives through the Moringa Tree", at: http://moringaconnect.com/.

Nisbett, N., 2017: "A Narrative Analysis of the Political Economy Shaping Policy on Child Undernutrition in India", in: *Development and Change*, 48, 2: 312–338, at: https://doi.org/10.1111/dech.12297 (4 June 2018).

Odeyo, J., 2014: "Leveraging Business for Social Change: Building the Field of Social Business Changemakers", at: https://www.changemakers.com/socialbusiness/entries/moringa-oleifera-nutritional-supplement (28 September 2015).

Ramos, J., 2015: "Organic Entrepreneurship and Moringa in A Case Study: Baca Villa Co.", at: https://www.researchgate.net/publication/317356339_Organic_enterpreneurship_and_Moringa_in_Cambodia_A_case_study_Baca_Villa_Co (19 November 2016).

Sengupta, S.; Singh, B.; Prasanna., R., 2016: "Breaking the Cycle of Malnutrition in India in Fulwari, Surguja; Chhattisgarh; Aman Persera, Meghalaya", in: *ASCI Journal of Management*, 45 (September): 20–35.

Shivarajah, P., 2015: "Moringa Cultivation Gets a Boost as TN Grants Rs 1", in: *Times of India*, 3 November, at: https://timesofindia.indiatimes.com/city/madurai/Moringa-cultivation-gets-a-boost-as-TN-grants-Rs-1-5-cr-to-popularise-it/articleshow/49640706.cms (15 May 2017).

Smallholder Farmers Alliance, 2015: "Moringa: Export Market Potential for Smallholder Farmers in Haiti".

Stolarz, S., 2015: "Could a Miracle Tree Help End Malnutrition in India?", in: *BBC News*, 2015, at: http://www.bbc.com/news/business-30504720 (16 April 2016).

Toensmeier, E., 2012: "Cuba Mass planting Moringa and Mulberry", at: http://www.perennialsolutions.org/cuba-mass-planting-moringa-and-mulberry (24 September 2017).

Chapter 17
Cotton Farming in India: Alternative Perspectives and Paradigms

Lipsa Mohapatra and Goutam Saha

Abstract In the present context, cotton farming in India is far from being a Sustainable Agricultural System. India is the second-largest producer of conventional cotton after China. More than 90% of the cotton is produced from genetically modified, pest-resistant, high-yielding-Bt seeds. Before 2002, cotton cultivation in India relied on natural farming techniques using indigenous seeds and some hybrid varieties. A report from the Textile Exchange (Cotton—India, material snapshot, 2016a, Organic cotton market report, 2016b, Organic cotton material snapshot material scenario common uses in apparel and footwear, 2016c) says that, after a series of protests against the Bt cotton, many NGOs and local farmers' associations in India have resorted to sustainable farming methods. Today, India is the largest producer of organic cotton and contributes about 70% of the world's supply. Through the review of literature on the impact of Bt cotton and organic cotton from both a farming and fashion and a textile industry perspective, the authors have tried to emphasise the importance of organic cotton farming which is sustainable, eco-friendly and generates a healthy livelihood for farmers.

Keywords Bt cotton · Organic cotton · Sustainable agricultural systems Organic standards · Sustainable fashion brands

17.1 Introduction

Cotton farming has been predominant in India since the Harappa and Mahenjodaro civilisations (approximately 3000 B.C.). Fine quality cotton cloth was exported to Greece in about 83-30 BC. The Mughals and British also imported fine quality cotton from various parts of India and Bangladesh. The English started importing cotton fibres from India to fuel their textile mills during the Industrial Revolution.

Lipsa Mohapatra, Associate Professor, NIFT, Bhubaneswar, Odisha, India; Email: lipsa.mohapatra@nift.ac.in.

Goutam Saha, Associate Professor, NIFT, Bhubaneswar, Odisha, India; Email: goutam. saha@nift.ac.in.

© Springer Nature Switzerland AG 2019   195
A. K. J. R. Nayak (ed.), *Transition Strategies for Sustainable Community Systems*,
The Anthropocene: Politik—Economics—Society—Science 26,
https://doi.org/10.1007/978-3-030-00356-2_17

The East India Company attempted to grow American cotton varieties in 1853 to create an alternative production source for its textile industry other than America. But despite all efforts, India used to produce 95% of desi variety (Boreum and Verbaceum) cotton and excelled globally in cotton farming. These seeds could be preserved for the next sowing season. After independence, there was a dominance of hybrid seeds and chemical fertilizers in cotton farming. Twenty major hybrid cotton varieties occupied nearly 40% of the total area and contributed to 60% of the total cotton production. From 1966–67 to 1996–97, cotton production increased from 5.6 million bales to 16.8 million bales. The area under cultivation increased from 7.8 million ha to 8.9 million ha (Santhanam/Sundaram 1997).

Andhra Pradesh, Maharashtra and Gujarat are currently the highest cotton-producing states in the country. Around 10 states with a total acreage of 118 lakh ha have produced a total of about 330 lakh bales in 2015–16. Maharashtra leads in both acreage (38.4 lakh ha) and production (76.7 lakh bales) (Vora 2016).

Mostly medium and long staple cotton is produced in most states of the country using hybrid or Bt varieties. The northern states like Punjab, Haryana and Rajasthan grow medium and short staple fibres and are mostly irrigated. Planting is done towards the end of April-May. Gujarat, Maharashtra, Madhya Pradesh, parts of Odisha, Andhra Pradesh, Karnataka, and Tamil Nadu grow medium to extra-long staple fibre and are largely rain-fed. Planting starts with the onset of the monsoon and continues until the end of August (Department of Agriculture and Cooperation GOI 2014).

According to the data compiled by the *Cotton Advisory Board* (CAB) and the *Ministry of Textiles* (MoT), while acreage is still increasing production has constantly declined since 2006–2007 from 554.5 kg/ha to about 506 kg/ha in 2014–2015. The outlook is far from promising, as both export prices of cotton and exports from India are decreasing, leading to a decrease in local cotton prices as well.

17.2   Bt Cotton (*Bacillus Thuringiensis*)

Bt seeds are *genetically modified* (GM) seeds that are high-yielding and pest-resistant in nature. The GM seeds are made by the terminator technology that strips the seeds of further reproduction capabilities. Thus once sowed in a season, the seeds generated by the new flowers cannot be stored for the next sowing period; which means that the farmer has to buy new seeds every year. These seeds will result in high yields only when they are well irrigated. The farming also requires a higher input of chemical fertilizers and pesticides (Yusuf 2010).

Also, Bt cotton is a cash crop and is mostly grown on a plantation scale. It is produced as a mono crop i.e. single crop plantation. This leads to excessive soil degradation, loss of surface water, ground water pollution and proliferation of insects, pests and disease-causing pathogens, like bollworms, aphids, jassids, thrips and whitefly (Gopalakrishnan et al. 2007). Thus if the crop fails, the farmer is left with neither any alternative source of income nor access to any other food crops to meet the minimum food requirements of his family and livestock (Mayee et al. 2011).

17.2.1 Bt Cotton in India

In 1995, Indian farmlands were used as testing grounds for Bt cotton seeds. In 2002, Maharashtra Hybrid Seeds Co. (Mahyco) entered into a joint venture with Monsanto, an American multinational agrochemical company, and started selling Bt cotton in India. Every stage of the industrialisation of cotton rested on violence, and India is no exception to this (Beckert 2015). Due to many issues related to Bt cotton, such as higher input cost, farmers' indebtedness, and poor pecuniary conditions, there was a steady movement among farmers for using *desi* cottonseeds and replacing hybrid Bt cottonseeds (Santhanam/Sundaram 1997). However, the seed market was flooded with Bt seeds and it was difficult to obtain the desi varieties. This not only led to a monopoly of Monsanto over the seed market but also over the prices they offered and the promises they made, leading to a state of "agricultural imperialism" (Orphal 2005).

17.2.2 Organic Cotton

Organic cotton farming is the process of growing cotton completely free from the application of any chemical pesticides, fertilizers, herbicides, etc. The crop is mostly rain-fed and does not pose any threat to human health, cattle, birds and insects or the environment as a whole. It represents an alternative method of farming that is self-sustainable. It uses many natural techniques of plant growth, pest and weed control, mulch and compost as fertilizer, intercropping for weed control and food crises management, and crop rotation for soil nutrition and soil moisture retention management (Textile Exchange 2016a; Dorugade/Satyapriya 2009).

17.2.3 Organic Cotton in India

India is currently the largest producer of organic cotton in the world, followed by China, Turkey, the USA and others (Textile Exchange 2015a, b). Commercial organic cotton is now grown in more than 22 countries worldwide, including the Sub-Saharan countries like Africa, Uganda, Tanzania and others (Afari 2010).

From a decent production of about 40,000 tonnes in 2005–2006, organic cotton production in India went up to 241,697 tonnes, accounting for 81% of global organic cotton production in 2009–10. But due to the decrease in cotton fibre prices, production decreased by 3.8% in 2015–2016. However, India is still the pioneer in organic cotton production, contributing around 74% of the world share in 2015 (Nagarajan 2012).

According to Statista and the US Department of Agriculture (2016), India still leads the world in organic cotton production at 5,748,000 metric tonnes, followed by China at 4,790,000 metric tonnes and USA at 2,680,000 metric tonnes (US Department of Agriculture 2018). It shows a positive trend towards the adoption of organic cotton farming majorly by Indian farmers. It also indicates that India is facing a major challenge in the world market due to the rapid production growth of China, USA, Pakistan, Brazil and other countries, leading to reduced exports from India (Textile Exchange 2017a).

17.2.4   Impacts of Bt Cotton Cultivation in India – Why Switch to Alternatives?

17.2.4.1   Bt Cotton Production Versus Organic Cotton Production (Higher Input Costs)

Initially, Bt cotton cultivation led to higher yields, but gradually the cost of cultivation was impacted by the high cost of seeds, pesticides and fertilizers. Monsanto completely controlled the seed market in India since its entry in 2002 and hence cheap indigenous varieties of seeds were unavailable on the market. The seed costs were Rs. 1,600 per packet of 450 g. The trait fees were as high as Rs. 1,200. A royalty fee or trait fee is the amount given by a company to its licence holders (Orphal 2005). The farmers were uneducated about the usage of pesticides for Bt cotton and indiscriminately used pesticides 20–30 times as against the requirement of 15 times. Thus the pests developed resistance and re-infested the Bt cotton, thereby lowering production. New pests started attacking the fibre, necessitating an ever great number of pesticide applications (Shetty 2004). The new variety of Bt II currently costs around Rs. 830–Rs. 1,100 per packet, and the royalty has decreased to around Rs. 164 (Directorate of Cotton Development Government of India 2017).

According to the data collected from farmers in Maharashtra and Madhya Pradesh by Textile Exchange in 2011, the following inferences were drawn and are depicted in Figs. 17.1 and 17.2.

From the above graphs it can be inferred that the profit generated from organic cotton production is the highest – at about 85% – in relation to the input costs, which are relatively low – about 49% of the output. However the gross income generated value was much less than that from Bt cotton. Figure 17.2 shows that although Bt cotton production gives the highest yield, its input cost is about 80% of the output, which is also the highest. Hence, the situation seems alarming, as the producers appear to have never really understood this huge difference.

Fig. 17.1 Input cost versus income. *Compiled by*: The authors

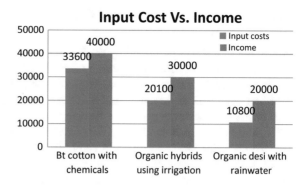

Fig. 17.2 Profits earned. *Compiled by*: The authors

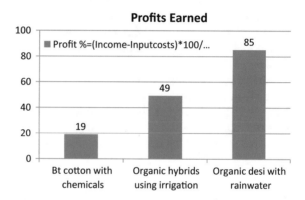

17.2.5   *Irrigation in India*

India is primarily a rain-fed country, and small farmers mostly depend on annual rainfalls. A study revealed that the micro irrigation technique can also be used, but as its initial cost of investment was too high, very few or only privileged farmers used the technique (Bhaskar et al. 2005). The northern states of Punjab, Haryana and Rajasthan are almost entirely irrigated, mostly using a furrow irrigation system (Chapagain et al. 2005). Other parts of central and south India, like Maharashtra, Madhya Pradesh, Andhra Pradesh and Karnataka, are mostly rain-fed (Kranthi 2014).

India is also one of the world's most water-challenged nations, as it has large acres of agricultural lands, industries and an ever-growing urban population, all competing for the limited surface water and groundwater resources. Also, much of India's cotton-growing lands coincide with these high and extremely high water stress areas (Dhawan 2017; Bhattacharyya et al. 2015).

17.2.6 Falling International Cotton Fibre Prices

According to a report published by the Textile Exchange, a non-profit organisation based in the US, the following data was retrieved: Indian cotton prices vary by staple length, from an average of $0.36/kg for short staple cotton below 22 mm, to $0.40/kg for standard varieties, and $0.58 for long staple cottons (USDA 2013). ICE benchmark costs for cotton in January 2015 were set at 26 cents/kg, a fall of 2.7%, and the International Cotton Index had also fallen 2.5% to 31 cents/kg This was a 27% reduction from prices in early 2014 (Textile Exchange 2016b).

As with other commodities, cotton prices can be volatile depending on supply and demand factors and on governmental subsidies and policies. Prices have declined in recent years due to increased cotton production (USDA 2013). The *International Cotton Advisory Committee* (ICAC) estimated an increase in cotton production in India of up to 6.5 million tonnes in 2014–15, but exports have gone down and this surplus production has caused local prices to dip further (USDA 2017).

According to a report published by the Textile Commissioner, Ministry of Textiles, in 2015, cotton exports were expected to decrease to about 6.8 million bales in 2015–2016 from about 13 million bales in 2012–2013, as China is consistently exporting less cotton from India each year. Hence production from conventional seeds may be considered as an alternative.

17.3 Impact on Environment – Carbon Footprint

Cotton production contributes about 0.3 to 1% of total green house gas (GHG) emissions. The increase in atmospheric CO_2 and temperature favours plant development and lengthens the growing season, but it also leads to a decrease in water availability for irrigation and an increase in pests (Seyfang/Haxeltine 2012).

17.3.1 Future Prospects

A technical report submitted by the International Trade Centre, Geneva, in 2011 states that the seasonal surface temperature in India is projected to increase by 2–4 °C by the 2050s. The number of rainfall days during monsoon will decrease over a major part of the country. There will be a significant change in the water cycle due to climatic changes, and an increase in the severity of droughts and floods in places like Gujarat and Rajasthan. All other parts of the country fed by rivers will experience seasonal or regular water-stressed conditions. Thus as the temperature will increase and the water availability will decrease, a resultant decrease in crop yields will also become significant (Ton 2011).

The *Central Institute for Cotton Research* (CICR) found that selected desi cotton varieties/hybrids will adapt well to such drastic conditions because cotton is resilient to high temperatures and drought due to its vertical taproot system. Productivity will increase according specific farm conditions in India. However, the pest problem will still exist and will be aggravated in future. By and large, the impact of climate change on cotton production and productivity will be favourable (Ton 2011; Kranthi 2001).

17.3.2 Impact on Environment – Water Footprint

Water footprint is an indicator of the amount of water used for plant growth and polluted by chemical fertilizers and pesticides. Cotton consumes about 2.6% of the water used globally (Chapagain et al. 2005).

The conventional method of cotton farming makes indiscriminate use of chemical fertilizers and pesticides. This causes excessive pollution of running water and ground water. The water footprint analyses the consumption of water at various stages of cotton cultivation and its environmental impact. The *Grey water footprint* (GWF) measures the volume of freshwater required to assimilate all the pollution created by the use of chemical fertilizers and pesticides and make it suitable for further use. In other words, the GWF reflects the farmer's contribution to the degradation of water quality (Franke/Mathews 2011).

Canada, along with the Water Footprint Network (2013), conducted a study across 480 cotton-producing farmlands in India. The study compared the GWF of organic cotton and conventional cotton and found that the latter creates as much as 200 times more pollution than organic cotton. Therefore the study concluded by suggesting alternative methods of farming and supporting organic cotton production for a cleaner and healthier society (Textile Exchange 2016b).

In 2014, a report by Textile Exchange mentioned that organic farming saved 226.7 billion litres of water, 300.2 million kW of energy and 96.2 million kg of CO_2 emissions.

17.3.3 Impact of Bt Cotton and Its Products on Consumers Worldwide

The *International Agency for Research on Cancer* (IARC) of the *World Health Organization* (WHO) classified glyphosate – a chemical found in herbicides used for transgenic cotton crops – as "probably carcinogenic" in 2015 (International Agency for Research on Cancer 2015). Researchers from across the globe have found traces of glyphosate in almost all products like cotton swabs, sanitary napkins, diapers, gauze and others. Scientists say that glyphosate can promote cancer in parts per trillion concentrations. It can be readily absorbed by the vaginal walls and enter into the blood stream and can cause cancer, infertility and death in women. Scientists from the UK government also found that the antibiotic resistance genes in cotton can make *gonorrhea untreatable* (Qaim/de Janvry 2005).

Secondary co-products of cotton ginning include cottonseed used for oil, animal feed, cosmetics, and fertilizer, and linters, which are very short fibres used in the production of rayon, acetate, cellophane, nail polish, and methylcellulose (Chen/Burns 2006). Cotton seed oil is extracted for human consumption, while the residue, cotton seed cake, is fed to animals. Hence the carcinogenic chemicals may enter human blood through the food chain as well as through direct usage of the cotton products (Allsop et al. 2015).

17.3.4 Impact of Pesticides and Herbicides on Farmer's Community

India has the highest acreage of land under cotton cultivation worldwide and consumes around 50% of the total pesticides applied to all farm products. This not only affects the land, water and air but also compromises the health of farmers and their families, women and children in particular. Indian cotton farmers apply pesticides with bare feet and bare hands, mostly wearing a vest due to the extreme hot climate of this country. Protective garments are not only expensive but generally also unavailable. Hence cotton farmers are directly exposed to the dangerous effects of these chemicals (EJF PAN 2007).

Many cotton-producing farmers surveyed in Andhra Pradesh, Karnataka, Punjab, and Haryana have been mildly or severely affected by pesticide poisoning, which poses physical, neurological and gynaecological health hazards. Endosulfan and Aldicarb, the most commonly used pesticides for cotton, can kill an adult if a single drop gets absorbed by the skin. Women workers are most vulnerable because prolonged exposure during pregnancy can lead to immediate/spontaneous abortions, stillbirths, birth defects, early neonatal deaths, disruption of the endocrine system, sterility, an underdeveloped brain in the child and a weakened immune system. Cases of cancer, such as leukaemia, lymphoma, brain cancer, colon cancer, rectal and prostate cancer, are not uncommon (International Agency for Research on Cancer 2015). Such pesticides also enter our food chain through cow's milk as the cotton seeds are used as fodder for the cattle. Cotton seed oil is also used for direct human consumption (Hashmi/Dilshad 2011).

17.3.5 Bt Cotton Production in India and Its Impact

To analyse the various impacts of Bt cotton from the day it was introduced to this date, the authors have discussed several cases from various parts of the country. Cases from Karnataka, Maharashtra, Rajasthan, Punjab, Haryana and Gujarat have been discussed.

17.3.5.1 Case Study I: Karnataka

A report released by the Institute of Economics in Horticulture, Germany and the *Food and Agriculture Organization of the United Nations* (FAO), Italy (2005) was based on a primary survey conducted in Karnataka. The study on *Comparative Analysis of the Economics of Bt and Non-Bt Cotton Production* was carried out on 100 farmers producing Bt as well as indigenous varieties of cotton in the 2002/2003 cropping season. The survey concluded with the findings that farmers simply assumed Bt cotton to be high-yielding and to not require any pesticides. However post cultivation they found that there was no significant difference in yield between the Bt cotton and indigenous varieties, even after proper irrigation. Pesticide cost were the same for Bt and indigenous varieties. Bt was resistant to pests only for a certain period of time, after which it was difficult for farmers to control pests like bollworms (Prasad/Dhar 2012).

The study concluded by considering Bt as an additional pest control seed but this depended much on the infrastructural and ecological conditions of the farmers adopting it. *Bt was not the best alternative.*

17.3.5.2 Case Study II: Rajasthan, Punjab, Haryana and Gujarat

In 2004–2005, the desi variety cotton (*G. arboreum*) was grown in Rajasthan, Punjab and Haryana. These areas were *frequently infested by bollworms, cotton leaf curl virus* (CLCuV) and water logging problems. Agricultural farming in these states is fully irrigated by using furrow or drip water techniques. So adopting Bt cotton was easy and resulted in high yields (Gopalakrishnan et al. 2007).

In Gujarat, *Herbaceum (Desi)* cotton was grown despite the challenges of drought and/or late heavy rainfall resulting in flood situations. With the use of 90% of hybrids of *G. hirsutum* group and different varieties of Bt cotton hybrids, Gujarat emerged as the largest cotton producer, with 100 lakh bales during 2006–07. It used rainwater harvesting techniques and farm pond irrigation methods to sustain its cotton production process and obtain higher yields (Jackson 2005).

But according to recent reports, these states have witnessed a downturn in cotton production due to untimely rainfall and pest attacks like whitefly, which has resulted in huge losses (Vora 2016).

17.3.5.3 Case Study III: Maharashtra

The 'cotton belt' of India, now referred to as "farmers' suicide belt", Maharashtra has seen more than 300,000 farmers die between 1995 and 2013. Most of these deaths have been attributed to Bt cotton cultivation (Bravi/Nilaeus 2016).

Maharashtra is a rain-fed area and farmers with small farm holdings found it very expensive to introduce modern irrigation techniques. The farmers mostly borrowed money from local moneylenders at high interest rates to buy seeds, pesticides and fertilizers. However, crops have failed due to low rains and lack of irrigation facilities and non-compatibility of the seeds with the local farming conditions. Due to mono cropping, such situations also lead to food scarcity for the farmer's family and leave no alternative source of income (Mishra et al. 2006).

The above three case studies indicate that *benefits from Bt cotton may be considered a myth.*

On the next page, Fig. 17.3, designed by the authors, illustrates the vicious cycle of life and death associated with "Magic seeds" (Bt cotton).

17.4 Alternative Farming Solutions

17.4.1 Integrated Pest Management and Nutrient Management Programme

According to the *CICR Annual Report 2012–13*, a nutrient management programme designed to develop fertilizers and pesticides the organic way was carried out in Nagpur and Coimbatore districts of Maharashtra and Tamil Nadu respectively. The experiment was carried out with four organic components, namely Farm Yard Manure (FYM), vermicompost, bio-enriched cotton compost and sun hemp. Two varieties of cotton seeds, namely Suraj (G. hirsutum) and JLA 794 (G. arboretum), were used. Three sprays of neem oil (at 300 ppm) were used as pesticides. Cover crops and crop rotation helped in weed control, improved the soil quality and fibre yield and retained about 2% more moisture in the soil than the monocropping of cotton. Similar experiments were carried out with Bt cotton and the following results were obtained (CICR Annual Report 2012–13 2013). Input costs with organic seeds were about Rs. 2,000 per acre, which included the cost of seeds, fertilizers, pesticides, growth regulators and weedicides, while input costs with Bt seeds were around Rs. 6,000 per acre, which was three times higher. Yield in the case of organic seeds was about 4.15 quintals per acre compared with 6.35 quintals per acre with Bt seeds, which is roughly 1.5 times the production output with organic seeds.

The above study significantly draws the fact that organic cotton cultivation can yield higher results with best farm management practices, reduced use of synthetic fertilizers and pesticides and crop rotation techniques. These experiments were

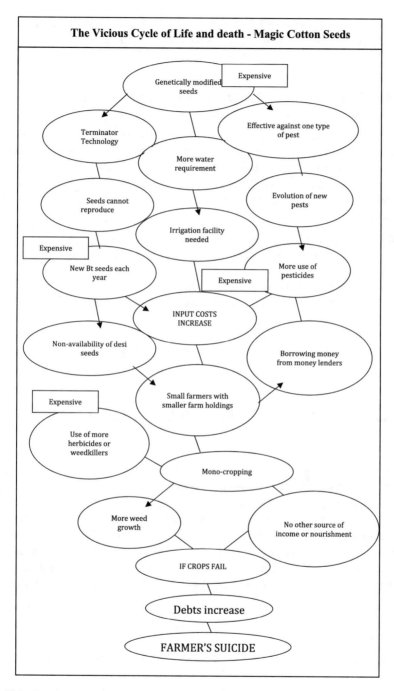

Fig. 17.3 The vicious cycle of life and death – magic cotton seeds

carried in two geographically different zones of the country and the results were similar, hence these methods may be applicable at other places too.

17.4.2   FAO-EU IPM Programme for Cotton in Asia

The Food and Agriculture Organization of the United Nations and European Union Integrated Pest Management Programme operates in six countries across the continent (Bangladesh, China, India, Pakistan, Philippines, Vietnam) to provide alternative farm practices for sustainable farming (WHO 2008). According to the programme, the following methods will reduce our dependency on harmful chemicals:

- Bird species as natural predators of cotton pests;
- *Crop rotation with crops like wheat, pulses, legumes, etc. that are less susceptible to pests*;
- Selecting seeds according to local ecology;
- Border crops (i.e. maize, sorghum, marigold and sunflower) around cotton fields to keep out pests;
- Intercropping (i.e. soybean, castor);
- Use of chemical pheromones that discourage pest growth.

17.4.3   High Density Planting (HDP) of Indigenous (desi) Varieties of Cotton

CICR is encouraging the farmers of Maharashtra to return to the age-old technique of *High Density Planting*, in which the farmers plant more plants per acre. Though the yield per plant is low, the overall yield will increase due to the greater number of plants, and the investment in seeds will be less because the cost of indigenous varieties are as low as Rs. 150 per kilo. Also the farmer can use the seeds from the first year's crop for subsequent years of farming. This was tested on 500 acres of land in Wardha district and yielded about 6 to 8 quintals per hectare at a fraction of the Bt cultivation cost (Venugopalan et al. 2014; CICR Annual Report 2012–13 2013).

17.4.4   The Progress Towards Growing Organic Cotton

Currently many farmers have switched to organic cotton farming. About 72,280 hectares of land has been used to plant the desi cotton varieties. Thus the acreage under desi cotton has taken a quantum leap in the last ten years and recently accounted for 7% of the total acreage in North India (Sood 2016).

The main reason for the switch was that the desi cotton varieties are resistant to leaf curl virus (CLCuV) and pests like whiteflies. Scientists expect the acreage to increase up to 25% in coming years (Forster et al. 2013).

17.4.5 Minimum Support Price (MSP)

The Government guarantees producers of organic cotton a *Minimum Support Price* (MSP) for their produce. The MSP of long staple cotton was raised from Rs. 2,030 to Rs. 3,000 per 100 kg in 2008–09, while that of medium staple cotton went up from Rs. 1,800 to Rs. 2,500 per 100 kg and private companies paid a premium of 25% on MSP to the organic cotton growers (NITI AAYOG 2016).

17.4.5.1 Multiple Income Sources from Organic Cotton Farming

Many farmers from Yavatmal, Khargon and Wardha (Maharashtra) and also some parts of Madhya Pradesh switched to organic farming a few years back. They have taken up multi-cropping and hence their farms' soil conditions have improved, and now hold more moisture, which is required for cotton farming. They have overcome the problem of food scarcity and alternative sources of income by harvesting lentils, jowar and vegetables. Also the organic cottonseeds are sold as animal fodder and fetch thrice the price of Bt seeds. Finally, the seeds can be conserved for the next season's sowing as well as sold to other farmers at a very low price of Rs. 30 per kg (Textile Exchange 2016b; Kranthi 2015; Sood 2016).

17.4.5.2 Rise in Demand for Organic Cotton Products Worldwide

According to *Pesticide Action Network* (PAN), UK, the demand for organic cotton is very high in the American and European countries. A report prepared by Ipsos MORI in 2005 stated that the UK market was steadily growing for products made from certified organic cotton. The Soil Association, UK also said in 2011 that the market for organic cotton was constantly growing, though more than 65% of textiles were made from conventional cotton (PAN UK/Solidaridad/WWF 2017).

However it indicates a way forward, as many leading global designers like Katherine Hamnett and brands/retailers like Walmart, Harrods, Marks and Spencer, Co-op Switzerland and others have successfully introduced organic cotton lines of apparel (Willer 2017). According to reports by Textile Exchange (2014, 2015a, b, 2016a, b, c), the trend is catching on fast owing to a rise in awareness amongst consumers and retailers, who are increasing their organic cotton lines.

17.4.5.3   Efforts of Indian Farmers

Desi Cotton Growers Association of Karnataka – a small group formed by 36 farmers from various districts of Karnataka – raised 45 indigenous varieties of cotton successfully. They sourced the seeds from across the country. The seed varieties include tree cotton from Bengal, Cernuum from Meghalaya, Dev-Kapass from south Karnataka, Pandarapura from North Karnataka and Maharashtra, Pondru from Andhra Pradesh, Karung Kanniparthi from Tamil Nadu, Bangal Desi from Rajasthan and Gujarat besides many other local varieties (Srivatsa 2012).

Input costs were very low and production per acre for was high (12 quintal/acre) in comparison to Bt cotton (8–10 quintal/acre). The major challenge faced by the group was sourcing the local varieties from across the country. Secondly, the local varieties are short staple fibres which are majorly consumed by the medical and surgical market in the country and can also be used to make denims (Directorate of Cotton Development Government of India 2017).

17.4.5.4   Textile and Fashion Industry Requirements

The product requirement analysis of retailers for 2025 was based on a preliminary survey of retailers to discover the future assumptions about consumer desirability of certain types of products in 2025. On the basis of the product analysis, the kind of staple length of fibres required was identified. It was found that the demand is mostly for the medium and long staple length cotton fibres i.e. from 24 mm to about 40 mm. Hence organic cotton or *desi* varieties can solve the requirement for short and medium-length fibres, whereas other local hybrids may be useful for long and extra long fibres. The demand for garments like knitted T-shirts, denims, shirts and pants will significantly rise by the year 2025, while that for chambray, towels, lawn dresses, chiffons and voiles may remain the same as in 2014 (SSTF, OCRT, and Textile Exchange 2015a, b; Pay 2009; Ton 2007).

17.5   Organic Standards and Certifications

To become eligible to obtain organic certification or organic status, farms must meet certain standards for two to three continuous years. Various organisations worldwide (listed below) certify farm products as organic according to specifications laid down by them and facilitate organic cotton usage through maintaining sustainable agriculture standards (Textile Exchange 2016c).

- *Indian Standard for Organic Textiles* (ISOT), India
- *National Organic Program* (NOP) standard by the United States Department of Agriculture (USDA)
- The *Global Organic Textile Standard* (GOTS)

- *Organic Content Standard* (OCS 100)
- Organic Content Standard (OCS blended)
- Soil Association Certification, United Kingdom
- Chetna Organic, India
- Fair Trade, America
- *Better Cotton Initiative* (BCI)
- ECOCERT, France
- *Japanese agricultural standard* (JAS).

17.6 Fashion and Textile Value Chain

Looking at the above trends and growth opportunities, many Indian and international companies have geared up to make organic clothing lines for men, women and children. Details of some such brands and organisations are provided in Table 17.1.

Table 17.1 Brands using organic cotton

Name of Indian apparel and home furnishings brands	Name of lingerie brands	Name of international apparel brands	Name of international apparel brands
Indigreen	Brook There, USA	Patagonia, USA	Volcom, USA
No Nasties	Najla, USA	Nudie Jeans, Sweden	Prana, USA
Samtana	Alas, India	Muka Kids,	Loomstate, USA
Tvach	Base Range, France	Remei AG, Switzerland	Stella McCartney
Forty Red Bangles	Only Hearts, Peru	Inditex, Spain	Continental, London
Anokhi	Kowtow, New Zealand	Earth Positive, Sweden	Organic Initiative, New Zealand
UV & W	Nico, Australia	C&A, Netherlands	Indigenous, USA
Bhu:sattva	Skin, Peru	H&M, Sweden	Cotonea, Germany
Do you speak Green	Pact, USA	Kathmandu, Australia and Netherlands	Skunkfunk, Spain
Ethicus		Nike, USA	
Paruthi from Upasana		Decathlon, France	
Tula		Lindex, Gothenburg	
AND		Stanley & Stella	
Fabindia			

Source Developed and compiled by the authors from the websites of the organisations

Many Indian and international brands currently manufacture organic clothing, such as women's wear, lingerie, denims and sportswear, kids' wear, women's hygiene products, home furnishings, etc. They have linked up with many farmers or organisations like Chetna in India, who produce certified organic cotton fibres (Textile Exchange 2017b; Alaya 2014).

17.7   The Way Forward

In all the above discussions, we have tried to put forward a perspective that the world understands the need to construct a green and sustainable business system. India is not an exception. The most widely used natural fibre is cotton and it generates livelihoods for billions of people across the world. It touches the lives of billions in various forms of textile material. Hence it may be positioned as a symbol of sustainability and risk-free life to construct a greener, healthier and fairer society for all. We have also tried to draw attention to the rising demand for organic cotton by a multitude of brands that wish to make a difference to our lives. Hence, small farmers' co-operatives, NGOs, government bodies, and other private fashion retailers may join hands to create sustainable systems based on deep ecological principles.

References

Afari, V., 2010: "Horticultural Exports and Livelihood Linkages of Rural Dwellers in Southern Ghana: An agricultural household modeling application", in: *The Journal of Developing Areas* 44(1): 1–23.

Alaya, P.N., 2014: "10 Indian Organic Clothing Brands that You Should Be Proud of Wearing." 2014; at: http://www.thealternative.in/lifestyle/10-indian-organic-clothing-brands-that-you-should-be-proud-of-wearing/ (5 March 2015).

Allsop, M.; Huxdorff, C.; Johnston, P.; Santillo, D.; Thompson, K., 2015: "Pesticides and Our Health, a Growing Concern", in: *Greenpeace*. Vol. 2.

Beckert, S., 2015: "Empire of Cotton: A New History of Global Capitalism—Book Review", in: *Studies in People's History* 1(2): 136–143; at: https://doi.org/10.1177/2348448915574932 (5 January 2016).

Bhaskar, K.S.; Rao, M.R.K.; Mendhe, P.N.; Suryavanshi, M.R., 2005: "Micro Irrigation Management in Cotton." Nagpur. www.cicr.org.in (16 April 2013).

Bhattacharyya, A.; Janardana Reddy, S.; Ghosh, M.; Raja Naika, H., 2015: "Water Resources in India: Its Demand, Degradation and Management", in: *International Journal of Scientific and Research Publication* 5(12): 346–356; at: http://www.ijsrp.org/research-paper-1215/ijsrp-p4854.pdf (11 December 2016).

Bravi, L.; Nilaeus, M., 2016: *Cotton Cultivation—An Exploratory Study of Agricultural Opportunities to Fight Poverty in India*. Sweden.

Chapagain, A.K.; Hoekstra, A.Y.; Savenije, H.H.G.; Gautam, R., 2005: "The Water Footprint of Cotton Consumption." *UNESCO-IHE*, 2005.

Chen, H.-L.; Burns, L.D., 2006: "Environmental Analysis of Textile Products", in: *Clothing and Textiles Research Journal* 24(3): 248–261; at: https://doi.org/10.1177/0887302X06293065 (10 April 2016).

CICR Annual Report 2012–13, 2013: "Mega Seed Project." Nagpur; at: http://www.cicr.org.in/cicr_ar_1213/5_production.pdf (19 October 2015).

Department of Agriculture and Cooperation GOI, 2014: "National Mission for Sustainable Agriculture (NMSA)." New Delhi.

Dhawan, V., 2017: "Water and Agriculture in India". *Background Paper for the South Asia Expert Panel during the Global Forum for Food and Agriculture,* Germany; at: https://www.oav.de/fileadmin/user_upload/5_Publikationen/5_Studien/170118_Study_Water_Agriculture_India.pdf (15 March 2018).

Directorate of Cotton Development Government of India, 2017: "Status Paper of Indian Cotton." Nagpur.

Dorugade, V.A.; Satyapriya, D., 2009: "Organic Cotton", in: *Man-Made Textiles in India,* June 2009.

EJF PAN, 2007: "The Deadly Chemicals in Cotton"; at: ISBN No: 1-904523-10-2 (13 May 2015).

Forster, D.; Andres, C.; Verma, R.; Zundel, C.; Messmer, M.M.; Mäder, P., 2013: "Yield and Economic Performance of Organic and Conventional Cotton-Based Farming Systems—Results from a Field Trial in India", in: *PLoS ONE* 8(12); at: https://doi.org/10.1371/journal.pone.0081039 (18 May 2015).

Franke, N.; Mathews, R., 2011: "Grey Water Footprint Indicator of Water Pollution in the Production of Organic Versus Conventional Cotton in India", in: *Water Footprint Network;* at. http://waterfootprint.org/media/downloads/Grey_WF_Phase_II_Final_Report_Formatted_06.08.2013.pdf (11 June 2016).

Gopalakrishnan, N.; Manickam, S.; Prakash, A.H., 2007: "Problems and Prospects of Cotton in Different Zones in India", in: *Central Institute for Cotton Research, Regional Station, Coimbatore* (December 2007); at: http://www.cicr.org.in/pdf/ELS/general3.pdf (6 August 2016).

Hashmi, I.; Dilshad, K.A., 2011: "Adverse Health Effects of Pesticide Exposure in Agricultural and Industrial Workers of Developing Country", in: Stoytcheva, M. (Ed.): *Pesticides—The Impacts of Pesticides Exposure.* (Croatia: InTech Europe); at: https://doi.org/10.5772/13835 (11 November 2015).

International Agency for Research on Cancer, 2015: "IARC Monographs Volume 112: Evaluation of Five Organophosphate Insecticides and Herbicides." *WHO,* 2015; at: https://doi.org/10.1016/j.chemosphere.2016.03.104 (17 July 2017).

Jackson, G.J., 2005: "Organic Cotton Farming in Kutch, Gujarat, India." Mumbai.

Kranthi, K.R., 2001: "Bio-Technologies for Cotton Production: Why Are the Yields Low in India?" CICR, Nagpur, India; at: http://www.caionline.in/download_event_publications/56 (11 January 2015).

Kranthi, K.R. 2014: "Indian Cotton Yield in 2014—Prediction Analysis", in: *Cotton Statistics & News, Cotton Association of India,* (November 2014).

Kranthi, K.R., 2015: "Desi Cotton—Returns ?" *Cotton Statistics & News, Cotton Association of India* (July 2015).

Mayee, C.D.; Singh, P.; Dongre, A.B.; Rao, M.R.K.; Raj, S., 2011: "Transgenic Bt Cotton", in: *CICR Technical Bulletin No: 22.*

Mishra, S.; Shroff, S.; Shah, D.; Deshpande, V.; Kulkarni, A.P.; Deshpande, V.S.; Bhatkule, P.R., 2006: "Suicide of Farmers in Maharashtra Background Papers", in: *Indira Gandhi Institute of Development Research, Mumbai.*

NITI AAYOG, 2016: "Evaluation Report on Efficacy of MSP on Farmers". *Development Monitoring and Evaluation Office, Govt. of India.* Vol. 231.

Orphal, J., 2005: "Comparative Analysis of the Economics of Bt and Non-Bt Cotton Production", in: *Pesticide Policy Project Publication Series.* Hanover.

PAN UK, Solidaridad, and WWF, 2017: "Sustainable Cotton Ranking 2017"; at: http://www.wwf.
  de/fileadmin/fm-wwf/Publikationen-PDF/Report_Sustainable_Cotton_Ranking_2017.pdf
  (25 October 2017).
Pay, E., 2009: "The Market for Organic and Fair-Trade Cotton Fibre and Cotton Fibre Products"
  (Food and Agriculture Organization of the United Nations).
Prasad, B.; Dhar, M., 2012: "Cotton Market and Sustainability in India", at: www.wwfindia.org
  (14 April 2015).
Qaim, M.; de Janvry, A., 2005: "Bt Cotton and Pesticide Use in Argentina: Economic and
  Environmental Effects", in: *Environment and Development Economics* 10(2): 179–200; at:
  https://doi.org/10.1017/S1355770X04001883 (19 June 2016).
Santhanam, V.; Sundaram, V., 1997: "Agri-History of Cotton in India—An Overview", in: *Asian
  Agri-History* 1(4): 235–251; at: http://www.cicr.org.in/research_notes/cotton_history_india.pdf
  (22 July 2015).
Seyfang, G., Haxeltine, A., 2012: "Growing Grassroots Innovations: Exploring the Role of
  Community-Based Initiatives in Governing Sustainable Energy Transitions", in: *Environment
  and Planning C: Government and Policy* 30(3): 381–400; at: https://doi.org/10.1068/c10222
  (25 May 2017).
Shetty, P.K., 2004: "Socio-Ecological Implications of Pesticide Use in India", in:
  Economic and Political Weekly 39(49): 5,261–5,267; at: http://www.jstor.org/stable/10.2307/
  4415873%5Cn, http://www.jstor.org/discover/10.2307/4415873?uid=3738256&uid=2&uid=
  4&sid=21101170117791
  (26 June 2015).
Sood, D., 2016: *India Cotton and Products* (USDA).
Srivatsa, S.S., 2012: "Farmers Here Have Cottoned on to Indigenous Varieties", in: *The Hindu*
  (11 July 2012); at: http://www.thehindu.com/todays-paper/tp-national/tp-karnataka/farmers-
  here-have-cottoned-on-to-indigenous-varieties/article3625744.ece (28 September 2016).
SSTF, OCRT, and Textile Exchange, 2015: "Seed Availability for Non-GM Cotton Production—
  An Explorative Study." Netherlands; at: http://bit.ly/1BpdRR2%5Cn, http://farmhub.
  textileexchange.org/upload/learningzone/RoundTable/Seed_Availability_for_non-GM_Cotton-
  Production_May2015.pdf (11 October 2016).
Textile Exchange, 2015: "Textile Exchange Annual Report 2015." Texas; at: http://
  textileexchange.org/sites/default/files/2010_Annual_Report_web.pdf (24 November 2016).
Textile Exchange, 2016a: "Cotton—India, Material Snapshot". Texas.
Textile Exchange, 2016b: "Organic Cotton Market Report 2016". Texas.
Textile Exchange, 2016c: "Organic Cotton Material Snapshot Material Scenario Common Uses In
  Apparel and Footwear." Texas; at: http://textileexchange.org/wp-content/uploads/2017/06/TE-
  Material-Snapshot_Organic-Cotton.pdf (8 June 2016).
Textile Exchange, 2017a: "Organic Cotton Market Report 2017". Texas.
Textile Exchange, 2017b: "Quick Guide to Organic Cotton". Texas; at: https://doi.org/10.1007/
  978-3-662-47714-4 (16 February 2018).
Ton, P., 2007: "Organic Cotton: An Opportunity for Trade", in: *International Trade Centre*.
  Geneva; at: http://www.intracen.org (9 April 2016).
Ton, P., 2011: "Cotton and Climate Change: Impacts and Options to Mitigate and Adapt", Geneva;
  at: www.intracen.org (24 May 2016).
US Department of Agriculture, 2018: "Leading Cotton Producing Countries Worldwide in 2016/
  2017." Statista; at: https://www.statista.com/statistics/263055/cotton-production-worldwide-
  by-top-countries/ (16 June 2018).
USDA, 2013: "Importing Organic Products to the U.S." USDA.
USDA, 2017: "Organic Cotton Market Summary." USDA.
Venugopalan, M.V; Kranthi, K.R.; Blaise, D.; Lakde, S.; Sankaranarayana, K., 2014: "High
  Density Planting System in Cotton—The Brazil Experience and Indian Initiatives", in: *Cotton
  Research Journal* 5(2): 1–7.

Vora, R., 2016: "Late, but Widespread Rains Bail out Cotton Crop in Gujarat", in: *The Hindu Business Line*, (9 August 2016); at: https://www.thehindubusinessline.com/economy/agri-business/late-but-widespread-rains-bail-out-cotton-crop-in-gujarat/article8964897.ece (8 January 2017).

WHO, 2008. "2nd FAO/WHO Joint Meeting on Pesticide Management." Geneva.

Willer, H., 2017: "European Organic Market Data 2015." Switzerland; http://orgprints.org/31200/31/willer-2017-european-data-2015.pdf (4 March 2018).

Yusuf, M., 2010: "Ethical Issues in the Use of the Terminator Seed Technology", in: *African Journal of Biotechnology* 9(52): 8,901–8,904; at: http://www.academicjournals.org/AJB/PDF/pdf2010/21DecConf/Yusuf.pdf (14 October 2015).

Chapter 18
MDCM Approach to Sustainable Agriculture Production for Odisha

Bishnu Prasad Mishra, Banashri Rath, Laxmidhar Swain
and Aamlan Saswat Mishra

Abstract Agricultural production has been never considered in a sustainable manner using mathematical modelling. Meeting the demand for food and nutrient security for the people living in an area is very important for researchers in the present context and climate change conditions. An increase in the population of 1.2% per annum has forced farmers to grow more using the latest techniques without considering soil health and sustainable crop production. Crop rotation, a common practice for sustainable crop production, is currently insufficient to make agricultural production sustainable. The right crop for a predetermined area according to the social and environmental requirement helps to address sustainability issues to a large extent. This complex problem can be resolved using appropriate mathematical tools like MCDM (Multiple Criteria Decision-Making), which has been formulated to address the requirements of the people. Suitable boundary conditions are used to simulate optimal production, keeping natural reserves intact for twenty-five crops, and taking into account five priorities in Odisha during the rabi season.

Keywords Sustainable crop production · Priority based optimisation
MCDM · Goal programming · Rabi crops · Major crops

Bishnu Prasad Mishra, Dean, School of Engineering and Technology (SOET), Paralakhemundi, Centurion University of Technology and Management (CUTM), Odisha, India; Email: bp.mishra@cutm.ac.in

Banashri Rath, Divisional Head, Odisha Industrial Infrastructre Development Corporation (IDCO), Bhubaneswar, Odisha, India; Email: banashreerath@gmail.com

Laxmidhar Swain, Network Director, Centurion University of Technology and Management (CUTM), Bhubaneswar, Odisha, India; Email: laxmidhar@cutm.ac.in

Aamlan Saswat Mishra, Std-XII, Dayanad Anglo Vedic (DAV) Public School, Unit-8, Bhubaneswar, Odisha, India; Email: aamlansaswat@gmail.com

© Springer Nature Switzerland AG 2019
A. K. J. R. Nayak (ed.), *Transition Strategies for Sustainable Community Systems*,
The Anthropocene: Politik—Economics—Society—Science 26,
https://doi.org/10.1007/978-3-030-00356-2_18

18.1   Introduction

According to limits to growth theory and subsequent revisions, it has been proved that per-capita growth combined with the population growth has accelerated the use of natural resources (Angehrn/Dutta 1992; Angehrn 1991). This has resulted in instability in the ecosystem and unsustainable agricultural production. The researchers have computed possible agricultural production with the help of mathematical modelling techniques, taking into account people's requirements and priorities. The limiting factor must be the use of natural resources in a regulated manner to mitigate the issue of sustainable agricultural production. The secondary published data which is collected from the field, verified and published by Odisha Agriculture Department, has been used for simulation purposes. The *Multiple Criteria Decision-Making* (MCDM) technique with priorities was chosen to formulate the equations to find the best sustainable solutions. The crop production parameters for 26 crops have been considered in this research. These 26 crops are most commonly grown in Odisha as per the basic, cultural, and social need of farmers. The crop production inputs for all these 26 crops have been collected from *Orissa University of Agriculture and Technology* (OUAT), Bhubaneswar, Odisha, and used in the development of model equations (Anonymous 2004a, b; Ata/ Sennaroglu 2008). Further, the priority rank has been obtained from the villagers of the district and ratified by the *Krishi Vigyan Kendta* (KVK) scientists working in the region. The QSB package developed by the researchers of *Indian Institute of Technology* (IIT), Kharagpur, West Bengal, India was used to find a solution to framed problems. The results are complicated to interpret. They were simplified for the purposes of computing the area to be cultivated under different crops, water to be used, labour employment, revenue generation and food security to make the crop production sustainable over a long time. The sensitivity analysis has been performed to see the production variation of the different crops which it is possible to cultivate in that region in a sustainable manner (Alston et al. 2003).

18.2   Literature Review with Comments

The crop production data have been collected, analysed and plotted for interpretation. The data on the proportion of different crops has been reviewed for Odisha over time. The analysis shows that the proportions of different crops during the years 1950–2000 are different (Anonymous 1979, 2000). The increase in population, changes in food habits, and environmental factors have influenced cropping practice to a large extent. The percentage share of the rice crop is stable at about 23–24% of the total crop production over the last fifty years (Anonymous 2004a, b). The percentage share of other crops has increased from 8 to 17% during the same time, which is need-based. The share of coarse cereal has decreased from 30 to 16% during the same study period. The adoption of modern farming and changes in economic status have replaced this course cereal area with cash crops. Sustainable planning must include these clauses when framing policies (Anonymous 1999, 2010).

The last fifty years' major crop production data were collected from the Agriculture Department, Odisha and reviewed. It has been observed that the total cropped area under major crops has undergone a major shift. This shift is positive for some crops and negative for others. The major crops in Odisha are paddy, wheat, course cereals, pulses, oil seeds, jute and mesta. Considering the population increase and rice as its staple food, it has been observed that the production of paddy in Odisha has remained stable at 22–24 million tons per year since 1995. This is enough to meet the staple food requirements of Odisha. Wheat production has increased from 7.5 to 13 million tons per year. Conversely, course cereals, which are the staple food of the lower income group, have recorded a sharp decline from 30 million tons to as low as 15 tons. Although pulses are frequently consumed in Odisha, the production of them has decreased from 16 to 10 million tons. The change in food regulations has prompted farmers to grow more oilseed crops, hence production of oilseeds has shown a steady increase from 7 to 14 million tons. The demand for sweets is steady. They are targeted at the elite class, and no change has been recorded in the production of sugar cane. Other crops include various cash crops. An increase in the economic status of farmers has been observed. More farmers have cultivated cash crops to get more and more profit and support their children to improve their social status. The change in crop selection and economic development, along with the population growth, has been linked to the change in crop production in Odisha state.

Further, production data for major crops grown in Odisha have been reviewed. Out of eight prominent crops grown in Odisha, there is a dearth of pulses crop production although they are part of a healthy diet. The pulses crops are grown as pyre crops all over Odisha, resulting in lesser yield. The production and productivity of pyre crops are much lower and no innovation has been reported in pyre crop management. However, according to some reports, the production of pulses has been increasing slowly since 1998 (Sinha et al. 1989). The production trend of millets, "a rich man's delicacy", has shown a decrease in production (Rath et al. 2003) even though it has huge national and international market potential. To increase the economic status of *below poverty line* (BPL) families, proper value addition is required for these crops, followed by export. These are short duration crops which require less water and short production periods. Agriculture Production Policy interventions for the promotion of these crops are there, but the efficiency of implementation is much lower. The spread effect of crop production has not been observed, although the programme has been implemented for many years.

The contribution from different crops to the total production has been further studied. The main crops are: rice (23%), wheat (13%), course cereals (16%), pulses (10%), oilseeds (13%), cotton and mesta (5%), sugar cane (2%). Other cash crops share the remaining 18%. An increase has been observed in the production of crops like oilseeds, other cash crops and wheat. The total crop percentage shares of other major crops have decreased. The reasons for such changes can be linked to social, political and economic developments.

The production share of pulses has decreased as the storage and processing methods have not been developed to prevent loss during storage. Similarly, due to less profitability in paddy production, there is diversification of crop production from paddy to other crops (Angehrn 1991; Anonymous 2004a, b). This implies that cash crops diversification planning can be readily implemented through agricultural policy. Once a cash crop is demonstrated to have added value, marketing it for the economic empowerment and nutritional security of all farming stakeholders can be addressed. The substantial growth in wheat production in Odisha has been noticed as a change in food habits. The area under the production of other crops has been increasing to develop the economic empowerment of stakeholders. Although sugar cane production is a small percentage of the total crop production, the growth trend is remarkable. It provides a vision that crops found to be remunerative are readily accepted by farmers. In view of global food requirements, a Fair and Remunerative Price (FRP) for some thrust crops has been tried in order to promote new commercial crops.

The food habits of Odisha people are clearly visible once the figures have been collated. The area and production of paddy in each district of Odisha have been computed and tabulated to find the less productive areas of the state. To address the sustainability issue in such districts, Balangir district has been selected. Paddy production is very low in Jharsuguda district and high in Bargarh district of Odisha. It has been inferred that the districts with irrigation have higher production of paddy. Deogarh and Gajapati have the least production of paddy according to the data obtained from the statistics cell of the Department of Agriculture. This sample relates to data from the 2010 financial year.

Inputs for crop production and its management are very important, as crops require timely interventions. The major inputs like land, water, seed, fertilizer, finance and machinery in all the districts of Odisha have been reviewed. The data collected and listed below can be used as the upper/lower limits of the MCDM problem model formulation. The use of these data is never reported for such simulation using MCDM techniques.

Data about the agricultural resources available in Odisha and used for paddy production have been collected for all thirty districts: Angul, Balangir, Balasore, Bargarh, Bhadrak, Boudh, Cuttack, Deogarh, Dhenkanal, Gajapati, Ganjam, Jagatsinghpur, Jajpur, Jharsuguda, Kandhamal, Kalahandi, Kendrapara, Keonjhar, Khordha, Koraput, Malkangiri, Mayurbhanj, Nabarangpur, Nayagarh, Nuapada, Puri, Rayagada, Sambalpur, Subarnapur and Sundargarh. Each has different potential for paddy production. Sustainable production of paddy was the objective of the study. A sustainable development model has been achieved through GOAL programming techniques using production, net area shown, rainfall, loan disbursed in paddy production, seed supplied, fertilizer and machinery marketed in the above districts. All these production parameters have been identified after due consultation with the farmers. The data have been obtained from the statistics cell of the Department of Agriculture, Government of Odisha for the year 2010.

18.3 Cropping Pattern

Cropping pattern (crop rotation/crop sequencing) is practised in Odisha for soil nutrient management as a means to arrive at sustainability in crop production. The eight most popular rice-based cropping patterns being followed in the state of Odisha are described below (Government of India 2002).

Rice-Rice-Rice: This crop rotation is most suitable for areas with high rainfall and assured irrigation facilities in the summer. Soils with a high water-holding capacity and a low rate of percolation are particularly suitable for such crop rotation. In some canal irrigated areas of Odisha, the Rice-Rice-Rice cropping pattern with 300% cropping intensity is followed. In such areas three crops of rice are grown year after year.

Rice-Rice-Cereals (other than rice): This cropping pattern is followed in the areas where the water is not adequate for growing a rice crop in the summer. The alternative cereal crops being grown are ragi, maize and jowar.

Rice-Rice-Pulses: In the areas where water scarcity precludes the cultivation of cereal crops in summer, short-duration pulse crops like green gram and black gram are grown as pyre crops. Nitrogen fixation is the added advantage of this cropping pattern.

Rice-Groundnut: This cropping pattern is being followed by some farmers in coastal parts of Odisha. After harvesting the rice crop, groundnut is grown in the summer.

Rice-Wheat: This crop rotation has become the dominant cropping pattern in northern parts of the India, but it is not found predominantly in Odisha.

Rice-Wheat-Pulses: In this cropping pattern, after harvesting the wheat crop, green gram or cowpea is grown as a fodder crop in the alluvial soil belt of northern states. Additionally, cowpea is also grown in the red and yellow soils of Odisha and black gram is grown in areas with black soils.

Rice-Toria-Wheat: This crop sequence is not widely followed in Odisha; however, it is adopted in northern parts of India. Among the above-mentioned cropping patterns followed in the country, the Rice-Wheat cropping pattern is adopted by many farmers. The Rice-Wheat cropping pattern has been practised in the Indo-Gangetic plains of India for a long time.

Rice-Fish Farming System: The fields with sufficient water-retaining capacity for a long period and free from heavy flooding (coastal areas) are suitable for the rice-fish farming system. This system is being followed by the small and marginal poor farmers in rain-fed lowland rice areas. These farmers are not able to invest much in agricultural development. They raise a modest crop of traditional low-yielding rice varieties. In order to improve the economic condition of these farmers, the Central Rice Research Institute, Cuttack, Odisha, India has developed this production technology for the rice and fish farming system. Steps have already been taken to popularise the rice-fish farming system in lowland areas to increase the production and productivity of crops and thereby improve the economic conditions and food/ nutritional security of the resource-poor farmers of these areas.

18.4   Linear Programming Simulation to Compute the Optimum Agriculture Production in a District of Odisha

A simulation considering no priority has been formulated and the output in terms of economic growth has been proposed below (Laxminarayan/Rajgopanan 1977; Sinha et al. 1988). The prevailing farm gate price has been considered. Water, agricultural labour availability, and seed and land availability have been considered for five major crops grown in a district (Bolangir) using linear programming (LP). After the LP formulation has been made, a suitable boundary condition can be incorporated to find optimal production in an area of Odisha. It is also possible to discover the exact amount of revenue that can be generated from the agricultural sector in a district. The district level planning must be done considering these ground realities. The present formulation is a starting point for identifying the resources used for production and the resources which could not be used due to other limiting input parameters. The staple foods for Odisha, like rice, green gram, black gram, groundnut and maize, have been considered in this simulation. The constraints to be used in simulating the agricultural production of Bolangir district have been obtained from the statistics department of the Government of Odisha, and the LP formulation has been framed accordingly. The three major constraints of water, labour and seed have been considered in the present LP formulation. The total arable land in Bolangir has been considered as a limitation with respect to area.

The productivity of some major crops in the Odisha context was computed and found to be: paddy: 1,888 kg; green gram: 350 kg; black gram: 400 kg; maize: 1,556 kg; and groundnut: 1,695 kg. The minimum support price for the year 2011–12 was also collected: paddy: Rs. 1,080.00; green gram: Rs. 37.00; black gram: Rs. 33.00; maize: Rs. 9.80; and groundnut: Rs. 27.00. The water requirement, labour, seed and area under all these crops have also been collected. The area under paddy, green gram, black gram, maize and groundnut cultivation was found to be 204,600 ha, 17,540 ha, 19,810 ha, 3,760 ha, and 10,480 ha respectively. Similarly, the water requirement to cultivate paddy, green gram, black gram, maize and groundnut was found to be 1,000 cm, 25 cm, 25 cm, 50 cm and 48 cm. Paddy, green gram, black gram, maize and groundnut cultivation were found to consume 25 kg, 8 kg, 10 kg, 6 kg, and 50 kg of seeds respectively. The labour requirement for cultivating the above crops was also obtained from the statistical cell of the department. Paddy, green gram, black gram, maize and groundnut cultivation was found to use 96, 45, 45, 84 and 70 man-days. A linear programme was framed and solved to discover the maximum production from one district as a sample case for cross-checking. The simulation result shows that production from the above five major crops is only 508.16 crores in the Kharif season (summer). The water required for crops like paddy, green gram, black gram, maize and groundnut has been collected from reference books and OUAT publications. The actual data about labour, seed requirement and area under the above crops were used for LP formulation and simulation purposes.

Further details were obtained during the simulation. The crops generally grown by the farmers of Odisha were noted and the resources used are stated below. Production inputs like water, labour, fertilizer, animal power, plant protection, finance, revenue and yield were taken into account in the calculations. The twenty-five crops commonly grown in Odisha for many years were used as the sustainable production model when writing the goal programming model. They are paddy, wheat, barley, green gram, cowpea, pigeon pea, groundnut, sesame, castor, cotton, tobacco, sugar cane, jute, mesta, niger, sunflower, mustard, lentil, field pea, horse gram, Bengal gram, black gram, ragi, maize, and jawar. The water requirement for these crops was found to be 100 cm, 40 cm, 50 cm, 15 cm, 15 cm, 40 cm, 48 cm, 25 cm, 58 cm, 62 cm, 55 cm, 170 cm, 48 cm, 45 cm, 25 cm, 40 cm, 50 cm, 40 cm, 15 cm, 20 cm, 20 cm, 15 cm, 40 cm, 55 cm, 55 cm respectively. Similarly, the labour required in man-days has been found to be 82, 45, 45, 30, 40, 40, 70, 36, 30, 109, 114, 262, 86, 62, 30, 30, 40, 30, 10, 27, 30, 27, 57, 84, 84 respectively. The fertilizer requirements of various crops have been obtained from OUAT. The data are 48 kg, 42 kg, 32 kg, 32 kg, 32 kg, 32 kg, 40 kg, 24 kg, 70 kg, 64 kg, 56 kg, 144 kg, 48 kg, 26 kg, 32 kg, 30 kg, 48 kg, 26 kg, 32 kg, 30 kg, 48 kg, 32 kg, 40 kg, 15 kg, 32 kg, 24 kg, 32 kg, 64 kg, 64 kg respectively. Animal power is used in rural areas, and the requirement of animal power has been obtained from OUAT. The data are 26, 18, 18, 8, 11, 10, 14, 10, 11, 15, 13, 16, 14, 13, 11, 11, 11, 8, 0, 7, 8, 8, 10, 16, 11 days respectively. The cost of cultivation has been computed by OUAT experts and published as a reference source. The costs of cultivating the above crops are Rs. 20,666, Rs. 24,470, Rs. 24,320, Rs. 6,549, Rs. 19,860, Rs. 19,450, Rs. 19,768, Rs. 18,890, Rs. 17,730, Rs. 24,608, Rs. 37,962, Rs. 63,587, Rs. 29,750, Rs. 24,090, Rs. 16,560, Rs. 16,790, Rs. 28,220, Rs. 15,330, Rs. 7,870, Rs. 8,164, Rs. 8,514, Rs. 8,050, Rs. 21,510, Rs. 17,534, Rs. 21,975 respectively.

The yields of the above twenty-five crops were obtained from the research trial of OUAT. They are found to be 20, 11, 12, 4, 7, 8, 6, 4, 4.5, 6, 5, 400, 11, 7, 4, 3, 6, 5.5, 8, 6, 4, 7, 10, 14, 9 tonnes per hectare.

The minimum selling prices (MSP) of the crops were obtained from 2011–12 data. Prevailing market prices were used for crops not covered by the MSP scheme.

GOAL programming was developed using the above data. For sustainable production, water, labour, production, revenue and food security were assigned different priorities during programming to identify the most satisfactory solutions.

18.5   Formulation of the Sustainable Model Using GOAL Programming

Taking into consideration the available natural resources used in crop production, the model has been formulated to determine which crops should be produced to meet social, economic and environmental needs (Anonymous 1991, 2001). Water use, labour use, production enhancement, revenue generation and food security have been considered as the conflicting goals to be achieved with priorities [15, 16]. The priorities have been changed to discover the matching results. It has been observed that priority 1 (water availability and use) and one or two more priorities have been achieved in the proposed model (Banashri et al. 2012; Banashri 2013). None of the solutions are optimal but they are fair in nature.

The detailed solution with the changing priority has been given below. The solution with water as the top priority is shown in Tables 18.1 and 18.2.

In model 1, water was used as priority number one. To solve this model equation, a 94 × 39 matrix was used with 77 iterations. In this solution optimal use of water and production has been achieved. No resources are left for agricultural production. This implies the optimal solution has been obtained with respect to water and production. The solutions with respect to labour, revenue and food security have been partially achieved.

In model 2, labour was assigned to priority number one. The solution shows that the optimal solution has been obtained with respect to labour and production. The solutions with respect to water, revenue and food security have been partially achieved. The available water may be used judiciously to address food security.

In model 3, production was considered to be priority number one, while other parameters were assigned different priorities. Changing the priority yielded at least six more permutations. The optimal solutions with respect to production and labour were obtained. Satisfactory results were obtained for three other goals: water, revenue and food security. Food security is a critical activity which has to be achieved using the best possible means.

In model 4, revenue was considered to be priority number one. Other priorities were given less importance. The optimal solutions were obtained with respect to revenue, production and labour. A satisfactory solution was achieved with respect to water and food security. The water left over could be used for growing other low-duty crops according to the requirements of the individual farmer (Table 18.3).

The sensitivity analysis presented in above table shows that in three cases (water, labour and production) two objectives have been fulfilled. In the other two cases (revenue and food security) three objectives have been fulfilled. The optimal solution in other cases has not been achieved. The realistic solution has been presented in the sensitive analysis table. The sustainable production system cannot use all the labourers available as agricultural labourers. This presents a compelling case for the migration of workers to other states or other employment areas, such as service sectors.

Table 18.1 Model 1: Solution with water as the top priority

	Goal	Value	(Min.) =	
G1	Goal	Value	(Min.) =	0
G2	Goal	Value	(Min.) =	20,350,080.00
G3	Goal	Value	(Min.) =	0
G4	Goal	Value	(Min.) =	1,151,854,080.00
G5	Goal	Value	(Min.) =	199,235,125,248.00

	Constraint	Left hand side	Direction	Right hand side	Slack or surplus	Alternate Allowable min. RHS	Solution Allowable max. RHS	Exists!!	Shadow-Price Goal 1	Shadow-Price Goal 2	Shadow-Price Goal 3	Shadow-Price Goal 4	Shadow-Price Goal 5
1	C1	3,834,457.25	=	3,834,457.00	0	3,417,229.00	28,651,628.00		0	−0.82	0	−0.26	−206.92
2	C2	3,417,229.25	=	3,417,229.00	0	M	3,834,457.00		0	0	0	0	0
3	C3	23,500,000.00	=	23,500,000.00	0	3,149,920.00	M		0	1.00	0	0	0
4	C4	873,709.06	=	873,709.00	0	M	1,843,732.75		0	0	0	0	0
5	C5	1,152,850,048.00	=	1,152,850,048.00	0	995,968.00	M		0	0	0	1.00	0
6	C6	10,200,000.00	=	10,200,000.00	0	229,889.00	M		0	0	0	0	1.00
7	C7	199,999,995,904.00	=	199,999,995,904.00	0	794,066,944.00	M		0	0	0	0	1.00
8	C8	20,000,000.00	=	20,000,000.00	0	781,964.00	M		0	0	0	0	1.00
9	C9	4,226.00	=	4,226.00	0	M	38,263.35		0	0	0	0	0
10	C10	18.00	=	18.00	0	0	85,111.38		0	−12.20	0	−7.60	−16,206.20
11	C11	15.00	=	15.00	0	0	M		0	0	0	0	1.00
12	C12	836.00	=	836.00	0	0	M		0	0	0	0	1.00
13	C13	51.00	=	51.00	0	0	M		0	0	0	0	1.00
14	C14	135.00	=	135.00	0	0	M		0	0	0	0	1.00
15	C15	248.00	=	248.00	0	0	M		0	0	0	0	1.00
16	C16	261.00	=	261.00	0	0	M		0	0	0	0	1.00
17	C17	15.00	=	15.00	0	0	M		0	0	0	0	1.00
18	C18	74.00	=	74.00	0	0	M		0	0	0	0	1.00
19	C19	8.00	=	8.00	0	0	M		0	0	0	0	1.00
20	C20	41.00	=	41.00	0	0	20,062.97		0	−122.60	0	28.20	−28,816.60
21	C21	9.00	=	9.00	0	0	78,920.15		0	−46.64	0	−1.52	−19,831.84
22	C22	16.00	=	16.00	0	0	M		0	0	0	0	1.00

(continued)

Table 18.1 (continued)

23	C23	93.00	=	93.00	0	0	M	0	0	0	1.00
24	C24	21.00	=	21.00	0	0	M	0	0	0	1.00
25	C25	93.00	=	93.00	0	0	M	0	0	0	1.00
26	C26	112.00	=	112.00	0	0	M	0	0	0	1.00
27	C27	10.00	=	10.00	0	0	M	0	0	0	1.00
28	C28	35.00	=	35.00	0	0	M	0	0	0	1.00
29	C29	244.00	=	244.00	0	0	M	0	0	0	1.00
30	C30	617.00	=	617.00	0	0	M	0	0	0	1.00
31	C31	42.00	=	42.00	0	0	M	0.00	0	0	1.00
32	C32	179.00	=	179.00	0	0	M	0	0	0	1.00
33	C33	253.00	=	253.00	0	0	M	0	0	0	1.00
34	C34	9.00	=	9.00	0	0	M	0.00	0	0	1.00

Source Output of the program

Table 18.2 Model 2: Constraint summary with water as the top priority

12-19-2013 19:50:02	Decision variable	Solution value	Basis status	Reduced cost Goal 1	Reduced cost Goal 2	Reduced cost Goal 3	Reduced cost Goal 4	Reduced cost Goal 5
18	X18	0	At bound	0	2.80	0	2.40	-7,060.70
19	X19	0	At bound	0	2.30	0	3.90	-4,776.20
20	X20	0	At bound	0	-10.60	0	-1.80	-4,032.60
21	X21	0	At bound	0	-13.60	0	-2.80	-4,381.60
22	X22	0	At bound	0	-14.70	0	-4.10	-4,955.20
23	X23	0	At bound	0	-24.20	0	0.40	-13,245.20
24	X24	0	At bound	0	-38.90	0	-1.70	-6,170.40
25	X25	0	At bound	0	-38.90	0	-1.70	-6,170.40
26	X26	0	At bound	0	-9.90	0	3.30	-10,606.40
27	X27	0	At bound	1 00	-0.82	0	-0.26	-206.92
28	X28	0	At bound	0	0.82	0	0.26	206.92
29	X29	417,228.00	Basic	0	0	0	0	0
30	X30	0	At bound	1.00	0	0	0	0
31	X31	0	At bound	0	1.00	0	0	0
32	X32	20,350,080.00	Basic	0	0	0	0	0
33	X33	970,023.81	Basic	0	0	0	0	0
34	X34	0	At bound	0	0	1.00	0	0
35	X35	0	At bound	0	0	0	1.00	0
36	X36	1,151,854,080.00	Basic	0	0	0	0	0
37	X37	0	At bound	0	0	0	0	1.00
38	X38	9,970,111.00	Basic	0	0	0	0	0
39	X39	0	At bound	0	0	0	0	1.00
40	X40	199,205,928,960.00	Basic	0	0	0	0	0
41	X41	0	At bound	0	0	0	0	1.00

(continued)

Table 18.2 (continued)

12–19–2013 19:50:02	Decision variable	Solution value	Basis status	Reduced cost Goal 1	Reduced cost Goal 2	Reduced cost Goal 3	Reduced cost Goal 4	Reduced cost Goal 5
42	X42	19,218,036.00	Basic	0	0	0	0	0
43	X43	34,037.35	Basic	0	0	0	0	0
44	X44	0	At bound	0	0	0	0	1.00
45	X45	0	At bound	0	−12.20	0	−7.60	−16,206.20
46	X46	0	At bound	0	0	0	0	0
47	X47	0	At bound	0	0	0	0	1.00
48	X48	15.00	Basic	0	0	0	0	0
49	X49	0	At bound	0	0	0	0	1.00
50	X50	836.00	Basic	0	0	0	0	0
51	X51	0	At bound	0	0	0	0	1.00
52	X52	51.00	Basic	0	0	0	0	0
53	X53	0	At bound	0	0	0	0	1.00
54	X54	135.00	Basic	0	0	0	0	0
55	X55	0	At bound	0	0	0	0	1.00
56	X56	248.00	Basic	0	0	0	0	0
57	X57	0	At bound	0	0	0	0	1.00
58	X58	261.00	Basic	0	0	0	0	0

Source Output of the program

Table 18.3 Sensitivity analysis of all five goals

Description	Sensitivity analysis summary					$x_1 \ldots X_{26}$	Optimal solution				
	Priority						Goal 1	Goal 2	Goal 3	Goal 4	Goal 5
	Highest	High	Medium	Medium low	Low						
Water	S_1+, S_2-	S_3-	S_4-	S_5-	$S_6-\ldots S_{26}-$		0	2.00E+07	0	1.15E + 09	1.91E+11
Labour	S_3-	S_1+, S_2-	S_4-	S_5-	$S_6-\ldots S_{26}-$		0	9.63E+08	0	9.01E+08	2.20E+03
Production	S_4-	$S+, S_2-$	S_3-	S_5-	$S_6-\ldots S_{26}-$		0	9.63E+08	0	9.01E+08	2.20E+03
Revenue	S_5-	S_1+, S_2-	S_4-	S_3-	$S_6-\ldots S_{26}-$		0	4.43E+09	0	0	3.40E+03
Food Security	$S_6-\ldots S_{26}-$	S_1+, S_2-	S_4-	S_5-	S_3-		0	4.01E+08	0	1.00E+10	0

Source Tabulated from all the above outputs

18.6   Conclusion

MCDM programming with priority has re-invented the requirement of inputs for the target of sustainable food production. Taking into consideration the soil health, food security and long-term sustainability, the inputs and area to be cultivated for the production of cereals, pulses, oilseeds etc. have been computed considering multiple goals with priority. Soft constraints, such as seed supply, mechanisation, finance, and number of crops in a year (cropping intensity), have been determined and specified under present conditions. When the agriculture production environment changes in a particular area, the simulation can be modified and the corresponding value of the soft constraints can be determined. This will help planners to determine the inputs to be made available for targeted production. The area under different crops can be computed before the actual agricultural production takes place. The Rabi and Kharif programme can be planned as a Specific, Measurable, Attainable, Realistic and Time-bound (SMART) production goal. This simulation has used 150% cropping intensity for paddy with other crops. The total area under production has been considered to be 73 lakh hectares. The production of twenty-five crops, such as paddy, green gram, black gram, maize and groundnut, has been simulated. The revenue generated from these twenty-five crops has been simulated to be 3,500 crores considering the cost of input and minimum support price declared by the Government for the financial year 2015.

References

Alston, J.M.; Larson, D.M., 1993: "Hicksian vs. Marshallian Welfare Measures: Why Do We Do What We Do?" *American Journal of Agricultural Economics*, 75: 764–769.

Angehrn, A.A.; Dutta, S., 1992: "Integrating case-based reasoning in multi-criteria decision support systems", in: INSEAD Working Paper 92/54TM.

Angehrn, A.A., 1991: "Designing humanized systems for multiple criteria decision making", in: *Human Systems Management*, 10: 221–231.

Anonymous, 2000: "Agricultural Modernization in Rural Odisha: First Impact on the Weaker Sections," in: *IASSI Quarterly*, 14(3–4): 85–93.

Anonymous, 1991: *Census of India* (Bhubaneswar: Department of Statistics, Government of Odisha).

Anonymous, 2001: *Census of India* (Bhubaneswar: Department of Statistics, Government of Odisha).

Anonymous, 2004a: *Economic Survey, 2003–2004* (Bhubaneswar: Directorate of Economics and Statistics, Planning and Co-ordination Department).

Anonymous, 2004b: *Odisha Reference Annual 2004* (Bhubaneswar: Information and Public Relations Department).

Anonymous, 1979: *Statistical Abstract of Odisha* (Bhubaneswar: Department of Statistics, Government of Odisha).

Anonymous, 1999: *Statistical Abstract of Odisha* (Bhubaneswar: Department of Statistics, Government of Odisha).

Anonymous, 2010: *Statistical Abstract of Odisha* (Bhubaneswar: Department of Statistics, Government of Odisha).

Ata, A.; Sennaroglu, B., 2008: "The usage of MCDM techniques in determining the constants of criteria's weight of warship design", in: *Journal of Naval Science and Engineering*, 4(1): 1–16.

Banashri, R.; Mangaraj, B.K.; Mishra, B.P., 2012: "Fuzzy logic-based simulation for modelling of sustainable marketing policy for modern rice mill in Odisha", in: *International Journal of Supply Chain Management*, 1(3): 34–42.

Banashri, R., 2013: "Decision support system for drafting of sustainable agriculture policy for Odisha", in: *International Journal of Supply Chain Management*, 2(2): 92–98.

De, P.K.; Yadav, B., 2011: "An Algorithm to solve Multi-Objective Assignment Problem Using Interactive Fuzzy Goal Programming Approach; in: *International Journal of Contemporary Mathematical Sciences*, 6(34): 1,651–1,662.

Government of India, 2002: *Odisha Development Report* (New Delhi: Planning Commission).

Laxminarayan, V. and Rajgopanan, S.P., 1977: "Optimal cropping pattern for basins in India", in: *Journal of Irrigation and Drainage Engineering*, ASCE (AMERICAN Society of Civil Engineering, IRI-Industrial Research Institute) 103: 53–59.

Rath, B.; Mohanty, R.P.; Mishra, B.P., 2003: "MCDM approach rural common resource planning for sustainable rural development", in: *Siddhant, A Journal of Decisionmaking*, 3(1): 135.

Sinha, S.B. et al., 1989: "Fuzzy Technique for Agricultural Planning", in: *Journal of Information and Optimization Science*, 10(1): 257–274.

Sinha, S.B.; Rao, K.A.; Mangaraj, B.K., 1988: "Fuzzy goal programming in multi-criteria decision systems: a case study in agricultural planning", in: *Socio-Economic Planning Sciences*, 22(2): 93–101.

Chapter 19
Sources of Productivity Growth and Livelihoods Resilience in Bihar in the Recent Decade: A District-Level Non-parametric Analysis

Surya Bhushan

Abstract This chapter applies the non-parametric *data envelopment approach* (DEA) to estimate *Total Factor Productivity* (TFP) growth rate for agricultural output during the period 2000–2012 across districts of Bihar. This approach finds that the shift in the frontier as well as improved efficiency play an important role as a source of productivity growth, suggesting that technological adoption and catching up may be a vitally important source for overall productivity growth. The approach identifies the frontier districts in terms of agricultural production. Using the geographically linked resources at district level, namely, bio-physical, social, economic and health resources, this chapter also develops a *Livelihoods Resilience Index* (LRI) at district level to explain the association with agricultural TFP growth. The positive association identified warrants further investigation of a smaller unit, say, household level, to explain the rural development dynamics in the predominantly agricultural and rural state.

Keywords TFP · Malmquist · DEA · Non-parametric · Agriculture
Bihar · Livelihoods resilience index

19.1 Introduction

Agricultural *total factor productivity* (TFP) growth, reflecting more efficient use of inputs in output, has been able to sustain a rapid transition out of poverty for society and create a possible strategic pathway towards a sustainable system (Timmer 2007: 7). Positive values of TFP growth present an indicator of how the output obtained from the system increases more quickly with respect to the input supplied on the

Surya Bhushan, Associate Professor, Development Management Institute (DMI), Patna, Bihar, India; Email: surya.bhushan@gmail.com The author is grateful to Professor K. V. Raju, Director, DMI, especially for developing the livelihoods resilience index, and providing helpful comments and valuable insights and suggestions on the earlier draft. The usual disclaimer applies.

one hand and conserving natural resources and promoting ecological integrity of the agricultural system on the other. The positive agricultural TFP allows economic entities to produce more food at lower cost, improve nutrition and welfare, and release resources to other sectors (e.g. Lewis 1954: 139; Rostow 1956; Ranis/Fei 1961; Timmer 1988), which increases income for the rural communities, which then promotes their spending on the non-farm sector and reduces poverty, especially in rural areas (e.g. Datt/Ravaillion 1998; Ellis 2000; Thirtle et al. 2003; Himanshu et al. 2011). Further, for the regions, with a dwindling supply of arable land per capita, population pressure acts as the driving force behind the adoption of steadily more intensive forms of agricultural technology, beginning with the abandonment of shifting cultivation and fallowing (Boserup 1965). In developing countries, juxtaposing agricultural TFP growth with the livelihood makes the case intriguing in the development thinking and practice, where a majority of families, both in the farm and non-farm sectors, derive their livelihoods from agriculture (e.g. Acharya 2006; Kumar et al. 2006). 'Sustainable livelihoods' is a multifaceted concept and refers to the maintenance or enhancement of people's access to food and income-generating activities on a long-term basis (e.g. Chambers/Conway 1992; Scoones 2009).

The main purpose of this study is to estimate the *total factor productivity* (TFP) growth of agriculture at district level in Bihar during the years 2001–2012 by applying the non-parametric data *envelopment analysis* (DEA) approach. By estimating input or output distance functions, non-parametric DEA can be used to decompose TFP growth into movements of the frontier (technical gains or innovation) and movements towards the frontier (efficiency gains or 'catching up') using the Malmquist index (e.g. Färe et al. 1994; Coelli et al. 2005 for further discussion). This might allow more insight into the main drivers of productivity. Many farmers in Bihar are still using low-yielding agricultural technologies, which lead to low productivity and production. Another relevant question for agricultural policy-makers is whether the agricultural sector can be made more efficient by achieving more output with the current input level, or by achieving the current output with less input usage than is currently observed. An important step in answering these questions is to understand the pathway of productivity and its importance in developing livelihood resilience. The resilience can be seen as a collective property – of the community rather than the individual (e.g. Ashley/Carney 1999; Krantz 2001; Donohue/Biggs 2015). The chapter also develops the livelihood resilience index (LRI), employing the methodology of the *Human Development Index* (HDI) estimation by World Development Reports of the World Bank at district level. This further analyses the relationship between the two, which might help policy-makers and planners to develop strategies for agricultural growth. The rest of this chapter is organised as follows: Sect. 19.2 presents an overview of agricultural productivity growth with reference to Bihar. The next Sect. 19.3, discusses the non-parametric approach to estimate the Malmquist TFP and LRI used in the study. The data and their sources are described in Sect. 19.3.3. Section 19.4 discusses the results and implications for the study. A summary and conclusions are included in the last Sect. 19.5.

19.2   An Overview

Bihar, the state located in the eastern part of India, is endowed with good soil, adequate rainfall and good groundwater availability, the best of natural resources for agriculture. Conversely, it is food deficit and houses the largest number of undernourished and poor people (>40%) in the country due to exceptionally low agricultural productivity (Joshi et al. 2011). Recently, Shah (2016) examined the paradoxes of Bihar's recent development experience, especially during 2004–05 to 2009–10. The state witnessed almost a double digit *gross state domestic product* (GSDP), but over 81% of its total population – the highest percentage of all Indian states (Alkire/Foster 2011) – is poor according to the *multidimensional poverty index* (MPI), and 56% of children under five are malnourished (compared to the national average of 42%). Bihar ranks bottom in terms of per capita income among the thirty states in India, and the primary sector still contributes more than the secondary sector of the state domestic product. The agricultural sector continues to play a crucial role in the development of the state, due to its large presence in both aggregate income and dependent labour force (Fig. 19.1).

Further, the recent decades of annual sectoral growth show that more needs to be done. While the non-agricultural sector of the economy is growing at almost double digits, the growth in the agricultural sector is below 4%, placing stress on the overall per capita GSDP growth (Economic Survey 2015–16: 6).

With the adoption of Green Revolution technology during the late 1960s, the tremendous agricultural growth in India during the past few decades has been of considerable interest to researchers, policy-makers and practitioners. With the availability of micro-level farm data, particularly of the *Commission for Agricultural Costs and Prices* (CACP) in India, a few crop-specific TFP studies have been conducted since the early 1990s (e.g. Kumar et al. 2004, 2006; Bhushan 2014, 2016). Bihar struggled to get into the high growth trajectory zone over the period for the majority of crops (Table 19.1).

Rada (2013) also estimated the TFP growth index for Indian states, where Bihar was ranked almost bottom during the last three decades (Fig. 19.2).

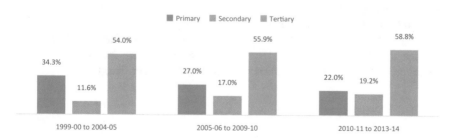

Fig. 19.1 Per capita income and structural decomposition of net state domestic product at constant prices 2004–05 (in Rs.). *Source* Compiled from Economic Survey, Bihar, 2015–16: 5–6

Table 19.1 Annual growth rates in input use, output, TFP and real cost of production (RCP) for different crops grown in Bihar: 1975–2005

CROPS	Input growth	Output growth	TFP growth	RCP growth	TFPG share in output growth
Rice	0.31	0.69	0.38	−0.71	55.7
Wheat	1.57	1.61	0.04	−0.04	2.6
Maize	0.26	1.33	1.07	−1.27	80.2
Gram	−8.58	−6.51	2.07	−2.19	(−)
Jute	0.63	1.9	1.27	−2.14	66.8

Source Compiled from Chand et al. (2011: 6)

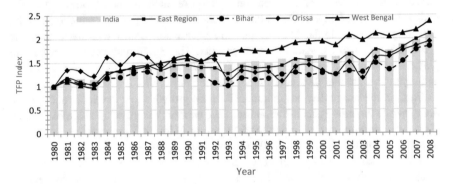

Fig. 19.2 National and eastern regional TFP growth rate indexes, 1980–2008. *Source* Compiled from Rada (2013)

Recently, using both parametric and non-parametric techniques, Bhushan (2016) estimated TFP growth of paddy in different states, including Bihar, during 1981–2010. The result for Bihar does not show a positive trend across the decades, regardless of the techniques used. The sluggish and negative productivity growth in Bihar may be due to low input use efficiency, partly because seed replacement rates are low and fertilizer use is imbalanced (World Bank 2006). The seed replacement rate, a metric to measure the quality of seeds for higher crop yield, has been around 15% for the majority of crops. However, it grew at an average annual rate of 10% in 2001–08, faster than the national average. In 2009, Bihar's fertilizer use intensity was 231 kg/ha, well below the level in Punjab (400 kg/ha) or Haryana (320 kg/ha). Against this backdrop, it would be interesting to examine the sources of TFP growth of agriculture in Bihar at district level in the recent decade.

19.3 Methodology

19.3.1 Malmquist TFP

In this chapter, we use a Malmquist index approach to examine inter-district agricultural efficiency and productivity. In contrast to conventional production function or other index approaches, the Malmquist approach can decompose the sources of productivity growth into two components (Nishimizu/Page 1982):

- The efficiency gains, i.e., the growth in the factor of production, indicating the movement along the best practice production frontier, and
- The technological gains, i.e., the shifting of the production frontier outward (inward) in the case of technological progress (regress).

The Malmquist Index, which Caves et al. (1982) introduced in its empirical usage and Färe et al. (1994) estimated non-parametrically, measures the radial distance of the observed output and input vectors in period t and t + 1 relative to two reference technologies: technology in period t and technology in period t + 1. Consider a sample of K districts, where each one produces M output using N inputs at time t, and define technology P^t as the transformation of all feasible inputs $[x^t = (x^1, \ldots, x^N)]$ into output pairs $[y^t = (y^1, \ldots, y^M)]$, i.e., $P^t = \{(x^t, y^t): x^t \text{ can produce } y^t\}$, t = 1, 2, …, T. We assume that P^t satisfies the axiomatic properties of non-emptiness, closedness, and convexity of freely disposable inputs and outputs, as defined by Shephard (1970). Let us define output distance function at time t as:

$$D_O^t(x^t, y^t) \equiv inf\left[\left(\theta : x^t, \frac{y^t}{\theta}\right) \in P^t\right] = [sup\{\theta : (x^t, \theta y^t) \in P^t\}]^{-1} \qquad (19.1)$$

where the inf (sup) operator finds the set's infimum (supremum), the greatest lower (upper) bound, scalar $\theta \leq 1$, which means scaling back the output by the least possible factor gives maximum proportional expansion of the output vector y^t, given inputs x^t and technology P^t. It is also equivalent to the reciprocal of Farrell's (1957) measure of output efficiency, which measures TFP "catching-up" of an observation (states in our case) to the best practice frontier technology. The value of $D^t (x^t, y^t)$ can be derived from the solution of the linear programming problem:

$$[D^t(x^t, y^t)]^{-1} = \max_{\theta, \gamma} \theta \; s.t. \; x^t - X^t\gamma \geq 0, -\theta y^t + Y^t\gamma \geq 0, \gamma \geq 0 \qquad (19.2)$$

The first constraint in the model states that to produce the observed output in the kth district, the actual use of input i for district k should be greater than or equal to the theoretically efficient input usage that is the weighted sum of all districts. The second constraint specifies that, given the actual amount of inputs used by the kth district, the maximum feasible output of district k should be less than or equal to the theoretically efficient output that is a weighted sum of all districts' outputs.

The Malmquist productivity index is the geometric means of the answers to these two questions (Caves et al. 1982; Hulten 2001): first, how much output could district A produce if it used district B's technology with its own inputs, and second, how much output could district B produce if it used district A's technology with its inputs?

Using the period t benchmark technology and Eq. (19.1), the period t output-orientated Malmquist productivity index is written as:

$$M^t = \frac{D^t(x^{t+1}, y^{t+1})}{D^t(x^t, y^t)} \qquad (19.3)$$

In order to avoid an arbitrary benchmark, Färe et al. (1994) rewrote the geometric mean yielding efficiency and technology and decomposed it into two parts:

Färe et al. (1994) further split the Malmquist index of Eq. (19.3) into technical change and efficiency change:

$$\text{Malmquist Index} = \frac{D^{t+1}(x^{t+1}, y^{t+1})}{D^t(x^t, y^t)} \left[\frac{D^t(x^{t+1}, y^{t+1})}{D^{t+1}(x^t, y^t)} \times \frac{D^t(x^t, y^t)}{D^{t+1}(x^t, y^t)} \right]^{1/2} \qquad (19.4)$$

The leftmost fraction captures the efficiency change or the 'catching up' effect. The rightmost term, in brackets, represents technological change or shifts in the technology frontier. Using Eq. (19.4), Malmquist Index $\gtrless 1$ entails productivity growth; stagnation or decline has occurred between periods t and t + 1. For instance, a Malmquist index greater than unity, say 1.25 (which signals productivity gain), could have a component of efficiency change smaller than 1 (e.g. 0.5) and a change in technology component greater than unity (e.g., 2.5). The distance functions are estimated by using linear programming problems, e.g., Eq. (19.2) (Coelli et al. 2005: 294–296).

19.3.2 Livelihood Resilience Index

The concept of livelihood encompasses secure ownership of, or access to, resources, assets and income-generating activities to offset risks, ease shocks, meet contingencies, and improve resilience. The *Sustainable Livelihoods Framework* (SLF) developed by DFID (2000) visualises five assets of livelihood, namely, human, social, natural, physical, and financial, which constitute the means of living. These are complex and dynamic in nature, with bi-directional causes and effects (Allison/Ellis 2001). The resilience setting within the livelihoods framework looks at the root causes of vulnerability in the systems of an individual, household, community or region. The approach looked into geographically linked resources that make systems resilient to exogenous shocks and endogenous trends. The exogenous shocks can be related to human, livestock or crop health; natural hazards, like floods or earthquakes; economics; and conflicts in the form of national or international wars. Endogenous trends can be demographic trends or resource trends (Cassidy/Barnes 2012). The *Livelihoods Resilience Index* (LRI) developed here includes four major components based on the availability of bio-physical, economic, social, and health resources.

- *Bio-physical Resources*: Critical to maintain and improve biophysical bases; for example, land, water and biological resources provide material resources for the livelihoods.
- *Social Resources*: Networks and connections, both formal and informal. Collective representation ensures a more broad-based distribution of economic benefits – both at present and in the future – in the form of secured livelihoods, especially for the weaker groups.
- *Economic Resources*: Indicate how resources both human and natural – for example, land productivity, marketable surplus, input-output ratio – are used under current technological conditions to meet the present developmental needs of the society.
- *Health Resources*: The health status of children, in particular, can be taken as an important indicator of the sustainability of well-being and livelihoods.

The LRI uses a balanced weighted average approach, and assumes each sub-component contributes equally to the overall index, even though each major component is comprised of a different number of sub-components (Sullivan et al. 2002). The intention was to develop an assessment tool accessible to a diverse set of users in resource-poor settings; therefore the LRI formula uses the simple approach of applying equal weights to all major components. This weighting scheme could be adjusted by future users as needed. The standardised index, adapted from that used in the Human Development Index, was computed and rescaled into a single scalar – a composite index (UNDP 2007).

$$Index_x = \frac{x - m_x}{M_x - m_x}, x - m_x = \text{distance variable}; M_x - m_x = \text{reference level(range)}$$

where x is the original sub-component for district x, and m_x and M_x are the minimum and maximum values, respectively, for each sub-component determined using data from the population.

19.3.3 Data Sources and Types

In order to estimate the TFP growth of agriculture at district level, panel data were needed. The data used for the analysis were state-level aggregates on input use and outputs collected under the *Comprehensive Scheme for Studying Cost of Cultivation of Principal Crops* at plot level by the CACP, Directorate of Economics and Statistics, Ministry of Agriculture for the period 2000–2012 (CACP, Various

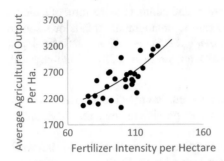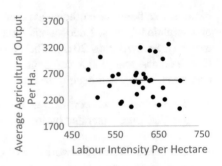

Fig. 19.3 Average resource intensity vis-à-vis agricultural output per hectare across districts for 2000–2010. *Source* Author's calculation

Issues). The detailed features of the CACP dataset are described in Bhushan (2014). Thirty districts were selected for the analysis.[1] The crops used to represent agriculture production per hectare in the districts were coarse cereals, pulses, oilseeds, and potatoes. The data were aggregated at average per hectare to present the overall agricultural potential and performance of the district. The quantity data were given priority over price data, wherever both available, to avoid the anomalies in price information due to market distortion: Output (in kg), Human Labour (in hrs), Chemical Fertilizer (in kg), and Machine (in hrs). The chemical fertilizers aggregate the nitrogen (N), phosphorus (P) and potassium (K) or NPK, and the Machine Hours include own, hired as well as owned and hired irrigation machines. All the data had been transformed by taking the centred moving average of three periods to avoid any short-term weather fluctuations, particularly rainfall effect. The following figures illustrate the resource intensity in the agricultural output across the districts in Bihar for the period 2000–2012 (Fig. 19.3).

There is a strong positive relationship between chemical fertilizer intensity and agricultural output over the decade, showing intensification of this input in agricultural production. However, the almost flat or non-existent relationship of labour intensity highlights an area for concern, as agriculture in Bihar employs more than 75% of the labour force in output production (Economic Survey 2015–16).

LRI is comprised of several indicators or sub-components (Table 19.2). The indicators were carefully identified and chosen at the district level to capture the essence of the livelihoods resilience as well as the practicality of available data. The data employs various Economic Surveys published by the Finance Department of the Government of Bihar. Some sub-components, such as the 'Child Survival Rate', were created because an increase in the crude indicator – in this case, the number of

---

[1]Bihar has thirty-eight districts, according to the latest data available. The districts notably absent from the analysis due to the non-availability of data are Araria, Kishanganj, Lakhisarai, Nawada, Sheikhpura, Sheohar and Sitamarhi. Arwal was carved out of the Jehanabad District in 2001.

Table 19.2 Livelihoods resilience indicators

Livelihood assets	Indicator	Variable construction	Explanation	Source
Biophysical	Wetlands	Wetland area to total area %	Wetlands are one of the most productive ecosystems and play a crucial role in the hydrological cycle. Utility wise, wetlands directly and indirectly support millions of people by providing services such as storm and flood control, clean water supply, food, fibre and raw materials, scenic beauty, educational and recreational benefits. Wetlands have been called the 'kidneys of the landscape' because of their ability to store, assimilate and transform contaminants lost from the land before they reach waterway	Wetland Atlas (2010)
	Cropping intensity	Gross cropped area/net sown area	This indicates the pressure on the land	Economic Survey 2015–16
Economic	Livestock	Livestock per household	This includes the livestock population of cows, goats, poultry etc.	Economic Survey 2015–16
	Irrigated area	% irrigated	Irrigation is an important component where the production process is dominated by agriculture	Economic Survey 2015–16
	Drinking water distance	Household with main source of drinking water within premises	Access to water for consumption is a key component both in the natural capital entitlements of households and of healthy ecosystems	NFHS-4 (2016)
	Electricity	Household with electricity	Expanding access to modern energy has been identified as a necessary condition towards the economic, social and environmental aspects of human development (IEA 2009)	NFHS-4 (2016)

(continued)

Table 19.2 (continued)

Livelihood assets	Indicator	Variable construction	Explanation	Source
Social	Female education	Female literacy	Apart from wider social benefits both locally and in terms of greater social change, female literacy has consistently been identified as an important factor in reducing poverty (e.g. Anand/Sen 1997; Maddox/ Esposito 2012)	NFHS-4 (2016)
	Caste	% of Dalits to total population	Dalits are at the bottom of the social hierarchy and are dominantly the poorest members of society	Census (2011)
	Market exchange	Urbanisation	The availability of markets for the exchange of goods and services	Census (2011)
Health	Child immunisation	Children aged 12–23 months fully immunised (%)	Immunising children protects individuals from infection and contributes to population-based immunity by reducing the circulation of infectious agents. This leads to community-wide health gains	NFHS-4 (2016)
	Acute undernutrition	Children under five years who are not wasted (weight-for-height) (%)	Wasting is an indicator of acute undernutrition, the result of more recent food deprivation or illness	NFHS-4 (2016)
	Child survival rate	(1000-Infant mortality rate)	Child life expectancy is a leading indicator of the level of child health and overall development of countries (WHO 2005)	NFHS-4 (2016)

Source Compiled from multiple sources, as indicated in the table

livelihood activities undertaken by a household – was assumed to improve resilience. In other words, we assumed that a household where the child survival rate is higher has better health resilience than a household which has a lower child survival rate. In the current study, the LRI is scaled from 0 (least resilient) to 1 (most resilient).

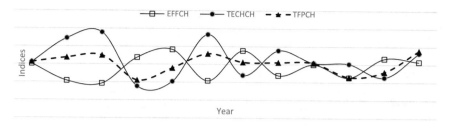

Fig. 19.4 Temporal decomposition of agriculture TFP in Bihar during 2001–2012. *Source* Author's calculation

19.4 Results and Discussion

A weighted average of TFP measures at Bihar level for a sample of thirty districts shows that annual growth between 2000 and 2012 was 0.30%. The decomposition of Bihar's TFP growth into efficiency and technical change shows that technical change contributed to productivity growth by 2.0%, while efficiency deterioration reduced productivity growth by 1.7% per year on average during the study period. Further, the input contribution, calculated as ratio of output change index to TFP change index, was positive at 0.3%, showing a mild 'input intensification' in the agricultural production. This average, however, hides significant variations across time, where two periods with contrasting results can be distinguished (Fig. 19.4).

For example, the poor performance and decline during 2004 stretches to 2005, during which productivity growth in Bihar was negative: 8.9 and 2.1% per annum. That period is followed by a period of recovery and improved performance that starts in 2006 and extends up to 2009. During this period, TFP grows at an annual rate of 2.3% (6% in 2006). Subsequently, TFP growth dips for the next two periods, 2010 and 2011, and recovers in the last period (2012) of the study. Another noticeable part in the graph is the presence of more fluctuations during the pre-2005 period. This phenomenon warrants further deep dive investigation.[2] Table 19.3 lists the results of TFP and its components for the period 2000–2012 at district level.

---

[2]Deokar/Shetty (2014) contend that Indian agriculture has been doing far better in the years since 2004–05 due to a series of policy initiatives. First, increased allocations for various departments concerned with the development of agriculture, animal husbandry and agriculture research and education. Second, during 2005–06, a National Horticulture Mission became operational, and it extended the programme beyond fruits and vegetables and embraced medicinal plants and spices. Third, a centrally sponsored scheme called the Support to State Extension Programmes for Extension Reforms was launched in 2005–06. Fourth, in 2005–06 a National Fund for Basic, Strategic and Frontier Application Research in Agriculture and a National Agricultural Innovation Project (in July 2006) were launched. Fifth, the terms of trade began to improve in 2004–05 after rapid increases in procurement prices followed by increases in the international prices of agricultural commodities. Sixth, the launch of the Bharat Nirman project in 2005–06 with a view to upgrading rural infrastructure comprising six components, namely, irrigation, electrification, roads, water supply, housing and telecom connectivity. Finally, a 'farm credit' package has continued uninterrupted thereafter and has provided a push to private investment in agriculture.

Table 19.3 Average TFP decomposition, input contribution, for 2000–2012

Districts	Efficiency (%)	Technical (%)	TFP (%)	Input contribution (%)
Rohtas	3.30	4.30	7.80	−7.13
Katihar	1.90	3.70	5.70	−6.23
Banka	0.40	4.30	4.80	−3.50
Bhagalpur	0.00	4.00	4.00	2.01
Purnea	1.60	2.40	4.00	−0.08
Aurangabad	2.80	0.10	2.90	−0.24
Begusarai	−0.10	2.50	2.30	−0.62
Gaya	−3.20	5.40	2.10	−4.07
Darbhanga	−2.20	4.10	1.80	−3.34
Madhepura	−0.30	1.40	1.10	−1.48
Kaimur	−1.10	2.10	1.00	2.60
Muzaffarpur	−2.30	2.80	0.50	2.51
Jehanabad	−1.60	2.10	0.40	0.51
Khagaria	−1.10	1.40	0.30	−1.60
Gopalganj	−1.90	2.10	0.10	1.20
Supaul	−2.60	2.70	0.00	−0.07
Bhojpur	−1.60	1.50	−0.10	−0.93
Nalanda	−2.80	2.70	−0.20	−1.27
East Champaran	−1.90	1.40	−0.50	1.61
Buxar	−1.60	0.80	−0.90	−1.41
Patna	−4.20	3.00	−1.30	−4.95
Samastipur	−2.40	1.10	−1.30	2.03
Saran	−2.30	1.00	−1.30	4.00
Saharsa	−2.90	1.10	−1.80	4.32
Madhubani	−4.40	2.60	−1.90	−0.45
Jamui	−4.40	1.80	−2.60	7.04
Vaishali	−3.50	1.00	−2.60	4.56
West Champaran	−5.10	1.60	−3.50	2.96
Siwan	−3.80	−1.50	−5.20	6.29
Munger	−3.80	−1.90	−5.60	5.57
Bihar	−1.70	2.00	0.30	0.26

Source Author's calculation

The last column represents average LRI, developed in accordance with the methodology described in the previous section. The technical change component captures shifts in the production frontier, providing a measure of innovation. The phenomenon of catching up is measured as an efficiency change component and captures the diffusion of technology. There exists a wide variation regarding the TFP growth and its components. Out of thirty districts, nearly half, i.e., fifteen districts had shown a positive TFP growth and the rest negative. Sasaram, Katihar, Banka, Purnea and Aurangabad had shown an impressive TFP growth due to the combination of improved efficiency and technical progress. In general, if we look at the other districts, the growth TFP is dominated by technical progress. Again, if we look at the input contribution,[3] the use of inputs in agricultural production has actually increased at overall level, showing 'input intensification' in agricultural production growth. This confirms the findings of Kalirajan/Shand (1997) that the output growth in agriculture is increasingly dependent on the input growth in the majority of states in India. The input-based growth is unsustainable in the long run. Kumar et al. (2004) also raised concerns over the indiscriminate use of natural resources in the intensively cultivated areas of Indo-Gangetic Plains.

The situation is worse in low-yield rain-fed states, where the use of modern inputs, like machine labour and chemical inputs, is still way below the national average. This has important policy implications, since it points to the crucial need for public investment in irrigation and water management in these states, as the majority of the farmers in these states are small and marginal and use input resources at sub-optimal level. However, at district level the input use actually declines for some high TFP growth districts. The input decline in some negative TFP growth districts, mainly Nalanda and Patna, raises concerns for the long-term sustainability of agricultural output production. These two districts recorded negative efficiency growth, despite positive technical progress. The TFP growth came mainly from progress in the best-practice technology in these districts.

Further, the decomposition of the LRI into various components – *Bio-physical Resilience Index* (BRI), *Economic Resilience Index* (ERI), Social Resilience Index (SRI), and *Health Resilience Index* (HRI) – presents a case for intervention strategy from the perspective of improving the resilience of livelihoods (Fig. 19.5).

The LRI, composed mainly of static components, on average, remains constant over the decade. The following scatter plot (Fig. 19.6) represents a positive correlation of LRI with TFP indices at the district level. Aurangabad and Saharsa are surprise entries in the top six, despite a low per capita income. They performed pretty well in terms of BRI and ERI. The low rank in terms of SRI for Aurangabad and Buxar warrants a special focus on urbanisation and female literacy to improve social resilience in the district. Further, the highest per capita income district, Patna, performed badly in BRI. This suggests the policy intervention needed in these districts is to restore the bio-physical resources, which may be over-exploited. The bottom six districts have no surprises, as these districts performed poorly in terms of all the four resilience indicators, namely, BRI, SRI, ERI and HRI.

Furthermore, as discussed earlier, the improved agricultural TFP growth may be associated with the geographically linked bio-physical, human, social, and health resources. This may help in improving the livelihoods resilience (Fig. 19.6). The positive association between these two confirms the need for policy-makers to

---

[3]Since productivity growth is defined as output growth divided by the input growth, the contribution of inputs to output growth can be calculated by dividing the output growth index by the Malmquist productivity index. If it is less than one (or in percentage terms negative), then total input actually declines. The output trend growth of agricultural yield growth is calculated by running log-linear regression on time.

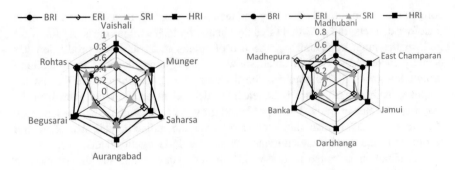

Fig. 19.5 LRI and its components for the top and bottom six districts. *Source* Author's Calculation

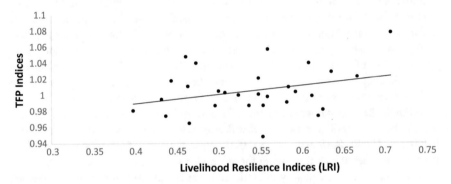

Fig. 19.6 LRI and TFP indices at the district level. *Source* Author's Calculation

improve the bio-physical, economic, social, and health resources, the components of LRI. This also calls for further exploration with more granular household level data.

However, this can be only a pointer towards understanding the static relationship; panel data regression can be helpful in exploring further dynamism at the spatial and temporal levels.

19.5 Concluding Remarks

The aim of this chapter is to understand the sources of productivity growth, namely efficiency and technical change, in agricultural growth patterns among Bihar districts by measuring and comparing agricultural total factor productivity or TFP growth, covering the period 2000–2012. The chapter applies non-parametric data envelopment analysis or DEA to estimate the Malmquist Productivity Index. The chapter also develops and provides a very crude measure to estimate the composite

Livelihoods Resilience Index (LRI), consisting of bio-physical, social, economic, and health resilience, at district level to understand the association with agricultural TFP growth. Several important conclusions emerge from these findings. First, unlike the recent findings by Bhushan (2016) at all India level, where technical progress was the primary source of wheat and paddy TFP growth, the agricultural TFP growth at district level in Bihar was mainly due to the combination of technical and efficiency changes. Overall, the agricultural production in Bihar provides evidence of technical inefficiency. The low technical efficiency may be due to the external shocks of flood and drought, management and incentive problems as well as information dissemination (e.g. Brooks et al. 1991; Foster/Rosenzweig 1996; Acharya 1997; Desai/Namboodiri 1997; Munshi 2004). Secondly, the total amount of resources used in agriculture by the majority of districts in Bihar increased during the decade under analysis. The technical efficiency offset the positive technical progress to prevent output improving. Finally, the approach enables identification of the "innovators" of Bihar's agricultural production, i.e., districts that have contributed to a shift in Bihar's overall production frontier.

Further, the exercise has attempted to address the complexity of agricultural TFP growth and livelihoods resilience in the districts. The LRI developed here encompasses bio-physical, social, economic, and health asset endowments, which are significant determinants in a district's ability to access and effectively use productivity-enhancing knowledge and technologies. The positive association, albeit weak, shows the importance of productivity to agricultural sector growth and demonstrates that its link to livelihoods is complex and depends on a variety of contextual factors, including the initial distribution of poverty, asset endowments, strength of market linkages and the extent and nature of the region's participation in the agricultural sector.

Finally, the chapter provides a quantitative understanding of the agricultural production system of Bihar districts over the recent past and, in the longer run, is a useful first step towards gaining a sense of what we can expect in the years ahead. An obvious extension to this study would be to estimate TFP at the crop and more granular unit, such as household, and also apply the *stochastic frontier approach* (SFA) by incorporating higher order policy variables, such as subsidies, government investment, variables representing resource endowment, infrastructure, groundwater extraction, etc., in the efficiency equation. Another interesting area of study would be to modify this approach by incorporating various sources of livelihoods dependent on agriculture in order to understand the linkages. Further research into the relationships between agricultural TFP growth and LRI is warranted to deepen our understanding of livelihoods approaches and their relative effectiveness. This might help to reinforce the importance of understanding the role of livelihoods resilience in achieving sustainable agricultural growth, especially in a region where agriculture provides the main source of livelihoods.

References

Acharya, S.S., 1997: "Agricultural Price Policy and Development: Some Facts and Emerging Issues", in: *Indian Journal of Agricultural Economics*, 52, 1: 1–27.

Acharya, S.S., 2006: "Sustainable Agriculture and Rural Livelihoods", in: *Agricultural Economics Research Review*, 19 (July–December): 205–217.

Alkire, S.; Foster, J., 2011: "Understandings and Misunderstandings of Multidimensional Poverty Measurement", *The Journal of Economic Inequality*, 9, 2: 289–314.

Ashley, C.; Carney, D., 1999: *Sustainable Livelihoods: Lessons from Early Experience* (Nottingham: Russell Press Ltd.): 1–64.

Bhushan, S., 2014: "Agricultural Productivity and Environmental Impacts in India: a Parametric and Non-parametric Analysis" (PhD Dissertation, New Delhi, Jawaharlal Nehru University).

Bhushan, S., 2016: "TFP Growth of Wheat and Paddy in Post-Green Revolution Era in India: Parametric and Non-Parametric Analysis", in: *Agricultural Economics Research Review*, 29, 1: 27–40.

Boserup, E., 1965: *The Conditions of Agricultural Growth. The Economics of Agrarian Change under Population Pressure* (London: George Allen & Unwin).

Brooks, K.; Guasch, J.L.; Braverman, A.; Csaki, C., 1991: "Agriculture and the Transition to the Market", in: *Journal of Economic Perspectives*, 5, 4: 149–161.

CACP, 1999–2016: *Comprehensive Scheme for the Study of Cost of Cultivation of Principal Crops in India* (New Delhi: Commission For Agricultural Costs and Prices, Department of Agriculture and Cooperation, Ministry of Agriculture, Government of India).

Cassidy, L.; Barnes, G., 2012: "Understanding Household Connectivity and Resilience in Marginal Rural Communities through Social Network Analysis in the Village of Habu, Botswana", in: *Ecology and* Society, 17: 4–11.

Caves, D.W.; Christensen, L.R.; Diewart, W.E., 1982: "The Economic Theory of Index Numbers and the Measurement of Input, Output, and Productivity", in: *Econometrica*, 50, 6: 1,393–1,414.

Census, 2011: *Ministry of Home Affairs* (New Delhi: Government of India).

Cervantes-Godoy, D.; Dewbre, J., 2010: *Economic Importance of Agriculture for Poverty Reduction*, OECD Food, Agriculture and Fisheries Working Chapters, No 23.

Chambers, R.; Conway, G.R., 1992: *Sustainable Rural Livelihoods: Practical Concepts for the 21st Century*. Discussion Chapter 296 (Brighton: Institute of Development Studies).

Chand, R.; Kumar, P.; Kumar, S., 2011: "Total Factor Productivity and Contribution of Research Investment to Agricultural Growth in India", Policy Paper No 25 (New Delhi: National Centre for Agricultural Economics and Policy Research).

Charnes, A.; Cooper, W.W.; Rhodes, E., 1978: "Measuring the Efficiency of Decision Making Units", in: *European Journal of Operational Research*, 2: 429–444.

Coelli, T.J.; Rao, D.S.P.; O'Donnell, C.J.; Battese, G.E., 2005: *An Introduction to Efficiency and Productivity Analysis* (Heidelberg: Springer).

Datt, G.; Ravallion, M., 1998: "Farm Productivity and Rural Poverty in India", in: *Journal of Development Studies*, 34, 4: 62–85.

Deokar, B.K.; Shetty, S.L., 2014: "Growth in Indian Agriculture Responding to Policy Initiatives since 2004–05", *Economic and Political Weekly*, Supplement: Review of Rural Affairs, XLIX, 26, 27 (28 June): 101–104.

Desai, B.M.; Namboodiri, N.V., 1997: "Determinants of Total Factor Productivity in Indian Agriculture", in: *Economic and Political Weekly* (27 December): A165–A171.

DFID, 2000: *Sustainable Livelihoods Guidance Sheets* (London: Department for International Development); at: http://www.eldis.org/vfile/upload/1/document/0901/section2.pdf (10 September 2016).

Donohue, Caroline; Biggs, E., 2015: "Monitoring Socio-Environmental Change for Sustainable Development: Developing a Multidimensional Livelihoods Index (MLI)", in: *Applied Geography*, 62: 391–403

Ellis, F., 2000: *Rural Livelihoods Diversity in Developing Countries* (Oxford: Oxford University Press).

Färe, R.; Grosskopf, S.; Norris, M.; Zhang, Z., 1994: "Productivity Growth, Technical Progress, and Efficiency Change in Industrial Countries", in: *American Economic Review*, 84, 1: 66–89.

Farrell, M., 1957: "The Measurement of Productive Efficiency", in: *Journal of Royal Statistical Society*, Series A, CXX, 3: 253–290.

Foster, A.D.; Rosenzweig, M.R., 1996: "Technical Change and Human Capital Returns and Investments: Evidence from the Green Revolution", in: *American Economic Review*, 86, 4: 931–953.

Himanshu, L.P.; Mukhopadhyay, A.; Murgai, R., 2011: *Non-farm Diversification and Rural Poverty Decline: A Perspective from Indian Sample Survey and Village Study.* Working Chapter 44 (London: Asia Research Centre, London School of Economics and Political Science).

Hulten, C.R., 2001: "Total Factor Productivity: A Short Biography", in: Hulten, Charles R.; Dean, E.R.; Harper, M.J. (Eds.), *New Developments in Productivity Analysis* (Chicago: University of Chicago Press).

Jannuzi, F.T., 1974: *Agrarian Crisis in India: The Case of Bihar* (New Delhi: Sangam Books).

Joshi, P.K.; Gautam, M.; Tripathi, G., 2011: *Constraints and Opportunities for Sustainable Agricultural Production in Bihar.* Workshop on Policy Options and Investment Priorities for Accelerating Agricultural Productivity and Development in India, 10–11 November 2011 (New Delhi: India International Centre).

Kalirajan, K.P.; Shand, R.T., 1997: "Sources of Output Growth in Indian Agriculture", in: *Indian Journal of Agricultural Economics*, 52, 4: 693–706.

Kishore, A., 2004: Understanding Agrarian Impasse in Bihar", in: *Economic and Political Weekly*, XXXIX, 31 (31 July): 3,484–3,491.

Krantz, L., 2001: *The Sustainable Livelihood Approach to Poverty Reduction: An Introduction*, Stockholm: Sida. Division for Policy and Socio-Economic Analysis.

Kumar, P.; Kumar, A.; Mittal, S., 2004: "Total Factor Productivity of Crop Sector in the Indo-Gangetic Plain of India: Sustainability Issues Revisited", in: *Indian Economic Review*, 39, 1: 169–201.

Kumar, P.; Singh, N.P.; Mathur, V.C., 2006: "Sustainable Agriculture and Rural Livelihoods: A Synthesis", in: *Agricultural Economics Research Review*, 19: 1–22.

Lewis, W.A., 1954: "Economic Development with Unlimited Supplies of Labour", in: *The Manchester School*, 22, 2: 139–191.

Mukherji, A.; Mukherji, A., 2015: "Bihar", in: Panagariya, Arvind; Rao, Govinda M. (Eds.), *The Making of Miracles in Indian States: Andhra Pradesh, Bihar, and Gujarat* (Oxford: Oxford University Press): 123–224.

Munshi, K., 2004: "Social Learning in a Heterogenous Population: Technology Diffusion in Indian Green Revolution", in: *Journal of Development Economics*, 73, 1: 185–213.

NFHS-4, 2016: "National Family Health Survey 4, 2015–16" (New Delhi: Ministry of Health and Family Welfare, India); at: http://rchiips.org/nfhs/factsheet_NFHS-4.shtml (13 April 2016).

Nishimizu, M.; Page, J.M., 1982: "Total Factor Productivity Growth, Technical Progress and Technical Efficiency Change: Dimensions of Productivity Change in Yugoslavia, 1965–78", in: *Economic Journal*, 92 (December): 920–936.

Prasad, P.H., 1975: "Agrarian Unrest and Economic Change in Rural Bihar. The Three Case Studies", in: *Economic and Political Weekly*, 10, 24 (14 June): 933–937.

Rada, N., 2013: "Agricultural Growth in India: Examining the Post-Green Revolution Transition", in: 2013 Agricultural and Applied Economics Association and Canadian Agricultural Economics Society Joint Annual Meeting, Washington, DC, 4–6 August.

Ranis, G.; Fei, J.C.H., 1961: "A Theory of Economic Development", in: *American Economic Review*, 51, 4: 533–565.

Rostow, W.W., 1956: "The Take-off into Self-Sustaining Growth", in: *Economic Journal*, 66: 25–48.

Scoones, I., 2009: "Livelihoods Perspectives and Rural Development", in: *Journal of Peasant Studies*, 36, 1: 171–196.

Shah, M., 2016: "Eliminating Poverty in Bihar: Paradoxes, Bottlenecks and Solutions", in: *Economic and Political Weekly*, LI, 6 (6 February): 56–65.

Shephard, R.W., 1970: *Theory of Cost and Production Functions* (Princeton: Princeton University Press).

Sullivan, C.; Meigh, J.R.; Fediw, T.S., 2002: *Derivation and Testing of the Water Poverty Index Phase 1. Final Report* (London: Department for International Development).

Thirtle, C.; Lin, L.; Piesse, J., 2003: "The Impact of Research-Led Agricultural Productivity Growth on Poverty Reduction in Africa, Asia and Latin America", in: *World Development*, 31, 12: 1,959–1,975.

Timmer, C.P., 1988: The Agricultural Transformation, in: Chenery, H.; Srinivasan, T.N. (Eds.), *Handbook of Development Economics*, Vol. 1. (Amsterdam: North-Holland): 275–331.

Timmer, C.P., 2007: "Agricultural Growth", in: Clark, D.A. (Ed.), *The Elgar Companion to Development Studies* (Cheltenham, UK: Edward Elgar).

UNDP, 2007: *Human Development Report* (New York: UNDP).

Wetland Atlas, 2010: *National Wetland Atlas* (cty, Bihar: Ministry of Environment and Forests, Government of India; Ahmedabad: Space Applications Centre, Indian Space Research Organisation); at: http://envfor.nic.in/downloads/public-information/NWIA_Bihar_Atlas.pdf (15 January 2016).

World Bank, 2006: *Bihar Agriculture: Building on Emerging Models of "Success"* (Washington, D.C.: Agriculture and Rural Development Sector Unit, South Asia Region Discussion Chapter Series Report No. 4).

Part IV
Organisations: Farmer Producer Organisations

Organisations have been the key engines of economic growth in human enterprise systems. However, today's organisational designs seem to greatly facilitate private financial capital creation rather than social wealth creation for the whole society. Although initiated on the principles of social capital formation, *Farmer Producer Organisations* (FPOs) gradually seem to adopt the design of organisations for private wealth creation. Therefore, there is a need to design FPOs such that they can evolve to be community enterprise systems rather than private enterprises. FPO, for our analysis, includes different forms of collectives: primary co-operatives, producer companies, farmers' clubs, *Self-Help Groups* (SHGs), producer organisations and producer collectives.

This section aims to highlight organisational design factors that can facilitate a higher frequency of interactions between the members/owners and a greater number of transactions throughout the year, and help members find greater value through these interactions and transactions. These design factors facilitate not only financial capital formation in the short run but also social capital formation in the long run to ensure a sustainable wealth creation process. The key design variables include *size, scope, technology, ownership* and *management*. Size refers to the number of members and geographical extent. Scope refers to the number and type of activities that an FPO can engage in. Technology refers to the process and product technology suitable for an FPO. Ownership refers to the shareholding structure in an FPO. Management refers to the governance and management structures, systems, processes, routines and type of managerial skills appropriate for an FPO. In addition to highlighting these issues, some chapters discuss the policies and challenges in making the producer organisations sustainable. In addition to the design issues, this section highlights the challenges and policy dynamics of facilitating legal provision for FPOs as producer companies.

Chapter 20
Effect of Firm Size on Performance Leading to Sustainability

Asish Kumar Panda and Amar K. J. R. Nayak

Abstract This chapter argues that enterprises intending to achieve sustainable performance need to restrict their firm size. It explains the indicators which can be observed in firms working towards sustainable performance. After examining several examples and case studies, it deduces that firms looking to achieve this performance must develop trust, co-operation, transparency and concern for others which can be achieved through better interdependence and interconnectedness between all stakeholders. This is possible if the firm stays small. This chapter tries to explain this aspect through examples and analysis.

Keywords Organisation design · Co-operation · Trust · Interconnection
Interdependence · Size · Sustainability · Social capital

20.1 Introduction

A firm survives and thrives by working through its external and internal context while adhering to its set of core values. Despite best intentions, many firms fail to survive and shut down their business. If financial success is the only criteria for long-term survival, 88% of fortune 500 companies have perished in the last sixty years (Perry 2014). The average lifespan of a company listed in the S & P 500 index of leading US companies has decreased by more than fifty years in the last century, from sixty-seven years in the 1920s to just fifteen years by 2012 (Gittleson 2012). In another study by BCG on a sample of 35,000 listed companies, they conclude that corporations are perishing sooner than before. Since 1950, the total lifespan of companies has significantly decreased (Reeves/Pueschel 2015).

These firms might have created greater products, captured a major market share and generated wealth for the shareholders, but could not sustain their advantages in

Asish Kumar Panda, Head-Corporate Interface, Centre for Management Studies, Nalsar University of Law, Hyderabad, state: Telengana, India; Email: asish@nalsar.ac.in

Amar K. J. R. Nayak, Professor of Strategy & NABARD Chair Professor, XIMB, state: Odisha, India; Email: amar@ximb.ac.in

© Springer Nature Switzerland AG 2019
A. K. J. R. Nayak (ed.), *Transition Strategies for Sustainable Community Systems*,
The Anthropocene: Politik—Economics—Society—Science 26,
https://doi.org/10.1007/978-3-030-00356-2_20

251

the long run. Strategic focus to stay competitive over a period of time has been skilfully summarised by Teece et al. (1997) by classifying them into four strands of strategic perspectives:

- Strategic conflict perspective: eliminate competition.
- Attenuating competitive forces perspective: reduce forces of competition.
- Resource-based perspective: gain advantage through critical resource positions.
- Dynamic capability perspective: gain advantage through capacity to innovate in a technologically fast-changing environment.

Firms are now looking inwards to develop their capability and alignment of resources to stay alive. The key internal design variables of size, scope, technology, ownership and management are getting recognised as influential in the long-term success of firms. In this chapter we shall consider one variable – firm size – and make a supposition about its effect with respect to firm performance leading to sustainability.

20.2   Firm Size Versus Performance

Several scholars have attempted to make an estimation of performance with respect to firm size. There have been mixed findings on this issue. Performance is taken as sales growth rate by some and technical efficiency, or productivity, by another group, whereas some take profit as an indicator. The table below summarises the findings of these chapters (Table 20.1).

Estrin et al. (2009) found that firm performance improves to a certain extent with respect to firm size but starts declining after a threshold point, making the relationship curvilinear, mostly in an emerging economy. All the above studies focus on the performance of firms purely in economic terms by analysing the annual rate of change in return on assets, sales revenue, profit over sales, and so on. Studies on sustainability performance with respect to firm sizes have not been undertaken by many. However, the sudden exit of many economically successful firms during the 1980s brought in the concept of the balanced scorecard, developed by Kaplan and Norton. This heralded a new era of looking at firms from different perspectives, i.e. those of shareholders, customers, employees and other stakeholders. This approach is used extensively by industries, governments and non-profit firms worldwide to align business activities to the vision and strategy of the organisation. However, the ultimate objective primarily remains the same as before, i.e. maximising shareholders' wealth. From the time of industrialisation, this has led to cut-throat competition, fighting for market share, disregard for the welfare of other stakeholders and the impairment of the environment and biodiversity. Then comes the concept of looking at performance through economic, social and environmental lenses, which launched evaluation systems such as the Global Reporting Initiative (GRI), the Dow Jones Sustainability Index and many more. But most of these are

Table 20.1 Firm size on rate of growth of sales

Positive relationship	Parker (2000); Polanec (2004); Vaona/Pianta (2008)
Negative relationship	Akcigit (2009)
No relationship	Simon (1964)
Dependent not only on size, but on strategy as well	Glaister et al. (2008); Escriba-Esteve et al. (2008)
Firm Size on Technical Efficiency and Productivity	
Positive relationship	John (1984); Acs et al. (1996); Cheng and Lo (2004); Biesebroeck (2005); Halkos/Tzeremes (2007); Castany et al. (2007); Leung et al. (2008); Jusoh (2010); *The Economist* (3 May 2012)
Negative relationship	Simon (1964); Diaz/Sanchez (2002); Truet/Truet (2009); Palangkaraya et al. (2009)
Dependent not only on size, but on strategy as well	Garicano et al. (2016)
Firm Size on Profit	
Negative relationship	Simon (1964); Diaz/Sanchez (2002); Truet/Truet (2009); Palangkaraya et al. (2009)
No relationship	Amato/Wilder (1985); Ha-Brookshire (2009); Muzir (2011)
Dependent not only on size, but on strategy as well	Liu (1995); Orser et al. (2000); Skypala (2005); Burson (2007); Beaver (2007)
Firm Size on Sustainability Efforts	
Positive relationship	Khaled (2006); Gallo/Christensen (2011)
No relationship	Blomback/Wigren (2009)
Negative relationship	Nayak (2015)

Source Compiled by the authors

either self-declared or audited by a third party with the aim of getting better results by making customers feel good. To understand the performance of sustainability intent, we need to understand the factors that can lead to simultaneous growth on all three fronts: economic, social and environmental.

20.3   Factors that Lead to Sustainability Performance

The compilations on indicators of sustainability adopted by various agencies and authorities are more on absolute terms and quantitative in nature while it gets very difficult to understand the strategic intent of the firm.

Sustainability is about building a society where firms address the triple bottom line instead of profitability as the only measure of performance. Firms moving towards creating a balance between economy, society and environment would be seen as approaching sustainable performance, which will make them maintain and expand economically, increasing shareholder value, enhancing corporate image, creating customer delight, improving the quality of products and services, following ethical practices, improving the quality of human resources, creating value for all stakeholders and also taking care of people who might lose their land and resources in the process of establishing and operating the firm. To achieve this, mere allocation of a certain percentage of economic profit to a Corporate Social Responsibility (CSR) fund will not be enough until it is not linked to the business strategy of the firm and not being driven by the vision and mission of the firm.

Firms would also gain from sustainability initiatives. These activities would reduce risks and waste, increase material and energy efficiency, and innovate and develop environmentally-friendly products. This makes an operation profitable and makes the firm stand out in the long run. Firms should therefore integrate economic, social and environmental objectives into their business strategy and strike a balance between these three (Szekely/Knirsch 2005).

Sustainability is not just a one-shot activity. Sustainability spreads across a larger space with many stakeholders spread over a very long period of time. It refers to a natural open system which is diverse and heterogeneous in character. The objective function is to balance and optimise multiple objectives of the ecosystem and manage with self control while helping to strengthen weaker stakeholders through an attitude of giving, loving and sacrifice (Nayak 2011). Love, sacrifice and co-operation are going to help achieve sustainability (Meadows et al. 2004).

Mathematical biologist Martin Nowak has worked for over forty years to reach a scientific conclusion that only co-operation can bring about long-term survival, contrary to the popular belief of competing and destroying the rival. This, he says, is especially applicable in communities where reciprocity would result in co-operation, which would lead to the well-being of everyone.

In *Built to Last* (2005), Jim Collins and Jerry Porras study hundreds of firms and observe that visionary and long-lasting companies never have profit as their primary motive. Rather, they are guided by a core ideology – core values and a sense of purpose beyond just making money. Instead of beating the competition, they primarily focus on improving their own mode of operation so that all the stakeholders will have a better tomorrow than they have today.

It is important to look for the strategic intent of the organisation. If it aims to improve the triple bottom line, the firm will strive to make a positive impact for all stakeholders. This is possible through better relationships in terms of interconnectedness and interdependence within all actors of the firm, co-operation, love, trust and concern for others' well-being. If we look at transaction cost economics (Williamson 1981), the cost gets reduced if transactions are more frequent. On similar lines, the more the interconnection and interdependence between actors, the greater the trust and co-operation. Taking this analogy, if the four elements are enhanced, this will lead to an increase in social capital, bringing sustainable performance for the firm. The following model has been developed using this concept (Fig. 20.1).

Fig. 20.1 Conceptual framework for sustainable performance. *Source* Panda/Nayak (2017)

While it is common to have a vision to be sustainable, design considerations and collective efforts are required to create the business environment which would lead to sustainability.

20.4 Firm Size as a Design Variable for Sustainability Performance

There are five major internal design variables of a firm which determine its behaviour and performance. They are size, scope, technology, ownership and management (Nayak 2010). All these variables are interlinked, and the design considerations are influenced by the strategic intent of the firm. A small firm can have a wider scope of products and services whereas large firms tend to be specialised to exploit economies of scale. Firms employing complex and indivisible technology tend to be large. Ownership in smaller firms can be collective or distributed, whereas in large firms it tends to be concentrated. Management can be more participatory in small firms, whereas large firms are governed by a set of rules.

Here we try to unravel the possible effect of firm size on performance. In *Small is beautiful* (1973), E. F. Schumacher explains how, out of the greed to amass more wealth and resources, and in the excitement of unfolding scientific and technological powers, firms are getting bigger, thereby creating a society which mutilates mankind, reducing trust and transparency.

Theoretically, all the co-operatives in the world should have survived, as their basic form of existence is to take care of all the members through interdependence and interconnectedness. But many of them have perished. The recent example of the Mondragon Co-operative Group, said to be world's biggest co-operative, could not keep its multinational Fagor group alive, as the employees lost the interconnectedness and trust due to its HR policies (Errasti et al. 2016).

When a firm starts getting bigger, governance becomes complex, creating layers of bureaucratic structure which reduce transparency, leading to mistrust within people. This is even true of political governance, where smaller political boundaries create a better understanding of the local issues and make the economic and social interventions effective.

Countries have increased their administrative units and sub-units. In Sub-Saharan Africa, almost half the countries have increased their number of administrative units by over 20% since the mid-1990s. The Czech Republic and Hungary have increased the number of their municipalities by about 50%. Brazil has also increased the number of its municipalities by about 50%. Indonesia has increased the number of its provinces from 26 to 33 and of its districts from 290 to 497, and Vietnam has increased its number of provinces from 40 to 64 (Grossman/ Lewis 2014).

In India we have seen the creation of new provinces, such as Uttarakhand, Telengana, Chhattisgarh and Jharkhand, from the earlier provinces of bigger size. In the province of Odisha, thirty districts were created out of thirteen in 1993, with the aim of improving governance.

In the corporate world, in 2009 Tata Consultancy Service broke up into twenty-three separate profit centres to be more agile and adaptive to change, which was not possible as one single big entity. In October 2016 another IT multinational, Infosys, announced that it was breaking up into fifteen smaller units. For better governance, firms should transform into either highly autonomous market-governed internal units or break hierarchical links from the earlier form of a large, single unit into autonomous firms (Zenger/Hesterly 1997).

Right in the heart of capitalism, Bo Burlingham studied fourteen small firms in the USA which have been there for several decades and have been doing extremely well. He analysed their success and has come up with some interesting revelations in his book, *Small Giants* (2007). Some of the firms he studied have existed since 1927 and are still doing well. He observed that almost all of them have deliberately resisted growing big and have rather focused on improving their connection with employees, customers and the community. Instead of competing with rivals, they concentrate on competing with themselves to improve their performance while creating a sense of belonging within their employees, who have never felt like mere employees, but instead have the feeling that the firm is their own. They have created such a good feeling within the community where they operate that the people feel proud to be associated with such firms. In spite of several temptations, they want to stay small to serve better. This has never resulted in a trade-off for economic results; rather it has shown greater growth.

Firms which have grown, commercially fighting it out with rivals with the basic intent to make more profits without concern for society or the environment, often perish in the long run or get merged or acquired to form a more infeasible entity. The recent acquisition of Monsanto by Bayer might possibly go this way.

We have studied some collective enterprises within India, where we checked the indicators for sustainability performance with respect to size and analysed the ground realities by speaking with different stakeholders of the enterprises.

The Amalsad Co-operative Society in Navsari district of Gujarat in India has existed since 1941 and has been doing extremely well in economic terms. In all these years it has kept itself small, with fewer than a thousand active producer member households. All the employees of the society are drawn from members through an open and transparent selection process. Instead of growing big, they have encouraged the setting up of another ten societies in the villages outside their area of operation, comprising of cluster of seventeen villages, and have helped them understand the basic needs of the people in their community, assisting them through continuous information exchange. The level of interdependence is very high and people in the society know each other well. Everyone, including the suppliers and non-members, feel very happy to be connected with this society. They collectively implement new activities which will be useful for everyone. The level of trust is so high that no member has ever questioned the procurement of produce, the prices set by the society from time to time, or the weight and gradation of the produce they supply. Everyone considers this society as their own. All the transactions are kept on record in IT systems as well as in hard copies so that any member can go through them whenever they need. The society has developed some fixed and limited market space for its produce where the goods get transferred by road and rail.

The Mahila Umang Producers Company was promoted in 2001 by the Pan Himalayan Grassroots Development Foundation (which has been working in the Himalayan area since 1972). This started as a non-profit venture to help women in the rural areas of the Kumaon region of Uttarakhand earn a living, and, seeing the impact of pilot projects, it converted itself into a producer company. It has spread within Kumaon region and also partly into Himachal Pradesh, with over 3,500 members working in the form of a cluster of self-help groups. Some of the employees are members, while some are directly recruited from outside. We found the level of trust and co-operation between members to be very high. However, due to the bigger spread of geographical area and larger number of members, the interconnectedness is relatively less on a daily basis. To mitigate this issue, the enterprise regularly organises cultural and social gatherings where every member can meet each other. This has still not created the sense of togetherness within members from distant locations. They attempt to sell their products through corporate store chains and also participate in international exhibitions, keeping exports in mind. Financially, the company is doing well, though relatively not as well as Amalsad.

Devbhumi, another producer company, has been operating in the Garhwal region of Uttarakhand for about seven years, promoted by an NGO, Appropriate Technology India. It has over 7,000 members from about 1,015 self-help groups. They market organic produce along with silk apparel, exploring wider markets within and outside India. Their management mostly consists of people who are non-members, and they aim to make this firm bigger to include more and more people. During discussion with members in the villages, it was observed that most of them were not aware of being co-owners of the firm and rather treated it as a customer with whom they can bargain for a better price. This may be mainly because the firm has not distributed profits to its members because it has not shown a book profit, though it does make an operating profit. The larger geographical spread and higher number of members reduces interaction within people, which also depletes the level of trust and transparency.

Nava Jyoti is a producer company located in the tribal area of Nuagada in Rayagada district of Odisha. To date it has limited its membership to below 700 and never intends to go beyond 1,000. It has limited its geographical spread to two Gram Panchayats (the smallest political units for local self-governance). The firm procures produce from members and supplies it to the market within the district.

Initially there were employees who were not members and there was a sense of mistrust within the people. This was furthered by a few mismanagement issues by some of the non-member employees. To rectify the situation, the board decided to employ only members with grass-roots-level management training to enable them to perform their duties efficiently. For the last two years, all the employees have been members, except the principal coordinator, who is a qualified manager to steer the firm to develop both economic and social capital. The members now get better prices for their produce than they would have received from other intermediaries. This has also impacted on the procurement price for other producers in the area, as they now look for better prices from other buyers who had been exploiting them for a long time. After the initial teething period, the firm has been making a profit for the last two years and there is a clear sense of belonging within members.

We now take the case of Gambhira Collective Farming Society (Naidu 2012), which has been operating since 1953 in four villages located on the banks of the Mahisagar River of Anand and Vadodara districts in Gujarat. There are just 291 members, cultivating collectively 526 acres of land by forming thirty groups of eight to fourteen members each. All the employees of the firm are from outside the system. With a set of rules and policies, a governance mechanism has been put in place to prevent free-riding, shirking and opportunism by members. Due to changes in the course of the river, some land and crops get destroyed every year. However, the society has created a mechanism to distribute the loss equally. Almost 80% of the land is used to cultivate tobacco, which is a major cash crop with high demand, despite its use being discouraged by the Government and NGOs. Because of the high profit margin achieved by this crop, the loss gets compensated. However, even after such a long period of time, the rules oblige all members of the group to walk out of the field together to avoid possible theft of the crop or equipment. The economic performance is good, but the main crop is not environmentally friendly and the society lacks the level of trust and concern for others inherent in other co-operatives. The reason for limiting the size of membership was limited land available for collective farming. In a meeting in 1960 it was decided that the society would not be open to new members. Therefore the legal heir (if more than one, this is decided by the family) becomes the new member upon the death of an existing member. However, in spite of being primarily engaged in mono-cropping and obliged to follow strict rules to operate, the society has been continuing to run their business because it has remained small.

From the cases discussed above and also from the examples given by Bo Burlingham, it can be conjectured that size has an effect on sustainability performance. Firm size can be seen in terms of the number of employees in a non-collective firm, the number of members and employees in a collective enterprise, or the geographical spread of the operation and reach of its market. With limited membership, a smaller geographical spread of operation, and a mission to work for the overall development of its members, Amalsad has been able to create a sense of belonging through trust and transparency. This is a bit diluted in the case of Mahila Umang, where the membership number is high with a larger geographical spread. In Devbhumi the members are not aware of their own position in the company, mainly because the membership is high and they also intend to expand over time. There is lack of trust and interconnection between members, and the management is still under the control of Appropriate Technology India, which looks after marketing issues, and hence uses this firm as a pseudo-supplier. Nava Jyoti, on the other hand, initially had local community mobilisation issues which it promptly overcame. However, appointing outsiders as employees did not build the requisite level of trust and transparency within the enterprise. The strategic intention to limit the firm size is aimed at increasing the level of interconnectedness and interdependence between members to develop trust, co-operation, bonding and transparency. This has shown positive results in the last few years now that the initial hiccups have been dealt with. Conversely, the Gambhira society, though small in size, has been working on the assumption of members being free riders and opportunists if left to themselves without tight control and monitoring. This firm has been running successfully for over seven decades, but the performance may not be tending towards sustainability.

From the examples discussed, we can state that the size of a firm plays a vital role in terms of sustainability performance and that firms which have a strategic intention to perform sustainably need to limit their size (employees, members, geographical spread etc.).

As this involves the development of social capital, Dunbar's number can also be referred to. Dunbar's number is a suggested cognitive limit to the number of people with whom one can maintain stable social relationships (Dunbar 1992). This has also been validated in further studies (Allen 2004). It says that an individual with the advanced human brain that he/she possesses can maintain a regular social relationship with a maximum of 150 people. This is applicable in collective enterprises where small groups can interact with each other. The organisation can reach a larger size, but not beyond a point where the effect of trust and transparency starts diminishing. As discussed and deliberated in the Round Table Discussion in 2016 at Xavier Institute of Management, the optimal size for producer organisations should be taken as a thousand members so that the indicators of sustainability performance can be visible (Nayak/Panda 2016).

20.5   Conclusion

As sustainability performance is not just numbers obtained from the financial statements of the firm, it is necessary to understand the strategic intent which can create an environment for trust, co-operation, transparency and concern for others. Firms working for sustainability performance need to develop adequate social capital to achieve this. To address this issue better, the firm needs to improve the interconnectedness and interdependence of its staff, which will not be possible if the size of the firm is large.

From the examples, it is seen that smaller firms with design intervention and proper strategic orientation perform sustainably. This is of great importance in collective enterprises where the ownership is dispersed. As the dispersion of ownership decreases, leading to only a few owners, the firm can get bigger until a point where its growth needs to be arrested to avoid governance complexities and complications. In any form of enterprise, there is an optimal size of firm beyond which the indicators of sustainability performance start to decline. Firm size therefore tends to have a definitive effect on sustainability performance.

References

Adams, M.; Thornton, B.; Sepehri, M, 2011: "The impact of the pursuit of sustainability on the financial performance of the firm", in: *Journal of Sustainability and Green Business*, 1, 1.

Akcigit, U., 2009: "Firm size, innovation and dynamic growth", Manuscript, 1–67.

Allen, C., 2004: "The Dunbar number as a limit to group sizes" (7 April 2004).

Amato, L.; Wilder, R.P., 1985: "The effect of firm size on profit rates in U.S. manufacturing", in: *Southern Economic Journal*, 52, 1: 181–190.

Amato, L.H.; Bursen, T.E., 2007: "The effect of firm size on profit rates in financial services", in: *Journal of Economics and Economic Education Research*, 8, 1: 67–81.

Beaver, G., 2007: "The strategy payoff for smaller enterprises", in: *Journal of Business Strategy*, 28, 1: 11–17.

Biesebroeck, J.V., 2005: "Firm size matters: growth and productivity growth in African manufacturing" (Chicago: The University of Chicago).

Blomback, A.; Wigren, C., 2009: "Challenging the importance of size as determinant for CSR activities", in: *Management of Environmental Quality: An International Journal*, 20, 3: 255–270.

Burlingham, B., 2007: *Small Giants: Companies that Choose to be Great Instead of Big* (London: Penguin).

Castany, L.; Lopaz-Bazo, E.; Moreno, R., 2007: "Decomposing differences in total factor productivity across firm size" (Barcelona: Research Institute of Applied Economics), Working Chapters 2007/5.

Cheng, Y.; Lo, D., 2004: "Firm size, technical efficiency and productivity growth in Chinese industry" (London: School of Oriental and African Studies, University of London), Working Paper No. 144.

Collins, J.C.; Porras, J.I., 2005: *Built to last: Successful habits of visionary companies* (New York: Random House).

Dallago, B., 2003: Small and medium enterprises in Central and Eastern Europe (Centre for Economic Institutions, Institute for Economic Research, University of Trento and Hitotsubashi University); at: http://cei.ier.hitu.ac.jp/activities/seminars/papers/Dallago_Dec.pdf.

Diaz, M.A.; Sanchez, R., 2002: "Firm size and productivity in Spain: a stochastic frontier analysis" (Valencia: University of Valencia, Department of Economic Analysis).

Dunbar, R.I., 1992: "Neocortex size as a constraint on group size in primates", in: *Journal of Human Evolution*, 22, 6: 469–493.

Elsayed, K., 2006: "Re-examining the expected effect of available resources and firm size on firm environmental orientation: an empirical study of UK firms", in: *Journal of Business Ethics*, 65, 3: 297–308.

Errasti, A., Bretos, I.; Etxezarreta, E., 2016: "What Do Mondragon Coopitalist Multinationals Look Like? The Rise and Fall of Fagor Electrodomésticos S. Coop. and its European Subsidiaries", in: *Annals of Public and Cooperative Economics*, 87, 3: 433–456.

Escriba-Esteve, A.; Sanchez-Peinado, L.; Sanchez-Peinado, E., 2008: "Moderating Influences on the firm's strategic orientation-performance relationship", in: *International Small Business Journal*, 26: 463–488.

Estrin, S.; Hanousek, J.; Kocenda, E.; Svejnar, J., 2009: "The effects of privatization and ownership in transition economies", in: *Journal of Economic Literature*, 47, 3: 699–728.

Gallo, P.J.; Christensen L.J., 2011: "Firm size matters: an empirical investigation of organizational size and ownership on sustainability behaviours", in: *Business and Society*, 50, 2: 315.

Garicano, L.; Lelarge, C.; Van Reenen, J. (2016). "Firm size distortions and the productivity distribution: Evidence from France", in: *American Economic Review*, 106, 11: 3,439–479.

Gittleson, K., 2012: "Can a company live for ever?", in: *BBC News* (19 January).

Glaister, K.W.; Dincer, O.; Tatoglu, E.; Demirbag, M; Zaim, S., 2008: "A causal analysis of formal strategic planning and firm performance: Evidence from an emerging country", in: *Analysis of Formal Strategic Planning, Management Decision*, 46, 3: 365–391.

Grossman, G.; Lewis, J.I., 2014: "Administrative unit proliferation", in: *American Political Science Review*, 108, 1: 196–217.

Ha-Brookshire, J.E., 2009: "Does the firm size matter on firm entrepreneurship and performance? US apparel import intermediary case", in: *Journal of Small Business and Enterprise Development*, 16, 1: 132–146.

Halkos, G.E.; Tzeremes, N.G., 2007: "Productivity efficiency and firm size: an empirical analysis of foreign owned companies", in: *International Business Review*, 16: 713–731.

John, M.P., 1984: "Firm size and technical efficiency: applications of production frontiers to Indian survey data", in: *Journal of Development Economics*, 16, 1–2: 129.

Jusoh, R., 2010: "The influence of perceived environmental uncertainty, firm size, and strategy on multiple performance measures usage", in: *African Journal of Business Management*, 4, 10: 1,972–1,984.

Keating, J., 2014: "State Division", in: *The World* (3 February).

Laforet, S., 2009: "Effects of size, market and strategic orientation on innovation in non-high-tech manufacturing SMEs", in: *European Journal of Marketing*, 43, 1: 188–212.

Leung, D.; Meh, C.; Terajima, Y., 2008: "Firm size and productivity" (Ottawa: Bank of Canada Working Paper); at: www.bank-banque-canada.ca (30 April 2016).

Liu, H., 1995: "Market orientation and firm size: an empirical examination of UK firms", in: *European Journal of Marketing*, 29, 1: 57–71.

Meadows, D.H.; Randers, J.; Meadows, D.L., 2004: *The Limits to Growth: The 30-Year Update* (White River Junction, Vt.: Chelsea Green Publishing Company).

Muzir, E., 2011: "Triangle relationship among firm size, capital structure choice and financial performance: Some evidence from Turkey", in: *Journal of Management Research*, 11, 2: 87–98.

Naidu, N.S., 2012: *Co-Operative Farming: A Case of Gambhira Collective Farming Society* (Dadri: Shiv Nadar University).

Nayak, A.K.J.R., 2010: "Optimizing Asymmetries for Sustainability", in: *Global Conference of the Jesuit Higher Education* (Mexico City, Mexico: Universidad Ibero Americana [UIA]).

Nayak, A.K.J.R., 2011: "Efficiency, effectiveness and sustainability", XIMB Sustainability Seminar Series, Working Paper 1.0 (December).

Nayak, A.K.J.R., 2014: "Logic, language, and values of co-operation versus competition in the context of recreating sustainable community systems", in: *International Review of Sociology*, 24, 1: 13–26.

Nayak, A.K.J.R., 2015: "Size and Organizational Design Complexity for Sustainability: A Perspective", in: *Vilakshan: The XIMB Journal of Management*, 12, 1.

Nayak, A.K.J.R.; Panda, A.K., 2016: "Technical Report: Optimal Design of Farmer Producer Organization (community enterprise system) in India.

Nazdrol, W.M.; Breen, J.; Josiassen, A., 2011: "The Relationship Between Strategic Orientation and SME Firm Performance: Developing A Conceptual Framework", in: *Regional Frontiers of Entrepreneurship Research 2011: Proceedings of the 8th AGSE International Entrepreneurship Research Exchange, Swinburne University of Technology, Melbourne, Australia, 1–4 February, 2011*, 713–723.

Palangkaraya, A.; Stierwald, A.; Yong J., 2009: "Is firm productivity related to size and age? The case of large Australian firms", in: *Journal of Industry, Competition and Trade*, 9, 2: 167–195.

Perry, M., 2014: "Fortune 500 firms in 1955 vs. 2014", in: *AEIdeas* (18 August).

Polanec, S., 2004: "On the evolution of size and productivity in transition: evidence from Slovenian manufacturing firms" (Florence: European University Institute; Ljubljana: University of Ljubljana).

Reeves, M.; Pueschel, L., 2015: "Die Another Day: What Leaders Can Do About the Shrinking Life Expectancy of Corporations", in: *BCG Perspectives* (2 July).

Shaffer, S., 2002: "Firm size and economic growth", in: *Economics Letters*, 76: 195–203.

Simon, H.A., 1964: "Firm size and rate of growth", in: *Journal of Political Economy*, 72, 1: 81–82.

Smith, K.G.; Guthrie, J.P.; Chen, M., 1989: "Strategy, size and performance", in: *Organization Studies*, 10, 1: 63–81.

Smyth, et al., 1975: "The measurement of firm size: theory and evidence for the United States and the United Kingdom", in: *The Review of Economics and Statistics*, 57, 1 (February): 111–114.

Szekely, F.; Knirsch, M., 2005: "Responsible leadership and corporate social responsibility: metrics for sustainable performance", in: *European Management Journal*, 23, 6: 628–647.

Truett, L.J.; Truett, D.B., 2009: "Firm size and efficiency in the South African motor vehicle industry", in: *Australian Economic Papers*, 48, 4: 333–341.

Vaona, A.; Pianta, M., 2008: "Firm size and innovation in European Manufacturing", in: *Small Business Economics*, 30, 3 (March): 283–299.

Williamson, O.E., 1967: "Hierarchical control and optimum firm size", in: *Journal of Political Economy*, 75, 2: 123–138.

Williamson, O.E., 1981: The economics of organization: The transaction cost approach. *American Journal of Sociology*, 87, 3: 548–577.

Zenger, T.R.; Hesterly, W.S., 1997: "The disaggregation of corporations: Selective intervention, high-powered incentives, and molecular units", in: *Organization Science*, 8, 3: 209–222.

Chapter 21
Organisational Design for Agriculture and Rural Development

Hemantbhai B. Naik

Abstract Agriculture is the backbone of rural India and most of the people directly or indirectly depend on agricultural-related work. After seventy years of independence, farmers are still having a problem getting the right price for their products in the marketplace. In India, agricultural produce marketing is a major issue. Farmer organisations have to find an agricultural produce market for their sustainability. This chapter demonstrates how, through suitable organisational design, integrated activities have been successfully undertaken for the last 50–75 years in many co-operatives of small farmer communities in Gandevi Taluka of Navsari district in South Gujarat. Individual farmers can reduce their operating costs by uniting to form co-operatives. Cross-subsidising their activities minimises risk factors and fosters sustainable development. Since most farmers do not have professional business training, input costs during the initial period can be high. Cross-subsidisation helps to minimise overheads and input costs. Cash rotation within the village through different activities should be a regular process. The profit of input supply of any activity should be utilised to minimise the cost of other activities. All these activities should take place in a limited geographical area so that they are easy to operate and monitor. The volume of daily transactions for consumptive purposes in a village is indeed very high. Therefore, farmer organisations should be involved in these activities. Farmer organisations must focus on creating a strong network credit co-operative. Deposits collected from members can be used by farmer organisations to operate more effectively and transact business on more favourable terms. This chapter attempts to elaborate how sustainable organisations can be designed to empower small farmers to protect their interests while promoting rural development.

Keywords Cross-subsidise · Diversified activities · Rural India market
Farmer organisation · Co-operative · Participation · Agriculture produce market

Hemantbhai B. Naik, Former Secretary, Amalsad V.V.K.S.K. Mandali Ltd. Amalsad, District Navsari, Gujarat, India; Email: hemantnaik.p@gmail.com

© Springer Nature Switzerland AG 2019
A. K. J. R. Nayak (ed.), *Transition Strategies for Sustainable Community Systems*,
The Anthropocene: Politik—Economics—Society—Science 26,
https://doi.org/10.1007/978-3-030-00356-2_21

21.1   The Need for a Suitable Farmer Organisation

This chapter discusses sustainable organisational designs for rural development, with special reference to the farming community. Even after nearly seventy years of Independence, 60% of the population of India resides in mostly backward villages with agriculture and allied activities as their main and still uncertain source of income. Poor and small and largely unorganised farmers are often squeezed on both input and output sides of the value chain, resulting in increasing but silent pauperisation in general, except in some tiny pockets, where farmers are better organised to corner the majority of government sops. Therefore, organising farmers to demand a fair deal in the marketplace is the primary and necessary condition for healthy agriculture and a prosperous rural economy. However, it is essential for these organisations to be suitably designed to sustain improvements in their lot. Only then can India boast of true development!

It is true that in post-independent India, especially in recent times, successive governments have placed a lot of emphasis on rural infrastructure development, besides announcing various schemes to benefit farmers. But these infrastructures happen to be more market-friendly – that is, more friendly to large corporations and traders, who command power in the marketplace, than to small unorganised farmers. The same is true of various government schemes – unless the farmers are well-organised to command bargaining power, these schemes either remain non-implemented or at best cornered by large and affluent farmers. This is why this chapter uses real-world examples to show how sustainable organisations can be designed to empower small farmers to protect their interests while promoting rural development.

The purpose of a farmer-friendly organisation is to protect the farmers from the vagaries of the market economy, which is usually dominated by large players – be they on the input/input sale services side or the output marketing side. The job of a farmer organisation is to solve several rounds of collective action processes among farmers and between farmers and outside stakeholders so as to make all necessary inputs and services (including household purchases) available to farmers at the right time, of the right quality and at the lowest price, while at the same time undertaking necessary value addition to farmers' produce to generate the highest possible consumer price for it. Only when the farmers own and are in command of such organisations, can the generated profits flow back to augment farmers' income on a continual basis. Besides ploughing part of the profits back into their organisation, thereby strengthening it, some of the profits are devoted to community welfare activities. Two underlying principles lie at the heart of true famer organisations – one, member centrality, and two, domain centrality. The former implies that the organisation would conduct competitive businesses to meet the needs of all members. Domain centrality, on the other hand, means that the farmer organisation would try to sell to premium markets whatever can be produced out of the resources available in the members' domain. Broadly, two types of organisations fit into the above scheme of thought – Farmer Co-operative Societies and Farmer Producer

Organisations. NABARD-promoted Farmer Clubs and aggregated SHGs can also approximate one of the two above-stated forms. Though some private corporations are trying to link farmers through contract farming, the nature of the contract is such that the farmers are invariably the 'vendors', not the principal, which means they can't appropriate the residual profits for sustainable growth and development.

Co-operatives have a strong international tradition, so much so that UN bodies nowadays look upon co-operatives as a better form of organisation than private corporations. They are also perfectly consistent with the Gandhian Philosophy. Gandhi believed that Indian economy lives and ought to live in villages. So he wanted rural development to take place around village co-operatives and through a participatory approach. This concept is absolutely perfect as long as a trustworthy professional is at the helm to run the co-operative business successfully, as the AMUL model shows.

This chapter demonstrates how, through suitable organisational design, integrated activities have been undertaken successfully for the last 50–75 years by many co-operatives of small farmer communities in Gandevi Taluka of Navsari district in South Gujarat. A deeper probe into the organisational design of these co-operatives is likely to provide important clues to developing sustainable rural development models for small farmers.

21.2 Organisation Design

The operating cost being too high for smallholder farmers, they must come together and join hands to meet all their requirements. To minimise risk, not only the farmers, but also their organisation must have a diversified but mutually reinforcing set of activities (e.g., marketing of farmer produce along with member supplies, one of which can cross-subsidise the other activity at the promotional stage). To achieve this diversification, a lot of prior preparation and entrepreneurial judgement is needed. It can't be achieved overnight. The following elements are crucial for judgement, as elaborated below (Fig. 21.1).

21.2.1 Size of Membership and Geographic Cluster

At the time of inception, activities should be restricted to a limited geographic cluster, so that a focused approach of intervention can be applied while mitigating the risk of failure. By restricting the area, operation will become smooth and monitoring will be easy. It should be fully democratic at the policy level and include all the important stakeholders involved in the process. Proper representation from each cluster must be ensured to make it participatory in nature, thus avoiding possible conflicts at later stages. Eventual investment should be the share capital or other sources, like bank loans or member deposits. Visionary leaders are needed to

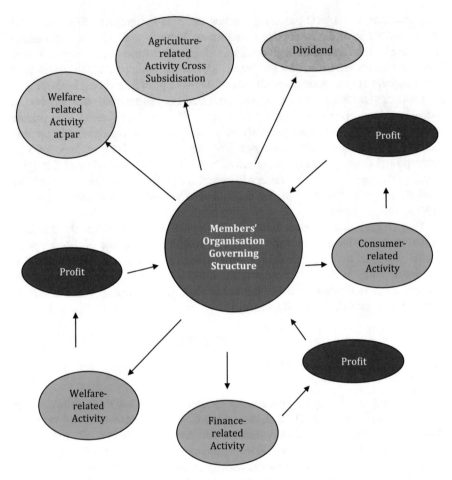

Fig. 21.1 Activity diagram. *Source* Author's conceptualisation

set the ball rolling in terms of designing suitable systems and processes, and must be promoted though training and practice to ensure that the necessary fine-tuning/ tinkering in the system design is done in a timely manner and that, through the continual 'relay race', the steady supply of leaders is never broken.

21.2.2 Scope for Consumer Activity

Seeing the potential, multinational corporations and big business houses have their eyes on the Indian rural market, though they are still struggling to enter it. Community-based organisations, on the other hand, have enormous opportunities if they can couple consumer business alongside other farmer activities and run the show through professional management, as the famous Warana Bazar in Maharashtra has shown. Not only does Warana Bazar supply all farmer needs for modern agriculture and modern living, besides selling all farmer produce in all its branches, but it has also come up with an Agro Mall in a big village called Vadgaon, forcing D-Mart to stop its big project. Besides encouraging the sale of local produce in branded packages, it supplies all farmer household and production needs at the right time and quality at moderate prices, thus avoiding exploitation by private business, retaining profit/commission on consumer activities in the village economy and especially empowering women. Warana Bazar is increasingly providing various consumer services around its consumer stores, thus improving member centrality. As consumer needs are everywhere, consumer activity is a lucrative opportunity for farmer producer organisations to tap in order to reinforce and even subsidise, if necessary, farmer producer activities. It is also a glorious opportunity to diversify business and, at the same time, expand member base in the community.

21.2.3 Financial Activity

Financial activity is one of the most important features of farmer organisations in a rural setting. Even after seventy years of Independence, more than 65% of rural areas don't have a formal banking network, resulting in large dependence on informal lenders, including moneylenders. It appears that the mission of the 1904 Co-operative Credit Society Act has remained grossly unfulfilled. Even formal institutions have a strong bias towards large lenders outside rural areas, as reflected in the abnormally low credit-deposit ratio of government banks. Moreover, poor farmers have transaction costs in accessing formal credit. In this situation, there is enormous scope for improving the rural network of credit co-operatives, even using advantages of NABARD refinancing policy, provided credit goes along with successful monitoring of credit-supported business activities. Just by collecting deposits from members, farmer organisations acquire the ability to transact business on more favourable terms while encouraging farmers to save.

21.2.4 Agriculture Activity

Supplying farmer inputs and marketing their produce after necessary sorting, grading and processing is the core activity. Every farm produce has its own characteristics, which determine its input demand as well as output marketing pattern. Though pooling is generally practised before marketing, depending upon customer needs the concept of separating equilibrium has to be applied alongside pooling. As huge fixed costs are involved in the use of machinery in agro-processing, branding and packaging, costs must be carefully imputed with the possible use of computer technology.

21.2.5 Service Activity

Every smallest member need should be fulfilled by the organisation. A holistic approach is necessary to maintain the interest of the farmers in the organisation. For example, the farmer organisation should own a tractor and provide it on a rental basis as per the requirements of members, thus putting a check on the behaviour of private tractor-leasing services. Such services may thus be sold at marginal profit. This is a device to augment member loyalty.

21.2.6 Welfare Activity

Keeping in mind that, unlike private companies, it is a community-based organisation and not for making huge profits, the organisation has to concentrate on community welfare too. The organisation may thus work in the fields of education, social welfare, health and community-based needs so that the rural standard of living goes up.

Gujarat, especially South Gujarat is the 'Mecca' of multipurpose-credit co-operative societies. In Navsari district, more than ten co-operative societies are running successfully on this pattern. Keeping member centrality and domain centrality in mind, diversified credit plus other activities are undertaken as credit itself can't stand alone. Another distinctive characteristic of these societies is that with huge deposits at their disposal, they don't need NABARD re-financing to run their show. As a result, the government stipulation that each loan has to be insured doesn't apply to them. Given the fact that each society knows by virtue of computerisation what each member is doing with his loan, monitoring is automatic, while peer pressure is effective in ensuring near zero default of loans for years together. The huge activity figures for the year ending March 2016 of these ten outstanding multipurpose co-operative societies of Navsari District bear testimony to their successful operation over a long period of time. They engage in multiple activities, including agricultural produce marketing along with backward linkages, consumer activities, finance activities and service/welfare activities. As each society has a different size of working area, they have different scopes for business, but they all maintain direct as well as indirect control over all their areas of operation. With a working area of seventeen villages, the Amalsad Co-Operative Society has the largest turnover in all aspects.

The most remarkable thing is that all these societies have to compete with private players, yet, running in profit since their inception, they have a market share of around 50% or more in all their business areas.

Though these co-operatives are working independently within their state-stipulated areas of operation, they are practising healthy competition among themselves in contestable markets, thus bringing out their best capabilities. Nevertheless, they are always ready to make strategic alliances among themselves in the spirit of 'co-operation among co-operatives' when it comes to a fight against the private sector or cornering some benefits through economies of scale. With a stable economic order established through community organisations, farmers are apparently happy and prosperous, and can find enough time for improved cultural activities.

Chapter 22
Producer Collectives: What Are We Missing Out?

Sankar Datta

Abstract It is widely recognised that small disaggregated producers need to come together and work as a collective enterprise. The performance of these organisations has been mired in a variety of constraints. This chapter argues that these collective enterprises work within some (i) Economic Conditions defining the business model; (ii) Organisational Design Conditions for efficient functioning and balancing governance; (iii) Socio-Economic Conditions and (iv) Institutional Conditions. An appropriate balance of all these conditions is necessary for these collective enterprises to function.

Emphasising the need for local entrepreneurs to take charge of these organisations, this chapter argues that leading a producer collective requires the ability to strike commercial deals rapidly and lead diverse producer groups. The chapter argues that this requires entrepreneurial ability in addition to leadership. Often in the process of 'promoting' FPOs, the ability of the local leader-entrepreneur is not allowed to flourish.

Keywords Co-operative · Entrepreneurship · Dynamic commercial world
Small and large farmers

It is widely recognised that small disaggregated producers need to come together and work collectively to:

- Get the benefit of economies of scale,
- Utilise technologies which require some scale to be used efficiently,
- Gain collective bargaining power, both economic and political,
- Utilise resources, either collectively owned, or that need to be shared, like human resources,
- Engage effectively with both the input and the output market.

Sankar Datta, Retired Professor and Head, Livelihood Initiatives, Azim Premji University, Bengaluru, Karnataka, India; Email: dattasankar@rediffmail.com

© Springer Nature Switzerland AG 2019   271
A. K. J. R. Nayak (ed.), *Transition Strategies for Sustainable Community Systems*,
The Anthropocene: Politik—Economics—Society—Science 26,
https://doi.org/10.1007/978-3-030-00356-2_22

But in spite of efforts at the highest policy levels, and profuse financial and intellectual support at all levels, experiences of promoting/supporting producer collectives, co-operatives or companies have not been encouraging at all India. In this chapter, I explore what we are missing out on.

22.1   Historic Background of Producer Collectives

Many collective initiatives started emerging in different parts of the world in the post-industrial revolution era, especially in the early nineteenth century.

- In 1761, the Fenwick Weavers' Society was formed in Fenwick, East Ayrshire, Scotland to sell discounted oatmeal to local workers. Its services expanded to include assistance with savings and loans, emigration and education.
- In 1810, Welsh social reformer Robert Owen, from Newtown in mid-Wales, and his partners purchased New Lanark mill from Owen's father-in-law, David Dale, and proceeded to introduce better labour standards, including discounted retail shops where profits were passed on to his employees. Owen left New Lanark to pursue other forms of co-operative organisation and develop co-op ideas through writing and lecturing. Co-operative communities were set up in Glasgow, Indiana and Hampshire, although ultimately unsuccessful. In 1828, William King set up a news chapter, *The Cooperator*, to promote Owen's thinking, having already set up a co-operative store in Brighton.
- The Rochdale Society of Equitable Pioneers, founded in 1844, was an early consumer (not a producer collective) co-operative, and one of the first to pay a patronage dividend, forming the basis for the modern co-operative movement.
- Raiffeisen conceived of the idea of co-operative self-help during his tenure as the young mayor of Flammersfeld. He was inspired by observing the suffering of farmers who were often in the grip of loan-sharks. He founded the first co-operative lending bank, in effect the first rural credit union in 1864.
- The Granger movement, or Order of the Patrons of Husbandry, a coalition of US farmers, particularly in the Midwest, fought monopolistic grain transport practices during the decade following the American Civil War. It was led by Oliver Hudson Kelley, an employee of the Department of Agriculture, who made a tour of the South in 1866. Shocked by the ignorance there of sound agricultural practices, in 1867 Kelley founded an organisation – the Patrons of Husbandry – which involved secret rituals and was divided into local units called 'Granges.'

Recognising the need for such collectives for dispersed producers, various policy actions were also taken. In India in 1904 the Co-operative Credit Societies Act was introduced to facilitate rural producers to come together to meet their own needs following the principles of self-help.

These problems (of peasantry in India) began to engage the attention of even the British colonial government as early as the 1870s: the practice of extending institutional credit to agriculture can be traced back to that period when farmers were provided with such credit by the Government during drought years. Thinking to do with credit co-operation began in the latter part of the nineteenth century. Finally, the Co-operative Credit Societies Act was passed in 1904 (Rakesh 2004).

Even before formal co-operative structures came into being through the passing of the Act, co-operation and collective action for economic purposes was prevalent in several parts of India. It was fairly common for village communities to collectively create permanent assets like village tanks or village forests, called Devarai or Vanarai. Similarly, there were instances of groups pooling resources, such as food grains after harvest, to lend to needy members of the group before the next harvest, or collecting small contributions in cash at regular intervals to lend to members of the group, namely Chit Funds in the erstwhile Madras Presidency, 'Kuries' in Travancore, and 'Bhishies' in Gujarat and Maharashtra. Other examples of co-operation include the 'Phads' of Kolhapur, where farmers impounded water by putting up bunds and agreed to ensure equitable distribution of water, as well as harvesting and transporting the produce of members to market, and the 'Lanas', which were yearly partnerships of peasants to cultivate jointly and distribute the harvested produce in proportion to the labour and bullock power contributed by their partners. But most of these collective actions were led by powerful people in the area, who also had a high level of social acceptability.[1]

After the first Co-operative Credit Societies Act was passed on 25 March 1904, more comprehensive legislation – the Co-operative Societies Act, 1912 – was enacted. Under this Act, co-operatives could be formed to provide non-credit services to their members. Among other things, this Act provided for the creation of the post of *Registrar of Co-operatives* (RoC). The Registrar was entrusted with the responsibility of both auditing and developing co-operative societies for various purposes. Under the assumption that small producers did not have adequate capital to run a business, the RoC was authorised to invest in these collective businesses on behalf of poor farmers. With the apprehension that others who actually made their fortune from agricultural trade would have a vested interest in stopping or distorting these collective businesses, the RoC was also empowered to make administrative decisions to protect the interests of the producers.

The McLagan Committee on Cooperation in India issued a report in 1915 advocating the establishment of provincial co-operative banks, which were established in almost all provinces by 1930, thus giving rise to the three-tier co-operative credit structure. Recognising the need for localised adaptation of the co-operative principles to the local cultural and legal environment, under the Montague-Chelmsford Reforms of 1919 co-operation became a provincial subject and the provinces were authorised to make their own co-operative laws. Under the Government of India Act, 1935, co-operatives were treated as a provincial subject.

---

[1]Datta (1994a) argued that there were several additional transaction costs of taking a collective action. Co-operative societies performed better where the co-operatives were led by farmers who had high social acceptability (in present-day terminology, maybe close to 'social capital', though the author does not use the term) and such additional costs could be absorbed.

The item 'Co-operative Societies' is a State Subject under entry No. 32 of the State List of the Constitution of India. In order to cover Co-operative Societies with membership from more than one province, the Government of India enacted the Multi-Unit Co-operative Societies Act, 1942.

After India attained Independence in August 1947, co-operatives assumed great significance in poverty removal and faster socio-economic growth. With the advent of the planning process, co-operatives became an integral part of the Five Year Plans. As a result, they emerged as a distinct segment in India's national economy. In the First Five Year Plan, it was specifically stated that the success of the Plan would be judged, among other things, on the extent to which it was implemented through co-operative organisations.

The *All-India Rural Credit Survey Committee* (AIRCS) Report, 1954 recommended an integrated approach to co-operative credit and emphasised the need for viable credit co-operative societies by expanding their areas of operation, encouraging rural savings and diversifying business. The Committee also recommended Government participation in the share capital of the co-operatives.

In 1958 the *National Development Council* (NDC) recommended a national policy on co-operatives. India's first Prime Minister, Jawaharlal Nehru, had strong faith in the co-operative movement. While opening an international seminar on co-operative leadership in South-East Asia, he said, "But my outlook at present is not the outlook of spreading the co-operative movement gradually, progressively, as it has done. My outlook is to convulse India with the co-operative movement, or rather with co-operation to make it, broadly speaking, the basic activity of India, in every village as well as elsewhere; and finally, indeed, to make the co-operative approach the common thinking of India... Therefore, the whole future of India really depends on the success of this approach of ours to these vast numbers, hundreds of millions of people."[2]

Recognising that 'Co-operatives have failed, but co-operation must succeed' (AIRCS Report, 1954; Misra 2010), various attempts have been made to improve on the design of these collective businesses. As the need for specialisation became stronger, co-operatives specialising in just a few activities, such as handloom weavers' co-operatives and milk producers' co-operatives, began to emerge. This helped their members build core competency in a select set of activities. Assuming that these businesses were not performing as expected, as most producers did not have sufficient accumulated capital to invest, provisions were made to supplement the capital of the co-operatives. Assuming that in thinly populated areas with a low density of economic activities, co-operatives could not reach a break-even volume, Large Area Multi-Purpose Co-operatives were created. To address the needs of some specific segments of producers, specialised collectives were developed under the *Drought Prone Area Programme* (DPAP), *Integrated Tribal Development Authority* (ITDA), *Small Farmer Development Authority* (SFDA), and so on. It was

---

[2]*Evolution of Co-operatives in India*, Government of India; at: http://pib.nic.in/feature/fe0299/f1202992.html.

not only these different types of co-operatives that were tried. People's collectives were also initiated for the management of common property resources: Water User Associations, Van Panchayats, Pani Panchayats, *Self-Help Groups* (SHGs), Rythu Mitra Groups, Mutual Benefit Trusts, to name a few.

In the wake of provisions for the Government to make investments in these organisations in order to protect the interests of investors, several social activists saw the powers enshrined in the Registrar of Co-operatives as a major impediment to the growth of co-operatives. Several policy advocacy efforts led to promulgation of the Multi-State Co-operative Societies Act, 1994, and the AP Mutually Aided Co-operative Societies Act, 1995, whereby the power of the Registrar was curtailed if the people capitalised their own venture and did not take any investment from the Government.

In 2002 another significant facilitative change for producer collectives was the insertion of Sect. 581 into the Companies Act 1956. This permitted producers to form their own business entity as a Producer Company, as long as they adhered to the principles of mutual assistance laid down in Sect. 581 G (2) of the Act. Though some of the statutory requirements have changed, what remains common is that these businesses also have to follow the principles of mutual help, which were earlier referred to as the Co-operative Principles.

The co-operative sector has played a distinct and significant role in the country's process of socio-economic development. There has been substantial growth of this sector in diverse areas of the economy during the past few decades. The number of all types of co-operatives increased from 1.81 lakh in 1950–51 to 4.53 lakh in 1996–97 and 6.11 lakh by 2009–10. The total membership of co-operative societies increased from 1.55 crore to 20.45 crore during the same period. The co-operatives have been operating in various areas of the economy, such as credit, production, processing, marketing, input distribution, housing, dairying and textiles. In some of the areas of their activities like dairying, urban banking and housing, sugar and handlooms, the co-operatives have achieved success to an extent but there are larger areas where they have not been so successful. The failure of co-operatives in the country is mainly attributable to dormant membership and lack of active participation of members in the management of co-operatives; mounting over-due payments in co-operative credit institutions; lack of mobilisation of internal resources; over-dependence on Government assistance; and lack of professional management. Bureaucratic control and interference in the management, political interference and over-polarisation have proved harmful to their growth. The predominance of vested interests, resulting in non-percolation of benefits to common members, particularly to the class of persons for whom such co-operatives were basically formed, has also retarded the development of co-operatives. These are the areas which need to be attended to by evolving suitable legislative and policy support.[3]

---

[3]See Mishra (2010); at: http://www.ncui.coop/pdf/indian-cooperative-movement-a-profile-2012. pdf for data on co-operatives in India.

In recent years there has also been significant progress in our thinking about what makes the collective efforts of people work or not work. Research in the domains of game theory and new institutional economics informs our understanding. Today we recognise that there are serious problems of Free Riding in any of these collective endeavours. We also recognise that there are serious problems of social choice,[4] which make it extremely difficult for collectives to make decisions. The prisoner's dilemma proposed by game theorists, whereby the outcome of an action depends not only on one's own actions but also on the actions of others, makes such collective action even more difficult. In addition to these complications, the issues of contract failure, market failure and coordination problems further make it difficult for such collectives to perform (Arrow 1950).

As Olson argued when looking at the logic of collective action, only when institutional arrangements are strong enough to resolve some of these tensions can collective action function (Olson 1971). This has been further developed by North (1990) and Ostrom (1991), who very concretely examined the role of institutions in economic performance. All of these arguments broadly indicate the need for every group to have a competent leader who facilitates the establishment of some of these mutually agreed norms. Looking into some of the specific characteristics of collectives that work, it was also argued by Doherty/Jodha (1971) that for a collectivity of farmers to perform, the following provisions are necessary:

1. Collective good is greater than the cost of collective action;
2. There is some organisational good, which cannot be obtained just by collectivising;
3. Individual profit before collective action must at least be protected;
4. Members must perceive some compensatory profit for the losses;
5. Members must have a functional identity within the cause of collectivisation;
6. The size of the group must be appropriate;
7. There are some structural guarantees, as this helps groups perform better.

22.2   What Makes These Enterprises of Collectives Different from Investor-Promoted Ones?

There are some very basic differences between these forms of collectives and the other commonly seen investor-owned firms. As identified by the International Co-operative Alliance, these are autonomous associations of "persons united

---

[4]See Kenneth J. Arrow's 'Impossibility Theory' (1985) and Amartya Sen's 'Theory of Social Choice' (1986).

voluntarily to meet their common economic, social, and cultural needs and aspirations through a jointly-owned and democratically-controlled enterprise."[5]

One defining characteristic is that they are democratic organisations, owned and controlled by their member-users. Only those who are engaged in the activity that the co-operative is also engaged in can buy their shares and become a user-owner. As has been discussed by Novkovic (2008),[6] this characteristic can have some serious implications for the day-to-day operations of such organisations.

Very importantly, co-operatives are business enterprises. But they do not come into existence when an investor sees an opportunity. They are usually formed when a group of members see that collective action will help them to meet one or more of their needs. Often such conditions arise out of the market failure conditions. Producers come together to deal better with the market and fulfil gaps left by the market and not adequately addressed by the State. Therefore, they often acquire the characteristics of third-sector organisations,[7] where basic belief in the core values plays an important role. These organisations, whether registered as a new generation co-operative[8] or a producer company (referred to as *Farmer Producer Organisations* [FPOs]) face some special challenges compared with any other investor-owned agri-business firm.

There have been various studies which have tried to understand what makes some co-operatives work but not others. Some of the early studies by Attwood and Baviskar indicated that it was the nature of socio-political conditions which favour collaboration between different producer groups that made a collective action successful (Attwood 1988; Attwood/Baviskar 1987). Datta (1997) observed that organisations which were led by and patronised by people who had higher social credibility (in today's parlance, maybe higher social capital), performed better. Meanwhile, Shah (1995) argued that it was the design of the co-operative that provided appropriate space for patronage-centric governance, and independent management structure that played a critical role in their success. Seetharaman/ Mohanan (1986) argued that the ability to simultaneously perform on both a social and commercial level determined whether these forms of producer collectives work or not. Emphasising the importance of governance in balancing the divergent goals and transparency to members in such a collective, Desai/Phansalkar (1992) highlighted the importance of a proper information system in such efforts. Building on a

---

[5]International Cooperative Alliance: "Statement of the co-operative identity"; at: http://www.wisc. edu/uwcc/icic/issues/prin/21-cent/identity.html (1995).

[6]See also the presentation by Chris Cooper, The UK Co-operative College: "Co-operative Values as a Driver for Business".

[7]Market failure leading to emergence of the need for co-operatives has been discussed extensively by Datta (1992); Pestoff (1992); Gunn (2004); Zuidervaart (n.d.). The economic rationale of emergence and therefore the characteristics of such third-sector organisations, especially the non-measurability of their outputs, and smallness of some of their outputs, requiring a different system of management has been presented in Hansmann (1987) and Datta (1994b).

[8]Here I refer to 'Multi-State Co-operative' or 'Mutually Aided Co-operative' as new generation co-operatives.

similar argument, Hansman has argued that striking a balance between managerial control and partnership between members and managers is critical for the performance of a producer collective. Two other systems requiring balance emphasised by him are a balance between a Representative Board and an Expert Board as well as working with internal and external stakeholders. Some of these problems faced by producer organisations have been presented below.

Reviewing various community mobilisation experiences in the world, the Pan African Christian AIDS Network (PACANet) has summarised that every community (not just a geographical community, but also a community of common interests or a virtual community) has (i) its own structures and institutions, (ii) leadership, (iii) norms, rules and regulations, (iv) shared resources, (v) culture, customs and beliefs, (vi) a mix of young and old, and men and women, (vi) a mutually accepted medium of communication, (vii) presence of some social deviants, (ix) a tendency to highlight 'wrongs' more than 'rights', (x) observed need for socialisation, entertainment and recreation, (xi) ability (or a mechanism) of managing diversity within the community, (xii) an inherent sense of individual and community responsibilities and (xiii) a shared set of experiences. It has been argued by them that, while building an institution of the poor, if the new institutional arrangements do not align with these elements, it fails to become a sustainable institution.[9]

22.3   Problems Faced by Producer Companies Today

22.3.1   Difficulties of Collective Action

Some of the basic problems of collective action, such as free-riders' problem, finding a balance between individual good and the collective good, and democratic choice versus. expert choice, dominate the performance of these producer organisations. Such a balancing function has also led to the emergence of the active role played by some of the leaders. Theories about why we need to cooperate; what sociology and economics have to say in theory and in practice; game theory and the rational individual; evolution and group selection; role and norm theories of sociology; social capital; sharing competencies co-operation; and imbalances of power shed some light on why such producer collectives often fail to function optimally.[10] Many of these problems continue to exist whether they are registered as co-operatives or producer companies. Many of these issues were extensively dealt with at the Symposium on Management of Rural Co-operatives, 7–11 December 1992, at the Institute of Rural Management, Anand (IRMA).

---

[9]"Training for Community Mobilization", by Pan African Christian AIDS Network; at: http://teampata.org/wp-content/uploads/2017/06/Training-for-Community-Mobilisation-for-ART-in-resource-limited-settings.pdf.

[10]Many of these issues arising from various theories that explain problems of collective action have been summed up by Gillinson (2004).

22.3.2 Owner Versus Supplier Duality

In an FPO the farmer member is the owner of as well the supplier to the company. The farmer, as a supplier, would ask for the maximum price for his crop, whereas, for the owner of the company, it would be beneficial for the price to be as low as possible. But if the price asked by the farmer as a supplier is high, then it may be attractive for him and his family, but not profitable for the company. Paying a higher price during the procurement season, when the marked is flushed with the commodity keeping the prices low, may make it attractive for the producer at that point of time, but hampers the organisation's overall profitability, its competitiveness in the market and capital formation. If this practice continues, the company will eventually shut down. So it is necessary to reach a compromise between these two conflicting interests of the producers.

22.3.3 Horizon Problem

Novkovic has also argued that there is a problem of time horizon in such collective efforts. This arises partly from the owner-supplier duality, which gets further aggravated if the collective expands to include diverse groups of producer members. Some of the results of this are the problems of collective action mentioned earlier. This also has implications for the inclusion of larger and smaller farmers and capital formation, mentioned later in this section.

22.3.4 The Promoting Agency Is Not the Promoter

In most companies, it is a group of entrepreneurs who come together to promote the company. They have complete ownership of the idea, including its details, and are willing to bear the risk. But in most producer companies, it is not the producer-owners who come together to promote the organisation. The idea is introduced by someone else, such as an external promoting agency, though a small number of producers who sign the initial Memorandum of Association legally become its 'promoters'. This has two significant implications for the sustainability of these companies. First, however the external agency tries to mimic an 'insider', it still remains an idea suggested by them. It takes many years and a series of 'favourable' events to transfer this sense of ownership to people. But, second, and more significantly, the producer-owners do not get the sense of the risk of their decisions. They have faith that if things go wrong, the promoting agency will take care of them. Ultimately, it is not 'their' money that is getting invested.

22.3.5 Shortage of Appropriate Human Resources

It is often assumed that the producers themselves do not have the capacity to manage a full agri-business firm. Building this capacity within the community is often so time-consuming that it does not fit in the 'project' mode of the promoting agency. Over time, with enhanced competition in the market, the margins in most agro-businesses have become very thin. This very often does not allow them to hire the services of professionals capable of managing such businesses, especially when the turnover is low. If such a professional is hired, because of their socially perceived status, they often take the upper hand in decision-making (at which they may be more competent). But this social hierarchy permeates down into the organisational hierarchy. If this person was part of the promoting agency, the role confusion becomes worse. This problem has been discussed in detail as a manifestation of the Principal-Agent Problem by Novkovic (2008).

In the Indian context we see that the farmer leaders of the Kheda District Milk Producers' Co-operative Union went out of their way to hire a professional manager to manage their milk business, which they realised they could not manage on their own. However, Warnanagar Sugar Co-operatives, Mulkanoor Co-operatives, and Sridharpur Farmers' Multipurpose Co-operative Society chose to groom their own human resources and grow slowly, keeping pace with the growth of the people who managed their business. But there are examples like Vasundhara Co-operative Society, Navsari, Shri Mahila SEWA Sahakari Bank Ltd., Ahmedabad, and Madhya Pradesh Women Poultry Producers Company Pvt Ltd (MPWPCL), Bhopal, where there has been engagement of professional managers fairly early in the life of these collectives, who have also played a critical role in shaping the institutional processes.

22.3.6 Small Farmer Versus Large Farmer

A farmer producer company will be viable if, and only if, there is the presence of both small and large farmers. It is important for the company to break even as early as possible, and for that it is necessary to have large as well as small farmers. Having only small farmers will lead to issues with respect to scale economies, and having only large farmers will lead to monopoly issues. It has been observed by Shah (1996) that there is usually a small number of active participants of the FPO, mostly large farmers, who substantiate the business of the FPO, whereas there is a large majority of silent participants who just sell their produce through it. Though using the principle of one-man-one-vote, it is the silent majority group who get to determine the business decisions of the organisation, while the active minority, who actually provide the large part of the business, have little formal say in these decisions. This poses a serious challenge for the governance of these collectives.

22.3.7 *Capital Shortage*

Generally, when farmers come together they do not have sufficient capital to start an FPO. When co-operatives started proliferating, they faced a similar issue, but at that point the State came forward to help the co-operatives by funding them. But the State funding was not unconditional. The Government started dictating terms. Now the farmers have transcended that stage; they are registering themselves as farmer producer companies and now there is no need to ask for – or any provision for – state funding. This puts them into a serious fund shortage, especially as most agricultural business is seasonal. To help members overcome the harvest-season-price-dip, most such FPOs are required to hold large quantities of stock. As these assets are usually very liquid, most bankers find it difficult to extend credit for such purposes. Therefore, management of the working capital becomes a serious challenge for these FPOs.

There are two sets of barriers to investment of capital in these forms of collectives. First, the legal forms of these organisations often do not permit capital investment from non-producers. Second, and more significantly, the risk perception of these organisations is unfavourable. In the absence of good-quality human resources in these institutions, especially when the collectivisation and institutional processes are inadequate, the risk perception with these institutions is very high for most external investors. And collective agri-businesses are often not in a position to provide an attractive risk-weighted return.

However, as this problem has persisted from fairly early in the history of co-operatives in India, external funding agencies, especially the Government, have taken the initiative to supplement the small capital of small and marginal producers. But scholars have observed that "during the first decades of the 19th century the Indian *credit co-operative system* (CCS) had an unparalleled great start: based on principles of self-management and self-financing, protected by a credit co-operative legal framework and effective supervision. However, during the second half of the 20th century, the states of India took over the management and financing of the sector. This has practically ruined the system, both institutionally as a self-help movement and economically" (Seibel 2013).

In India, there are quite a few examples of co-operatives which were organised by some of the people without promoting intervention by the government or a non-government organisation. They allowed infusion of capital from outside only at a later stage, when many of their institutional processes had stabilised. Examples include Kheda District Milk Producers' Co-operative Union, Warnanagar Sugar Co-operatives, Mulkanoor Co-operatives, Shri Mahila Sewa Sahakari Bank Ltd., Kolhapur Shetkari Sahakari Sangh Ltd., Sridharpur Farmers' Multipurpose Co-operative Society, Shri Mahila Griha Udyog Lijjat Papad.

22.3.8 The Promotion Process: An Organisation or an Institution

It has also been argued by scholars that when people came to form their own collectives, recognising the need for collective action, they ensured that those collectives achieved their goals. But in most cases when they were promoted as a need recognised by the Government or a Non-Government Agency, they failed to perform.[11] These producer collectives often lack the 'self-creating, self-propagating and self-preserving' characteristics of successful co-operatives observed by Shah (1996).

There has also been substantial work in recent years on the institution-building processes, as such. These involve, apart from identification and defining the problem being faced by people, defining the community of people; concretisation of the idea of a solution; coming to an agreement about the idea; defining the limit of tolerance to variations within the community; developing their norms, including norms of review and revision of norms; assigning specific roles to people; initiating operations; and reviewing it from the perspective of the mutually agreed purpose and solution using the agreed norms of review among other steps. It has also been observed that some of these processes, if not followed adequately, lead to the collapse of the institution.

22.3.9 A Proposed Model

From this review of the literature on producer collectives engaged in economic activities, it can be seen that four different sets of factors have been identified as critical for their performance:

1. *Economic Conditions*: Whether there is a net economic gain for the members to share or benefit from the collective action;
2. *Design Conditions*: Whether the design of the organisation, including its governance, structure, operating systems and management systems, is conducive to the success of the collective venture;
3. *Socio-Economic Conditions*: Whether the socio-economic conditions in the area are conducive to collective action; and

---

[11]There are several different ways in which performance of a co-operative has been assessed. Some looked at their economic performance only, while some others at their socio-political purpose as well. These have been reviewed in Datta (1994a). But for the purpose of this research, I will stick to a simple definition of performance: any co-operative which has been able to generate profit ten years after its inception is considered to be a performing co-operative. If it could not deliver on either the economic or socio-political purpose of the organisation, it would have been closed either by the members or by the bad business itself.

Fig. 22.1 This car needs all
four wheels to run. *Source*
Visualised by the Author

u19296670 fotosearch.com

| Economic Viability | Process of Institution | Design of Organisation | Political-economic Context |

4. *Institutional Conditions*: Whether the institutional conditions, including the processes of forming-storming-norming-performing to arrive at a mutually agreed form of behaviour, are favourable.

In other words, a Producer Collective functions well only when there are favourable *Economic Conditions: Socio-Economic Conditions and appropriate Design Conditions, and enabling Institutional Conditions* (Fig. 22.1).

22.3.10   Characteristics of the Leadership that Is Required

Producer collectives, whether co-operatives or producer companies, are institutions, where the ownership lies with only those who are actually producing the commodity (ies) that the enterprise is dealing with: aggregating, storing, processing and trading (or marketing). They are run by a principle of one-man-one-vote, using democratic principles of governance and not governance in accordance with the capital invested. This brings in certain critical requirements for the leadership of these organisations. First, they should be able to understand some of the critical variables of the commercial functions involved – aggregating, storing, processing and trading – in order to govern these functions. Second, they must be able to mobilise support of a large number using democratic means.

Even in an investor-owned business, the investor needs to understand the nature of the business only to a limited extent. Wasserman (2008) observed that most founders/business leaders surrendered management control long before their companies went public. By the time the ventures were three years old, 50% of founders were no longer the CEO; in year four, only 40% were still in the corner office; and fewer than 25% led their companies' initial public offerings. Most entrepreneurs studied by Wasserman believed that "I'm the one with the vision and the desire to build a great company." They knew how to hire the right person to place their vision on the ground. The founder creates the organisational culture, which is an extension of his or her style, personality, and preferences. From the get-go, employees, customers, and business partners identify start-ups with their founders, who take great pride in their founder-cum-CEO status (Wasserman 2008; Myers/ Majluf 1984). Similarly, the leader in a producer collective must have the vision that prosperity for a large number of other producers could be attained through the collective business. But not only that vision; s/he must also have the ability to mobilise the financial, material and human resources necessary to give shape to this vision. Further complications arise from: (i) the difficulties of collective action, (ii) owner vs supplier duality, (iii) the horizon problem, (iv) the promoting agency not being the promoter, (v) the promotion process: building an organisation or an institution, (vi) shortage of appropriate human resources, (vii) small farmer vs large farmer and (viii) shortage of net owned fund, which is specific to producer collectives, as discussed above.

This makes the case for a combination of skills and knowledge on the part of the leaders who lead producer collectives, especially during the formative stage of the institution. Though many of the programmes trying to promote such collectives do try to supplement some of these requirements, such provision of quasi-equity to supplement net-owned-fund, deputing manpower to take care of the day-to-day management of these organisations, many of these steps actually aggravate the problem. Unlike an investor-owned entity where the investor establishes his ability to mobilise resources by actually investing his own funds, here, without testing this ability, the leader is selected on the basis of some other parameters, most probably ability to produce more or ability to articulate. Even the idea of deputing staff to take care of the day-to-day operations exacerbates the problem of the promoting agency not being the promoter, and the promotion process focusing more on building an organisation not an Institution. Therefore, the agencies engaged in promoting producer collectives need to look for producers who, apart from being good producers, also demonstrate the abilities of a social entrepreneur.[12]

Very often the promotional team is thoroughly trained to recognise the business potential of the activity. Some of the teams also display an ability to understand the socio-economic context. More often than not, the design of the organisation is pre-decided. The team understands the specific commercial conditions, and the socio-economic reality has little space. But the ability, and also often the mandate, to identify entrepreneurs is wanting. Successful entrepreneurship requires a blend of analytical, creative, and practical aspects of intelligence, which, in combination, constitute *successful intelligence* (Sternberg 2003). In the particular case of leading

---

[12]For a discussion on the characteristics of people who have built institutions, see Ganesh/ Padmanabh (1985). Many of these characteristics of the institution-building process have also been discussed in Menard (2000).

a producer collective, the person also requires the ability to lead a democratic institution.

In the process of making the work of the incumbent leaders easier, often the promoting agencies respond by creating guidelines, rules and so on, whereas one of the strengths of the entrepreneur, as Davda (2016) has argued, is that entrepreneurs are like a workplace doctor: entrepreneurs can make their own rules to achieve the goal they have set out to achieve.

Robert Owen, who is known as the father of modern co-operatives, having purchased a spinning mill in England to run on a co-operative basis and who also inspired the first twenty-eight members of the Rochdale Pioneers; Oliver Hudson Kelley, who steered the Grange Movement in United States; Raiffeisen, who started the farmers' credit co-operatives in Germany; Hanseshwar Banerjee, who founded Sridharpur Co-operative Bank, a primary multipurpose credit society, in 1918; the communist leader A.K. Gopalan, who initiated the Indian Coffee House in 1936; Tribhuvandas Patel, under whose leadership the Gujarat Co-operative Milk Marketing Federation Ltd, popularly known by its brand name AMUL, took roots in 1946; Aligireddy Kasi Vishwanatha Reddy, who founded the Mulkanoor Co-operative in 1956; and Tatya Saheb Kore, who set up the Warana Sakhar Karkhana Ltd in 1959, were not just leaders, but were entrepreneurs with a high level of concern for the benefits of co-operation not only to themselves but also many others.[13]

Another important element that becomes essential for effective management of a producer collective is the dynamism of the mindset of the large number of people. As these organisations are controlled with the principle of one-man-one-vote, apart from the regular business uncertainties, rapid changes in the sentiments of the people play a critical role in the functioning of these organisations. Unlike the regular capital market, which has developed a mechanism for price discovery and capturing the sentiment in the market, in producer collective domains no such mechanism has yet been developed. Though thinkers about the well-being of such institutions, such as Shah (1995), have proposed a patronage-based voting system, whereby the voting power of a producer-owner depends upon the usage of the services of the collective, except in parts of milk business this model is yet to be tried.

With so many uncertainties, managing such an institution commercially is a complex process. With their conviction about the vision, and the zeal to make things work, entrepreneurs usually find practical solutions to these problems and communicate them to the large number of member-owners (Bonnstetter 2012). The ability to persuade others and stellar leadership qualities have been recognised as important traits of serial entrepreneurs. Rarely, if at all, do we look for these

---

[13]As defined in the online business dictionary, an entrepreneur is someone who exercises initiatives by organising a venture to take advantage of an opportunity and, as the decision-maker, decides what, how and how much of a good or service will be produced, supplies risk capital as the risk-taker, and monitors and controls the activities on the enterprise. http://www. businessdictionary.com/definition/entrepreneur.html.

characteristics when promoting a producer collective. As most of them 'play by their rules', the socially concerned among them are more likely to be rebellious types and not conformists to the ideas of the external promotion team.

22.4   Conclusions

The discussion above indicates that leading a producer collective, co-operative or company requires the ability to not only comprehend commercial deals very rapidly, but also balance the interests of the different types of producer groups whose interests are affected by different commercial decisions, especially when such a collective has to be an inclusive endeavour.

In attempts to promote more FPCs, such people need to be identified consciously. Efforts should be made not to 'build their capacity' but to share and develop a vision which is jointly owned by the entrepreneur leader and the promoting agency. Once a shared vision has developed, freedom must be given to design and operate the organisation based on the local context.

References

Arrow, K.J., 1950: "A Difficulty in the Concept of Social Welfare", in: *Journal of Political Economy*, 58: 328–346.

Attwood, D.W., 1988: "Socio-Political Preconditions for Successful Co-operatives: The Co-operative Sugar Factories of Western India", in: Attwood; Baviskar (Ed.): *Who Shares* (Delhi: Oxford University Press).

Attwood, D.W.; Baviskar, B.S., 1987: "Why do some Co-operatives Work but Not Others? A Comparative Analysis of the Sugar Co-operatives in India", in: *Economic and Political Weekly*, 22, 26: A-38–A-56.

Bonnstetter, B.J., 2012: "New Research: The Skills That Make an Entrepreneur", in: *Harvard Business Review*, 90, 12 (7 December 2012); at: https://hbr.org/2012/12/New-Research-The-Skills-That-Makes-an-Entrepreneur (16 April 2017).

Bromley, D.W., 1989: *Economic Interests and Institutions: the conceptual foundations of public policy* (New York: Basil Blackwell).

Datta, S., 1992: "*Economics of Co-operativization: A Comparative Analysis of Milk and Oilseeds*"; Paper presented at the Symposium on Management of Rural Co-operatives, 7–11 December 1992, IRMA, Anand.

Datta, S., 1994: "Management of Development Organisations" (IRMA: Anand).

Datta, S., 1994: Doctoral Dissertation: *Factors Affecting Performance of Village Level Organizations: Oilseed Growers' Co-operatives in Madhya Pradesh'*, Ballabh Vidyanagar, Gujarat: Sardar Patel University.

Datta, S., 1997: "Factors Affecting Performance of Village Level Organisations", in: *Finance India*, XI, 1: 79–81.

Davda, A., 2016: "Workplace Doctor: Entrepreneurs can make their own rules", in: *The National Business* (22 April); at: http://www.thenational.ae/business/the-life/workplace-doctor-entrepreneurs-can-make-their-own-rules (10 August 2017).

Desai, S.T.; Phansalkar, S.J., 1992: *"Role of Information Sharing in the Functioning of Board of Directors: Experience in co-operatives"*, Paper presented at Symposium on Management of Rural Co-operatives, 7–11 December 1992 (Anand, Irma).

Doharty, V.S.; Jodha, N.S., 1971: *'Conditions for Group Action Among Farmers'*, Occasional Paper 19 (Hyderabad: ICRISAT).

Ganesh, S.R.; Joshi, P., 1985: "Institution building: Lessons from Vikram Sarabhai's leadership", in: *Vikalpa*, 10, 4 (1 October): 399–414.

Gillinson, S., 2004: *Why Cooperate? A Multi-Disciplinary Study of Collective Action* (London: Overseas Development Institute, February).

Government of India, 2012: *"Evolution of Co-operatives in India"* (New Delhi: Government of India); at: http://pib.nic.in/feature/fe0299/f1202992.html (25 October 2013).

Gunn, C.E., 2004: *Third-Sector Development: Making Up for the Market* (Ithaca: ILR Press Books).

Hansmann, H., 1987: "Economic Theories of Nonprofit Organizations", in: Powel, Walter (Eds.), *The Non-Profit Sector: A Research Handbook* (New Haven: Yale University Press).

International Cooperative Alliance, 1995: "Statement of the co-operative identity"; at: http://www.wisc.edu/uwcc/icic/issues/prin/21-cent/identity.html (5 January 2011).

Menard, C. (Ed.), 2000: *Institutions, Contracts and Organizations* (London: E. Elger Pub.); at: http://web.missouri.edu/~cookml/CV/MENARD.PDF (17 May 2014).

Misra, B., 2010: *Credit Cooperatives in India: Past, Present and Future* (London: Routledge).

Myers, S.C.; Majluf, N.S., 1984: "Corporate Financing and Investment Decisions when Firms Have Information that Investors Do Not Have that year", in: *Journal of Financial Economics*, 13, 2: 187–221; at: https://doi.org/10.1016/0304-405X(84)90023-0 (15 April 2002).

Nelson, R.; Krashinsky, M., 1973: "The Major Issues of Public Policy: Public Policy and Organization of Supply", in: Nelson, Richard; Young, Dennis (Eds.): *In Public Subsidy for Day Care of Young Children* (Lexington, Mass: D.C. Health and Co.).

North, D.C., 1990: *Institutions, Institutional Change and Economic Performance* (Cambridge, Cambridge University Press).

Novkovic, S., 2008: "Defining the co-operative difference", in: *The Journal of Socio-Economics*, 37: 2,168–2,177.

Olson, M., 1971: *The Logic of Collective Action* (Cambridge: Harvard University Press).

Ostrom, E., 1991: "Rational Choice Theory and Institutional Analysis: Towards Complementarity", in: *The American Political Science Review*, 85, 1: 237–250.

Pan African Christian AIDS Network: "Training for Community Mobilization"; at: http://teampata.org/wp-content/uploads/2017/06/Training-for-Community-Mobilisation-for-ART-in-resource-limited-settings.pdf.

Pestoff, V.A., 1992: "Third sector and co-operative services—An alternative to privatization", in: *Journal of Consumer Policy*, 15, 1: 21–45.

Rakesh, M., 2004: "Agricultural Credit in India: Status, Issues and Future Agenda", in: *Reserve Bank of India Bulletin* (November).

Seetharaman, S.P.; Mohanan, N., 1986: *Framework for Studying Co-operative Organization: A Study of NAFED* (Delhi, Oxford and IBH).

Seibel, H.D., 2013: "Financial Cooperatives—What Role for Government? The rise and fall of the credit cooperative system in India", in: Oluyombo, Onafowokan O. (Ed.), *Cooperative and Microfinance Revolution* (Lagos: Soma Prints Ltd.): 93–105.

Sen, A., 1986 "Foundations of Social Choice Theory: An Epilogue", in: Elster, Jon; Hylland, Aanund (eds.), *Foundations of Social Choice Theory* (Cambridge: Cambridge University Press), pp. 213–248.

Shah, T., 1995: *Making Farmers' Co-Operatives Work: Design, Governance and Management* (New Delhi-Thousand Oaks, Ca: Sage).

Sternberg, R.J., 2003: "Successful intelligence as a basis for entrepreneurship" (New Haven, CT: Yale University, Center for the Psychology of Abilities, Competencies, and Expertise); at: https://doi.org/10.1016/s0883-9026(03)00006-5 (2 April 2003).

Wasserman, N., 2008: "The Founder's Dilemma", in: *Harvard Business Review*, (February); at: https://hbr.org/2008/02/the-founders-dilemma.

Williamson, O.E., 1975: *Markets and Hierarchies: Analysis and Antitrust Implications* (New York: Free Press).

Zuidervaart, L., n.d.: "Short Circuits and Market Failure: Theories of the Civic Sector"; at: https://www.bu.edu/wcp/Chapters/Soci/SociZuid.htm (15 May 2015).

Chapter 23
Companies of Farmers

Yoginder K. Alagh

Abstract Farmer Producer Companies (FPCs) originate in a Committee I chaired on preparing a Draft Bill for it, which led to the Second Amendment to the Companies Act 2002 (GOI, The companies second amendment bill, 2001, 2002) which gave a legal status to Farmers Producers Companies. There were, in fact, just a handful of such companies initially, and the first big critique came from the corporate sector. Homi Irani said that Farmer Producer Companies were, as he had convinced the Federation of Indian Chambers of Commerce and Industry (FICCI), not corporate companies. I agreed but took the position that we could experiment with organisational forms. They wouldn't agree so we lobbied and the FPCs were hurriedly kept in the 2012 Companies Bill, in a footnote, as it was to be followed by an 'appropriate Bill'. Initially there were only a few FPCs. In some States the Registrar of Companies was favourable. In others not, but National Apex Bank for Agricultural and Rural Development (NABARD) set up a fund to finance Farmer Producer Companies. Now FPCs are abundant. The corporate sector has taken to them. They face problems but have a bright future.

Keywords Companies act origin · Alagh committee · Collateral
Corporate critiques · Late acceptance by corporate sector · Recent problems of FPOs

23.1 Introduction

In March 2018 a senior official of the Department of Agriculture asked for comments on the proposed Bill on Farmer Producer Companies. This farmer organisation system originates in a Committee I chaired on preparing a Draft Bill for it, which led to the Second Amendment to the Companies Act 2002 (GOI 2002) which gave a legal status to Farmer Producers Companies. But my advice to the Ministry of Agriculture is not to proceed with the Bill, even though I am supposedly the

Yoginder K. Alagh, Chancellor of Central University, Gujarat & Emeritus Professor; India;
Email: yalagh@gmail.com

© Springer Nature Switzerland AG 2019
A. K. J. R. Nayak (ed.), *Transition Strategies for Sustainable Community Systems*,
The Anthropocene: Politik—Economics—Society—Science 26,
https://doi.org/10.1007/978-3-030-00356-2_23

granddaddy of Farmers Producer Companies, and therein lies a tale. But as Wodehouse would say let's begin at the beginning. I was a friend of the legendary Varghese Kurian. I would go to Anand for my work with H.M. Patel, I.C.S., my boss in the Sardar Patel Institute of Economic and Social Research. I was appointed a full-time Professor in that Institute at the age of twenty-eight with a one-line appointment letter sent to me on an Inland Letter Card, when I was the Chair of the Postgraduate Programme at the Indian Institute of Management, Calcutta, which I had joined after finishing my doctorate in the Economics Department of the University of Pennsylvania, which, of course, was also in the Wharton School. I wanted to get back to the University system, so I accepted and never resigned that job, always going to Delhi, Rome, New York, Waterloo in Canada, Tokyo and Helsinki on deputation. I am today Professor Emeritus at the Institute. But let us get back to Shri H.M. Patel and Farmer Producer Companies. Kurian, Vijay Vyas and I were reportedly Shri Patel's blue-eyed boys. In 1999 we were worried about the bureaucratisation of co-operatives. My friend V.N. Dandekar had caricatured them in a Presidential address to an Economics Association. Kurian was getting frustrated with the governmental system's abilities at Delhi and State Capitals to let co-operatives thrive in autonomy and function in the AMUL NDDB style. The Agriculture Ministry at Delhi and the Departments at the States would play politics and supersede them.

The corporate sector, it seemed, provided an answer. AMUL was the brand name of the Gujarat Co-operative Milk Marketing Federation (GCMMF). You had to have an Annual Audit of Accounts and election of Directors, both avoided by Politician 'Çoopérative Leaders', whom Kurien detested. But the heart of a co-operative was missing: one share: one vote. But we decided to build this into the company structure. I was chosen to draft the Bill. At the beginning of the century, Kurian organised a big meeting at Anand and there was the Anand Declaration. It wanted the reform of co-operatives. Later on Varghese Kurian was to fall out with Shri H.M. Patel's daughter, Amrita Patel, but in those days, she was his lieutenant and was in the loop with us. I worked on the idea and, as Chair of a Committee we got Shri Sharad Pawar to set up, I drafted the Draft Amendment to the Companies Act (Government of India 2001). From this emerged the 2002 Amendment to the Companies Act. The legislation got stuck up in Parliament in a Committee chaired by Shri Pranab Mukherjee, and I had to go and lobby with Madam Sonia Gandhi to get it nudged from there. She did and it became Law.

There were, in fact, just a handful of such companies initially, and the first big critique came from the corporate sector. When the Companies Act was to be reframed, FICCI argued that Farmer Companies were not companies. Pradan, the largest Indian self-help NGO, was experimenting with them and asked me to lobby for them. I did, and argued with Homi Irani, a colleague because by that time I was a fellow Director in a Tata company. Irani said that Farmer Producer Companies, were, as he had convinced FICCI, not corporate companies. I agreed with the argument that FPCs were not corporates in the traditional sense and I also argued that we were India and not the US or UK and, given our freedom movement and ethos, could experiment with organisational forms. In fact, AMUL-style co-operatives were an experiment in newer forms of organisational structures. FICCI wouldn't agree, so we lobbied and the *FPCs* were hurriedly kept in the 2012 Companies Bill, in a footnote as it was to be followed by an 'appropriate Bill'.

The present Bill is that Bill. Now FPCs are abundant. Initially there were just a few. Members of Primary Agricultural Co-operatives (PACs) got loans with the security of a standing crop but no other collateral. But the system would not finance Farmer Producer Companies. Pradan organised a seminar in Delhi and asked me to come. I did, and we said the situation was ridiculous. Deep Joshi, the Pradan founder, and his colleague Narendran were our points-men. We set up a committee under Nitin Desai to examine and solve these problems. He had returned from his Deputy Secretary General's position in the UN and went to a few States. He found that the practice varied. In some States the Registrar of Companies was favourable. In others the Registrar would make the farmer promoters of an FPC wait outside his office. It was a slow walk. But we eventually got NABARD to set up a fund for financing Farmer Producer Companies. This was also done indirectly. The Boston Consultants were reviewing NABARD's structure and in that context I was consulted for advice and suggested this as a new area. They put it in and NABARD was to list it as a new Functional Area for them and set aside a fund for it. Dear Reader do you see how policy reform is pushed in a closed circuit. Incidentally, I got Deep Joshi to succeed me as IRMA (Institute of Rural Management Anand Chairman), but that is another story. That story includes how I succeeded Kurien there only after I met him and he agreed that I should go there.

In practice, Farmer Producer Companies would still face many problems. The System knew co-operatives, but not this, as it was a new animal. I think the turning point came in the corporate sector. When I was a Director of Tata Chemicals, we acquired Rallis and the company nominated me to the Rallis board. We decided to get into increasing the productivity of pulses. I believed that was possible and showed the way in a Government of India (GOI) Committee I had chaired (GOI 2002). We started an experiment in Puddocotai with a brave Collector in Tamil Nadu. We plugged for FPCs and it worked. Rallis was to repeat it elsewhere. Others got into the Act and they started mushrooming. By now there is literature on the subject (see, for example, Singh/Singh 2018). Equally important enlightened corporate sector chiefs are advocating them (Gopalakrishnan/Thorat 2015). They say:

> The FPO will offer a variety of services to its members. It can be noted that it offers almost end-to-end services to its members, covering almost all aspects of cultivation (from inputs and technical services to processing and marketing). The FPO will facilitate linkages between farmers, processors, traders and retailers to coordinate supply and demand and to access key business development services such as market information, input supplies and transport services. Based on emerging needs, the FPO will keep adding new services from time to time. The set of services include financial, business and welfare services.

An indicative list of services includes:

- Financial services: The FPO will provide loans for crops, purchase of tractors, pump sets, construction of wells and laying pipelines.
- Input supply services: The FPO will provide low-cost and quality inputs to member farmers. It will supply fertilisers, pesticides, seeds, sprayers, pump sets, accessories and pipelines.
- Procurement and packaging services: The FPO will procure agricultural produce from its member farmers; it will do the storage, value addition and packaging.
- Marketing services: The FPO will do the direct marketing after procurement of the agricultural produce. This will enable members to save in terms of time, transaction costs, weight losses, distress sales, price fluctuations, transportation and quality maintenance.
- Insurance services: The FPO will provide various insurances like crop insurance, electric motor insurance and life insurance.
- Technical services: The FPO will promote best practices in farming, maintain marketing information systems, diversify and raise levels of knowledge and skills in agricultural production and post-harvest processing, which adds value to products.
- Networking services: Make channels of information (about product specifications, market prices) and other business services accessible to rural producers; facilitate linkages with financial institutions, build linkages of producers, processors, traders and consumers, and facilitate linkages with government programmes.

23.2   FPOs Could Play a Major Role in Transforming Indian Agriculture

Traditionally, Indian farmers have gained knowledge and skills by sharing within the community, through government programmes and through private sector involvement in inputs, processing and trade. The collective strength of farmers could enable them to increase their competitiveness through easier access to credit and technology, reducing the costs of distribution and providing greater marketing power and negotiation capacity. FPOs could emerge as one of the most effective pathways to address agricultural challenges. Through adequate policy and infrastructure support, these aggregators can become not only the 'connective tissue', linking supply and demand and bridging a major missing link, but also become instrumental in faster deployment and acceptance of modern agricultural technologies (including mechanisation). Adoption of modern technologies would result in demand for skilled technicians in agriculture. (Gopalakrishnan/Thorat 2015: 55–56)

This was, then, acceptance at a high level of corporate leadership of the organisation of FPOs. For some reason, I keep on getting in the news as their granddaddy. But much as I love the Agriculture Ministry and Agriculture Ministers – Shared Pawar and Nitish Kumar are friends and I have the highest respect for Radha Mohanji; ditto for the Agriculture Secretaries – the Agriculture Ministry at Delhi and more so the Departments at the State level do not have a corporate culture. Remember, in colonial days the Collector, the Superintendent of Police, and the District Agricultural Officer ran a District, and I am the first to admit that the Alagh Committee on Training and Reform of the Higher Civil Services has been implemented, but has not wholly succeeded (GOI 2012a). So please, let's keep farmers producers registration of FPCs, and NABARD must fund them, but don't throw away the baby with the bath water.

References

Gopalakrishnan, R.; Thorat, Y., 2015: *What India Can Do Differently in Agriculture* (Mumbai: Sarthak Krishi Yojana, The Information Company).

Government of India, 2001: "Committee of Experts for Legislation to Incorporate Cooperatives as Companies" (also called the 'Alagh Committee').

Government of India, 2002: *The Companies Second Amendment Bill, 2001*, Bill no. 88 of 2001, as passed in 2002.

Government of India, 2012a: "Ministry of Agriculture, Expert Group on Doubling Production of Pulses", Chairman: Yoginder K. Alagh.

Government of India, 2012b: Committee on Training and Reform of the Higher Civil Services of India (New Delhi: UPSC).

Sharma, V.P.; Kumar, P.; Ghosh, N.; Tuteja, U. (Ed.), 2018: *Glimpses of Indian Agriculture* (Delhi: Oxford University Press).

Singh, S.; Singh, T., 2018: "Producer Companies in India: A Study of Organisations and Performance", in: Sharma, V.P. et al. (Ed.): *Glimpses of Indian Agriculture* (Delhi: Oxford University Press).

Part V
Governance

Governance at the community level is the focus of this section. Community in our analysis consists not only of a certain number of families in a cluster of village(s) but also a certain extent of geography and its ecosystem. Accordingly, micro-watershed, the basic unit of an ecology, forms the technical base of a community. However, for optimising the size of a community for its commercial viability under diverse production systems and easier identification of its boundary by people of a community, a Gram Panchayat (GP) that may consist of two to four micro-watersheds, depending on whether it is in hilly, plain or coastal region, is taken as community for analysis and design.

This section covers facts of governance, namely the *efficiency principle* adopted in community governance, *problem-solving approach, structural orientation, decision-making method* and *transparency and accountability*. While the desirable direction of each of these variables may have been readily understood, the discussions need to figure out ways to overcome the challenges of community governance that can facilitate sustainability in community enterprise system and producer organisations, foster sustainable agricultural systems, enhance institutional interactions, deepen relationships between members within the community (GP) and strengthen co-operation and solidarity within the community.

Chapter 24
Community, but for Its Absent Presence

Nyla Coelho

Abstract It has taken humankind a brief span of twelve millennia to bring itself to the doorstep of an epoch of the Anthropocene where, on the one hand, as we are positioned today, there is spectacular technological prowess while on the other there is the unprecedented wearing-down of its social fabric. The situation is alarming as it unfolds at the backdrop of multiple crises of planetary dimensions with the climate crisis at the top of the list. The question pressing for an answer is: Will humankind live to tell the tale? Is a course correction possible? This essay skims over the evolution of human bands for an historical perspective on the matter, going on to then argue in favour of possible solutions that may be found in what we refer to as *community*. Further, it suggests actions that expect us to rise above the mundane to follow the diligent directives set down for us by Nature.

Keywords Community · Nature · Crises · Innate · Unique · Human being Beingness

Community has never been fully and finally achieved in the past, and the prospects of doing so today seem no more likely. In fact one cannot but conclude that nothing that human beings, individually or collectively, can do will stave off the collapse of the natural world, including the world of human society. To assert anything else at this juncture of planetary turmoil seems to be wishful thinking. The transition towards establishing some semblance of predictable stability in our lives might be possible if we are able to collectively replace our doggedly self-centred myopic ways with radically new ways of living, thinking and doing things – a rather tough call for humankind.

People everywhere and in all walks of life today find themselves living with a nagging feeling of unease, confusion, uncertainty, listlessness. This could be a subconscious response to the planetary level multiple crises that we have collectively brought upon ourselves and on all of Nature. Being unable to point clearly to the cause and thereby find ways to address it is leading to a sense of impotence

Nyla Coelho, Belagavi, Karnataka; E-mail: nylasai@gmail.com

© Springer Nature Switzerland AG 2019
A. K. J. R. Nayak (ed.), *Transition Strategies for Sustainable Community Systems*,
The Anthropocene: Politik—Economics—Society—Science 26,
https://doi.org/10.1007/978-3-030-00356-2_24

and despair. As individuals we see no way of dealing with these rather difficult-to-pin-down vague feelings of hopelessness. Of course, these are normal human reactions in the face of impending threats to one's wellbeing. In the collective memory of the human race these, in the past, had definite foci, such as a big cat straying into the neighbourhood forested area, a natural calamity, or an unusually bad season of insect attacks on crops. Definite practical courses of action suggested themselves and absorbed our attention. The crises we currently face are more vast and complex, beyond our individual and collective capacity to comprehend them in their entirety.

Today climate change, degradation of natural landscapes, depletion of fresh water, fossil fuel and mineral resources, pollution of our air, waters and soils, the precariously tilted ecological balance (primarily brought about by the overarching disproportionate predominance of a single species – Homo sapiens – on the face of the planet, and its careless ways), our narrow sectarian loyalties, the uncontrolled mindless push for economic growth all threaten us. We are unable even to imagine what the consequences will be, let alone what we might do to head them off. Or, if we can imagine what might be done, any action by me alone seems utterly inadequate given that the majority are either unaware, or refusing to even acknowledge these as threats, or are paralysed by the immensity and complexity of the challenges.

A related phenomenon is the pervasive sense of aloneness. We suffer from a lack of opportunities for mutual heart-to-heart communion with other people. Without this communion and without such bonding our lives lack meaning. This is an inevitable feature of our present-day atomised global culture. The acceptance of a life of unrelenting competition and a winner-take-all attitude is ruining our emotional and physical well-being. In this broken-down state the clarity of thought and the resolve to act are severely compromised.

24.1 Is There Any Way Forward Out of This Seemingly Dead-End Situation?

Looking back through human history, we learn that disquiet and anxiety are unique to periods of prolonged distress faced by populations, the conditions usually being restricted to geographic pockets. The reasons were factors such as scarcity, impositions, invasions, epidemics and natural calamities. However, this time round, the situation is an all-encompassing universal phenomenon, with enough indicators to place the blame on our contemporary global culture. What has happened and how? These are the most important questions that we must ask ourselves. Our very survival depends upon asking them, mustering up the courage to answer them honestly and then getting down to tackling them in earnest.

People in the past have banded together in groups and gone about their lives within more or less set norms, as is the instinctive strategy of all beings both within and outside their particular kind. Feelings of helplessness and insecurity were considerably allayed by banding together for mutual protection and support. Being members of a clearly defined group gave meaning to their lives. Further, such bands developed cosmogonies that indicated to them the place of human beings in the larger scheme of things, and the associated myths and rituals provided day-to-day direction in organising their lives.

Today we come across many disturbing instances which indicate that people have become increasingly atomised, displaying individualism and functionality in everyday interactions. The social bonding that took millennia to evolve is deteriorating rapidly. Technological breakthroughs have made the need for ordinary everyday communion with fellow human beings almost redundant. Robotics and artificial intelligence now threaten to take over our lives in yet newer ways.

24.2 Why Are We so Self-absorbed in Our Pursuits Today?

A naturally-evolved human band – community – is an assemblage of individuals inhabiting a common landscape with multiple interconnected relationships to other human bands (communities), and to communities of other beings in their immediate locality as well as further away. These bands are subsumed by larger communities, and these in turn are subsumed by still larger communities. The largest of these is the Earth community – a vast, dynamic whole. Each one at every level within it is a dynamic entity staying true to its innately unique *beingness*. This holds true for a grain of sand as much as it does for a person or a planet. Or, we may say, each and every entity is enabled to cognise and conform to its very own *beingness,* wholly and completely functioning as an entity pulsating with a unique consciousness. Any deviation from this results in a disturbance, setting in motion events which are Nature's way of course correction. For Nature, which for us Earth-beings encompasses the entire Earth community, is at all times true to its innately unique *beingness.*

We present-day human beings seem to have gone astray, moving ever further from our innately unique *beingness.* This is because we have come to believe in the narrative that we exist apart from and outside of it or can operate outside of our innately unique *beingness.* We have been educated and socialised into this assumption by contemporary society. The results of our misdemeanour are all around us to see. The first step on the road to recovering our collective *beingness* is therefore to abandon this assumption and accept that we are integral parts of Nature, enabled and contained by Her ways. Having done so, we must then, in all earnest, set about mending our wayward ways such that we realign with Nature's way, thereby slowly moving back on-course.

Nature's Way[1] is too vast and complex for us to grasp in its entirety. No part of a whole, at whatever level it functions, can know the whole of which it is a part. The part can come to know the structure and functioning of the whole only in terms of its most immediately surrounding larger part. The most obvious part for the community of human beings is the local landscape in which we find ourselves. Traditionally this included all the people with whom we share the space, as well as all the other beings that coexist with us there.

As we have moved further and further from our landscapes in pursuing our daily activities, their features have become indistinct; our sense of 'grounding' is lost.

24.3 How Did This Come About?

With the domestication of plants and animals, beginning about twelve millennia ago, hunter-forager communities settled permanently in particular landscapes. Interactions with other settlement communities were minimal. Each community was composed mostly of families who tended the land and animals, while others provided the community with a variety of allied skilled services. When the human population of a community settlement increased beyond the capacity of the landscape to provide it with adequate sustenance, it split, and sub-communities migrated to landscapes that were, until then, uninhabited by communities of human beings. Gradually, bartering and trade among settlement communities increased, and some communities began to specialise in the production of particular goods or in offering certain skilled services, foregoing the tending of plants and animals altogether. In a few settlements practices of localised governance were established. All these specialised later settlements, from this point of view, may be termed secondary or derivative settlements. They gradually grew in population size and, as they did so, community cohesiveness weakened, until, with the largest, only functionality remained. Further, the communities of other beings that had originally inhabited these derivative community settlements were driven to lead a life on the fringes. In today's world, half or more of the world's human population lives in derivative settlements, and the proportion is steadily increasing.

Meanwhile, given the steady decline in the extent and health of forested areas, the erosion of soil, the depletion of groundwater, and increasing human population in primary (land and animal tending) community settlements, people are migrating in increasing numbers to derivative settlements – larger towns and cities, as we know them today. A strong motivation for this migration is the so-called conveniences of life these places promise to offer and the lure of the flashy lifestyle in these settlements pictured in the popular media. Needless to say, most migrants are

---

[1]For a fuller description of Nature's Way refer to Jackson, M.G.; Coelho, Nyla, 2016: *Tending Our Land: A New Story* (co-published by Belagavi: People's Books, Kolkata: Earthcare and Secunderabad: Permanent Green, an imprint of Manchi Pustakam).

disappointed; they manage only to eke out a meagre living. Finding themselves uprooted from their moorings, they tend to be insecure, lonely, distraught and emotionally frail; even worse, they miss their own original community and landscapes and are forced to make do with functional communities. In this state of alienation the *beingness* that keeps each one anchored is greatly compromised.

In a community, in the true sense of the word, the members know one another based on everyday face-to-face interactions. They are at ease in the presence of other members of the community and trust one another implicitly. This trust can only be created by living together in a shared landscape day in and day out, night and day. All its members intuitively know that their individual and collective well-being ultimately depends on one another. In the absence of such arrangements, we create part-time or functional communities based on shared interests or mutual needs, where subconsciously its members are on guard all the time; unable to live their lives fully and completely.

Only when we relate to our fellow beings with trust and respect do our lives gain meaning. To wander desolately in a meaningless universe is the ultimate punishment for a wayward humanity.

The desperate efforts people everywhere are making to meet the felt need to re-connect with one another in meaningful ways (more and more with the assistance of modern technology today) are now universal in their prevalence. There is also a plethora of clubs, professional associations, platforms, sectarian religious groups and many more special-interest groups. At best these are impermanent. They are part-time, casual in approach, and members can drop out at the first occurrence of discord or personality clashes. There is no incentive to persist. They are palliatives devised by a fading civilisation. They are make-believe.

A critical issue that demands our attention is the schisms that have appeared in our world today. One is that of class, colour, race (also caste in India). Increasingly during the twentieth century (now spilling over into the twenty-first) this has been an unfortunate aberration, and considerable progress has been made in doing away with it. Another is gender bias, the subordination of women to men, and here too there has been considerable progress to get rid of this bias. Of late the rigid classification of individuals into male and female is also drawing flak from many quarters and LGBT (lesbian gay bisexual and transgender) groups. Doing away with these schisms to strengthen communities is not only necessary, but can most effectively be done only as a part of a more general effort to re-create empowered communities.

A way forward in recreating empowered communities could be to focus on nurturing the well-being of our local landscapes in every respect. In doing this we will simultaneously be recreating community, for rehabilitating and managing our local landscapes (neighbourhoods) is only possible by concerted community effort. Landscapes, once restored to health, can remain healthy only by continual effort by their resident community members. This applies to all communities everywhere. Only in this way can the future of the larger landscape units, and ultimately the future of the entire Earth landscape, be secured.

Box 24.1: Rules of Thumb for Sustaining Community. *Source* **Author**
In the day-to-day management of our personal and collective community
lives the following guidelines may be useful. These can be grouped into three
categories. They are only indicative and in no way exhaustive. Each community could create such guidelines of their own based on their particular
requirements.

1. Settlement livelihoods and economy:

- Make a conscious effort to understand individually and collectively the
  nature of the livelihood we wish to pursue. Whose purpose does it serve?
  Does the work, skill, knowledge create livelihood opportunities and
  directly serve the needs of one's community in terms of the service,
  outcome or end product?
- Ask and seek answers to the critical issues of equality, equity, justice,
  inclusiveness, equal opportunity, fair-play and fair-opportunity,
  co-operation, utility, risk to individual and community health with regard
  to the livelihood to be pursued.
- Treat any proposed change, new idea, scheme or innovation with great
  caution. Ask the questions: Is it at all necessary? In what ways does it
  benefit our present way of living? How will it affect our collective
  well-being and security? What are the prospects of its long-term
  usefulness?
- Shun the temptation to use 'labour-saving devices' or unnecessary
  automation if they come at the cost of denying work to human hands,
  cause unemployment even if their use enhances monetary profit, cause
  pollution or contamination of the landscape, are ethically inappropriate
  and alien to one's cultural traditions, are a long- or short-term health
  hazard, come with temporary short-term benefits but extract disproportionate costs in the long-term.
- Introduce a local currency or barter scheme so that revenues generated
  locally circulate locally, thus making the local community the prime
  beneficiary of all its endeavours.

2. Local needs come first:

- Make meeting local needs the first and foremost priority. Explore other
  communities, clients, users, customers only after local needs have been
  met completely. Any surplus can then be made available to neighbouring
  communities, nearby towns and cities.
- The matter of scale, scalability, replication, expansion, although holding
  great attraction, must be approached with caution. These tend to suck an
  enterprise into larger economies whose aims and policies might pose a
  challenge to local autonomy at a later date.

- Develop appropriate-scale enterprises that support local livelihoods and skills and whose output is largely utilised/consumed by the local community itself.
- Acquire surplus only to the extent that it assures the community of a modest way of life, well-being and security for all.

3. Resilient governance: a matter of the heart rather than the head:

- Resolve all matters of difference and conflict in amicable ways through dialogue. Communities disintegrate quickly if differences are not openly voiced and resolved in time.
- Provide care and assistance to community members who are ill, physically, materially or otherwise challenged, and integrate them into every-day community life.
- Do not allow people to face or resolve their calamities and tragedies by themselves; empathise and help one another to overcome adversity.
- Follow governance regulations diligently while at the same time collectively challenging any measures being brought in that can affect the long-term well-being of the community.
- Always remember that a community is a living entity that needs constant nurturing if it is to sustain itself and its constituent members.

It is equally, if not more, essential for us to actively participate in the nurturing of our young, caring for the elderly and the less fortunate amongst us. As social beings, this brings to the fore the better side of our nature, which otherwise, is best left unstated.

The larger implication of these imperatives is that we must abandon our contemporary development and modernisation agenda in favour of an agenda of permanence.[2] The all-encompassing ground rule for any of it to truly fructify is for each of us to give of ourselves generously.

---

[2]This is only possible through strict adherence to one's *beingness*. For further insights, see: Coelho, Nyla; Jackson, M.G., 2017: "Towards an Ethics of Permanence", at: https://www.ecologise.in/2017/05/news-update-141/ (10 May 2017).

Chapter 25
Attributes of Community Champions Volunteering for Leading Institutional Change in a Social Context

Amar Patnaik

Abstract With arguments for and against an agency embedded in the same institutional field being in a position to trigger large-scale institutional change endogenously still continuing, discussions on personality variables (related to the institutional entrepreneur or change agent or Community Champion, as this chapter argues to be more apt to use in a social context) have been relegated to the background. What sparks such entrepreneurial activity? Why does only one out of so many actors volunteer to lead such change? Based on literature from the disciplines of institutional change, volunteerism and transformative leadership and based inductively on four cases of successful transformational institutional change by change agents or Community Champions in Odisha, this chapter proposes a framework of nine attributes for voluntary action by these Community Champions. Unlike the variables identified in the literature on institutional change and volunteerism, these attributes are more relevant for a social context. Of particular interest is the attribute level of selflessness, which is strongly linked to altruism.

Keywords Institutional change · Community champion · Voluntary action
Attributes · Level of selflessness

Sustainable community systems are founded on strong relationships between the various individual actors within the community and also with nature and the overall environment, Mother Earth. These relationships influence the kind of institutional arrangements that take root in the community. Broadly, these relationships could be guided by the self-centred, interest-maximising '*homo economicus*' (the edifice on which neo-classical economics rests) or the boundedly rational (Simon 1928) '*social man*' whose actions are dictated (both constrained as well as created) by norms, values and belief systems. The latter, many times, could be non-reciprocal, unconditional one-way interactions.

Based on a literature survey and grounded inductive study of four detailed cases from different successful institutional change stories in Odisha, India, this chapter

Amar Pattanaik, Former Principal Accountant General, West Bengal, India;
Amar_Patnaik@post.harvard.edu; amar_patnaik@yahoo.com

© Springer Nature Switzerland AG 2019
A. K. J. R. Nayak (ed.), *Transition Strategies for Sustainable Community Systems*,
The Anthropocene: Politik—Economics—Society—Science 26,
https://doi.org/10.1007/978-3-030-00356-2_25

examines the extent to which this social man can provide foundational succour to sustainable systems in the community and the attributes of such actors that can enable a sustainable system to take root and thrive. This chapter further suggests which of those attributes contributes to promoting volunteerism within these actors.

25.1 Institutional Change Versus Ordinary Change

Change as a notion has come to mean different things to different people and has almost become a cliché due to a façade of being the 'in thing' to talk about, particularly in management literature. Is 'institutional change' any different from 'ordinary change' as used in literature and practice? Battilana (2008) argues that a change which is widely divergent, transformative in nature and of such a large-scale magnitude that it has the potential to permanently alter the very ecosystem or the environment in which the change itself was initiated or triggered is called an 'institutional change'. Hence, in short, large-scale and long-term change is 'institutional change'. It is "big change". Institutional change transforms institutions and the institutional fields in which they are located permanently, while ordinary change is rather incremental, may not be everlasting, and quite often does not change institutional structures and functions to a great extent. While institutional change literature in the discipline of economics and organisational theory recognises that such change can come about due to a sudden external shock or policy change, it has also been argued by organisational theorists and political scientists that such change can take place endogenously through an inspired and committed agency indulging in entrepreneurial/convening activity.

25.1.1 Institutional Change

In this chapter, it is clarified that organisation and institution are not synonymous terms as is commonly understood. While organisation is a structure, institution is the rule, principle and procedure that governs the running of the organisational structure – norms that govern the behaviour and action of all employees working within the organisation.

In the economic school, Acemoglu et al. (2004) argued that the elites in a society would want inefficient economic institutions to persist so that they could extract rent from them. They use their *de facto* (arising out of their economic wealth) political influence to subvert the *de jure* (legal) institutional power structures. Virmani (2005) holds that in a development context, institutions at a very macro level require long periods of time to change. He provides a "micro structure" of institutions so that change can take place faster. He argues for the need for an appropriate "change agent" to spearhead institutional change or reform. In the field of institutional economics, Williamson (1985) expanded the transaction cost concept of Coase (1923, 1930) to suggest that institutions would exist as long as they provide more benefits than the transaction costs of creating them. North (1992) argued that institutional persistence occurred due to path dependence arising out of network externalities, economies of scope and other complementarities. This means the group with a higher bargaining power tries to perpetuate the same. While North still relies on competition giving rise to appropriate incentive mechanisms to effect institutional change, he does admit that cognitive models or belief systems also act as influencers in institutional change. North (1994) discussed the role of learning processes and changes in perceptual/mental models in institutional change over time. He advocated six stages in which this model changes – initial dissatisfaction, some people ("institutional pioneers" or the "change agents") looking for alternatives, strengthening of personal knowledge of the pioneers (to strengthen their own conviction and logic) about the new model, diffusion to decision-making hierarchy through their active involvement and, finally, overcoming the fears of the neutral opinion leaders. Thus changing an economic or social institution appears to be an extremely political exercise – the act of mobilising supporters in support of a new paradigm.

In political science, institutional theory is based on the rational choice theory and views institutions as governance or rule systems that are designed to promote the 'interests' of individuals. Ostrom (1986) looked at institutions as 'prescriptions' for 'required, promoted or prohibited actions'. Most political scientists look at institutions as reflecting the 'preference' and 'power' of the units constituting them, but at the same time 'shaping' those preferences (in Farashahi et al. 2005). This is in contrast to economist North's argument that institutions are created to solve coordination and information asymmetry problems.

In sociological sciences, institutions are considered to be stable structures, path dependent and extremely sticky. They are homogeneous structures, practices and procedures to achieve *legitimacy* in preference to economic efficiency and performance (Meyer/Rowan 1977; DiMaggio/Powell 1983, 1991). The earlier neo-institutionalists from the disciplines of sociology and organisational theory asserted that actors' actions are not just guided by instrumental rationality, but by certain socially-constructed taken-for-granted norms (Meyer/Rowan 1977; DiMaggio/Powell 1983). These studies emphasised ways in which institutions constrained organisational structures and activities, and thereby explained the convergence of organisational practices within the same institutional environment, called field. This was referred to as "institutional isomorphism". This behaviour arose out of concerns for social (including organisational) legitimacy. It was explained that this arose from the logic of appropriateness (Levi 1990) rather than the logic of consequences due to the institutional pressures (Oliver 1992) to which they were subjected. However, the phenomenon of *institutional change* was not addressed at all by these sociologists.

But subsequent studies contributed by organisational theorists have dealt extensively with institutional change in both mature and emerging fields. They have essentially emphasised uncertainty, innovation, and legitimacy as the enabling conditions for institutional entrepreneurs to act and carry out a process of re-framing and re-legitimising the new institutional logic by mobilising allies and seeking collaborations to effect and sustain such institutional change in the field. They, however, largely failed to account for 'power' as a possible factor responsible for creating and perishing institutions. On the other hand, work on institutional change by political scientists who have taken 'interest' and 'power' into account in their research has been sparse and much more historical recountal in nature than providing any scientific study of an actual institutional change taking place because of power plays. The latter plays out extensively in a social arena which economists, political scientists and organisational theorists have left completely left.

25.1.2 Institutional Entrepreneurship and Institutional Change

It was DiMaggio (1988) who, for the first time, presented institutional entrepreneurship as a promising way to account for institutional change. According to him, institutional entrepreneurs are actors who have an interest in a particular institutional arrangement and mobilise resources to create new institutions or transform existing ones. The change occurs endogenously and not due to any exogenous influences. As for the process, he explained that institutional effects occur because actors engage in structural politics. In fact, it was he who brought agency and interest back into the study of institutions in order to explain institutional change long after the old institutionalists flirted with this notion. However, this perspective has its share of controversies arising out of the ability of actors, who are supposed to be institutionally embedded, to distance themselves from institutional pressures and to act strategically to change the very institutions which are constraining them to act along the charted path. This question alludes to the 'paradox of embedded agency' (Holm 1995; Seo/Creed 2002). To overcome the paradox of embedded agency, it is necessary to explain under what conditions actors are enabled to act as institutional entrepreneurs. Battilana (2006) argued that the social position of individuals is a key factor in understanding how they are enabled to act as institutional entrepreneurs despite institutional pressures. She argued, predicated on Berger/Luckmann (1966: 60), that the objectivity of the institutional world is a humanly produced, constructed objectivity. Before being 'objectivated' (i.e. experienced as an objective reality) by human beings, institutions are produced and re-produced by them. She also argued that the reason for this paradox is the neglect of analysis on an individual level in contrast to the organisational and societal levels of analysis. As stated by Friedland/Alford (1991), an adequate social theory must work at all three levels of analysis (i.e. individual, organisational and societal).

After DiMaggio and Powell, several organisational theorists and political scientists tried to explain how institutional change took place in the context of both *mature* and *emerging fields*. While the former has well-entrenched institutions, the latter has opportunities still available for the creation of new institutions, or new institutions are in the process of being formed. In all this work, the emphasis was on exploring what pushed an institutional entrepreneur to change the existing the institutional structure and what facilitated his action. But literature is scant as regards what prompts an actor to become an institutional entrepreneur when many other actors like him prefer to buy peace with the existing order/arrangement. What makes an actor volunteer to take up the mantle of entrepreneurship or championship? The existing literature does not throw adequate light on the attributes or characteristics of the institutional entrepreneur either. After all, not all individuals can become institutional entrepreneurs, nor do they end up as such!

25.1.3 Institutional Change in a Social Context

A nuanced study of the literature on institutional change and entrepreneurship reveals that most of the work is set against market contexts, where the rationality of the human actor is guided more by economic efficiency to maximise business profits as opposed to interests such as altruism, desire for social reform or just "doing something good for the society". The institutional change agent is motivated more by a concern to corner resources for itself/himself so as to stay in business (a resource-based view); and/or become more business competitive by improving efficiency through technological innovations, new skill-sets or managerial excellence (an economistic view); and/or attain institutional acceptance (legitimacy over efficiency and performance). Besides, the organisations referred to in most of these studies are set in the commercial world, where entrepreneurs act primarily to increase their profit margin or find new markets. None of the cases talked about an institutional change in the social sphere or, as this author would like to call it, *institutional social change*. The cases are also mostly drawn from the West where the business and social milieu is different on many counts from agrarian and rural-based economies such as India's. Most literature has dealt with entrepreneurship at organisational level; individual level entrepreneurship studies are very few. Similarly, most of the literature is not contextualised against state-sponsored, public-orientated interventions, which present a different set of challenges and enabling conditions for entrepreneurship to take place. Such contexts may require different kinds of skills from the institutional entrepreneur or champion to mobilise people within the community for collective action. There is also an implicit assumption about the absence of informal institutions in those contexts and the absence of the possibility that these could also impact the growth of entrepreneurial environment, the actor-skills and the entrepreneurial process itself. The underlying business logic in such kinds of institutional fields is competition and survival of the fittest. In a social context, particularly with reference to India, institutional change would necessarily have to encounter the vast array of deep-seated informal institutional arrangements, non-state and non-business actors and follow the logic of social appropriateness (Levi 1999). For social change of a completely divergent nature to occur, the institutions within the field may have to be re-organised, new institutions may have to be created and competition may have to give way to co-operation for collective action to take place. Since DiMaggio's conceptualisation is premised on self-interested actors, it is open to doubt if such an agency would succeed in bringing about institutional change in a social eco-system. The language and logic of competition and co-operation are diametrically opposite (Nayak 2008). Also, the literature on institutional change has largely not delved into issues such as injustice, inequity and absence of fair play – all significant concerns in a typical social arena in a developing country context in general and in the Indian rural context in particular. These aspects themselves could be serious fault lines that may spur Change Agents or Champions or Social Entrepreneurs to initiate action for change in the institutional field endogenously. This kind of institutional change in the social arena also finds coverage in the social change, social entrepreneurship and social movement literature in varying degrees and with different methodological approaches, often borrowing heavily from the disciplines of political science, organisational theory and leadership literature. However, the line of argument in this approach is different from the approaches followed in social entrepreneurship or social movement literature, though admittedly overlaps exist.

The endogenousness aspect in this change process suggests that this concept is likely to be linked to the concept of collective action (Ostrom 1994), which has already been extensively used to explain the management of common pool resources by communities without the involvement of the Government or the market. However, how does collective action start? Who sparks the collective to act? It could very well be the Change agent or Champion that this chapter discusses. The existing literature on institutional entrepreneurship from organisational and institutional theories is silent as to how entrepreneurship and collective action could result from an act of altruism or volunteerism by this Community Champion.

25.2   Who Are These Community Champions?

DiMaggio (1983, 1991) defined the institutional entrepreneur as an individual or actor who mobilises resources to exploit an opportunity identified by him or coming his way which would fulfil his interest. As mentioned in the preceding pages, such a definition of an institutional entrepreneur does not fit well into a social context where opportunistic behaviour by the institutional entrepreneur may prove to be counter-productive to achieving institutional change. In the social arena, the institutional entrepreneur needs to have more people-orientated skills by becoming an example of the change that he/she wants to usher in (Gandhi in Sethi 1978). Some of these people's characteristics could be selflessness, honesty, and commitment to the cause or to solving the problem faced by the people. Any implicit or explicit evidence of opportunistic behaviour for individual gain by the institutional entrepreneur may result in loss of credibility in front of the community and people moving away instead of thronging to him or her. The people have to perceive a sense of personal sacrifice being made by the institutional entrepreneur for the cause of the larger group. It is, therefore, argued that the use of the term 'institutional entrepreneur' in a social context may be inapt and misplaced. Hence this chapter espouses the term Institutional Champion or Community Champion, to describe an actor or a group of actors who trigger a long-term institutional change. Unlike DiMaggio's institutional entrepreneur, such people are not guided by any profit motive, even remotely. They have no personal interest to be served, but the larger community-wise interest. They volunteer their time and energy and mobilise people to steer this change process. In fact, it is their selflessness which attracts people to be convened and led by them, so as to bring about transformational institutional change. To be an institutional champion or exercise entrepreneurship, it is necessary to have certain attributes or characteristics that would facilitate the 'entrepreneurial (convening) work' to be done. Lounsbury (2004) identifies two such traits, *level of embeddedness* and *level of involvement*. Battilana (2006) argued that the *social position of the entrepreneur* was very important for guiding and spearheading the change process. But just as the bulk of literature in institutional change is located in a context of markets in the Western World, these works of Lounsbury and Battilana are posited against a background of markets. Needless to say, the very notion of markets presupposes an atomistic, self-serving and self-interested individual exhibiting opportunistic behaviour for maximising individual gains/interests. However, in a social context, these characteristics in an Institutional Champion would not only retard the process of convening or mobilisation, they may even give rise to negative emotions about the Champion. In this chapter, some more equally significant attributes of Institutional Champions that make them what they are and put them in a position to convene and mobilise others are identified from secondary sources as well as in a grounded manner through four case studies.

Even in participation literature, Chambers (1997) argues that imposing ownership and hence participation as an appendage to an existing programme framework for implementation will ultimately disillusion the people. He argues for the need to identify project Champions – *key individuals or alliances of individuals who battle institutional inertia* – and nourish them to take the project forward by slowly securing the participation and ownership of the targeted project beneficiaries. This brings into focus the primacy of actor-orientated agency to bring about institutional change in the social space. Once one is grappling with agency issues, the need arises for an agent with the characteristics and attributes that spark and spurs on suo-motu volunteering to bring about institutional change in a social context. How does one identify such individuals or groups of individuals? Also, why does one individual out of so many volunteer to lead such large-scale (almost transformative) change?

25.2.1 Volunteerism

Volunteerism is believed to have roots in the religious teachings of philanthropy and giving alms in most world religions like Christianity, Buddhism and some sects of Hinduism. Wilson/Musick (1999) suggest that volunteering increases societies' social capital, trust and reciprocity. However, it may be considered that even a reverse relationship in these variables is possible, as more social capital and trust engenders more collective action (as held by Ostrom 1990), and hence, more opportunities for actors within the community to volunteer. Lyons et al. (1998) suggested that volunteering may mean different things in different countries. For those working in the 'non-profit' space, volunteering could be seen as an activity based on altruism or philanthropy. This is mostly found in organisations that deliver some kind of public service. The 'civic society' paradigm regards at it as a member-benefit association which serves the interests of only the members of that association. Bussell/Forbes (2002) suggested that the study of volunteerism has always been multi-disciplinary. It is difficult to define, as there is no standard practice or defined criteria. However, broadly, it "involves contributions of time without coercion or remuneration."

  Smith (1994) states that more volunteer participation comes from larger contexts (territory and organisations), social background and role variables, personality traits, attitudes and situational variables. According to him, though background and attitude variables have been reasonably well studied, literature on contexts, personality and situations is thin. Berger (1991) and Schiff (1990) posited that the longer the length of residence in the community, the greater the probability of voluntary participation by an individual. Allen/Rushton (1983), in the case of community mental health volunteering, observed that voluntary participation was higher for individuals with more efficacy (internal locus of control), empathy, morality, emotional stability and self-esteem or ego-strength – all of which are indicators of social orientation. They admitted that though literature on these lines was thin, the area was promising. They averred that not much work seems to have been done on this issue by sociologists as they consider these topics to be closer to the discipline of psychology, hence there has been no or very little inter-disciplinary work. Rochester (1999) argued that while the literature on volunteering concentrates mostly on four principal themes – defining volunteering, measuring the overall extent of voluntary action, understanding the motivations of those who volunteer and looking at the organisations and management of the work of volunteers – it does not address the organisational contexts in which volunteering takes place. He suggested four models of volunteering based on different organisational contexts. Cnaan/Golberg-Glen (1990) identified twenty-eight different reasons for volunteering action by individuals, based on 200 articles on this topic. Though the contexts in these articles were mostly located in the United States, an acceptable view emerging is that the motivations of volunteers can be a complex mixture of the instrumental and the expressive and that they can change over time. The authors recognised that it was most demanding to understand the organisational behaviour of volunteers in different institutional arrangements.

Though altruistic attitudes have often been regarded as the central motive for volunteering, the prospect of a reward is, in fact, intrinsic to the act of volunteering and a major incentive for participation in volunteer groups. Smith (1981) posited that there could be no 'pure' altruism as even the volunteer gets some kind of pleasure or benefit out of such participation. Moore (1996) and Tihanyi (1991) have posited that younger volunteers are more interested in the value that this experience brings to their career, while older people take up volunteering as a response to a change in their lives such as loss of a partner. There is still little literature on the importance of the organisational contexts in determining why people volunteered. Cnnan et al. prefer to use a continuum from complete free will to obligated volunteer, meaning volunteering almost out of altruism to volunteering for a stipend. According to them, individuals and organisations may volunteer for reasons other than purely altruistic motives since there are many different contexts in which one can donate one's time. They also felt that little research has been done on local variations in volunteering and the process of volunteering. Hence, the identification of community champions could also be down to serendipity rather than design. However, some commentators (Kirk/Shutte 2004; Langone 1992) believe that a community leadership development programme would work.

Literature on volunteerism proposes various determinants of voluntary action. These have been categorised as contextual, social background, role-based, personality-based, attitudinal and situational variables. Over time, there has been a shift away from personality variables to attitudinal, background and contextual variables, primarily due to sociologists and political scientists finding difficulty in tackling personality variables, which draw heavily from the discipline of psychology. But literature on transformative leadership based on literature from psychological sciences still places a lot of premium on the personal attributes of this transformative or charismatic leader.

25.2.2 Volunteering and Leadership

While studies on leadership have, over a period of time, moved away from personalised, charismatic leadership traits or characteristics to understanding the various processes of leadership, leadership literature speaks of transformative leadership vis-à-vis a transactional leadership. It has been argued that while the latter leads to incremental changes only, the former brings wholesale, large, systemic change. The latter kind of change is something similar to institutional change, as described earlier. Hence, for institutional change to take place, a transformative leadership role has to be assumed by the Institutional Champion or Change Agent, who has been called a Community Champion in this chapter to reflect the fact that the change is taking place in a social space. Kark/Van Dijk (2007) observed that any kind of leadership is deeply tied to the individual's (both the leader's and the followers') internal motivational systems. Leadership literature has also brought out various traits of such transformational leadership. These are psycho-social variables which indicate the capacity of the transformational leader to motivate people and join the cause. These traits could be the same as those required in a Community Champion who volunteers to lead a change process.

25.2.3 Attributes of an Institutional (Community) Champion

In a social context, not everyone volunteers to lead a transformational change and be an Institutional Champion. Even if such a Champion is willing to lead the change, he or she needs to have authority, formal or informal, to be able to convene people in the community. While formal authority comes often from bureaucracy and Government hierarchy, informal authority is derived from a number of other sources, including an individual's own personality or psycho-social traits. But not all individuals with authority, either formal or informal, can convene people. There are some attributes or characteristic that set an Institutional Champion apart from others enjoying similar or even more authority. Lounsbury (2004) had identified two such characteristics, namely, *level of embeddedness* and *level of involvement*. However, based on secondary case studies from both Indian (Anna Hazare's Ralegan Siddhi and Popatrao Pawar's Hiware Bazar) and non-Indian contexts (Korten's eight cases [1980]) and also on the transformative leadership literature and four carefully structured purposively sampled case studies conducted in four different geographical regions of Odisha, India during 2012–14, it is argued that there could be some more significant attributes that the literature has so far not captured. The cases studying the ultimate gains to the community due to the intervention/experiment spanned different sectors such as education, health, environment, economic, cultural, capacity building, technology and skills gain. Care was also taken to select some recent governmental flagship programmes, such as the *Total Sanitation Campaign* (TSC), *Swarnajayanti Gram Swarojagar Yojana* (SGSY), and *Total Literacy Campaign* (TLC). The cases are as follows:

- Women Empowerment Programme in Tambakhuri village of Balasore district (supported by Unnayan, an NGO)
- Total Sanitation Campaign in village Bahalpur in Ganjam district (supported by GRAM VIKAS, an NGO)
- Programmes on Child Rights in Balaniposhi village of Keonjhar district (supported by PECUC, an NGO)
- Prevention of alcoholism in Dasingbadi village of Khandamal district (supported by Jagruti, an NGO)

The field study cases followed a structured multiple case methodology protocol and followed Eisenhardt's (1989) mixed and iterative methodology. A semi-structured questionnaire was prepared for interviews and a series of focus group discussions (FGDs) were held with different actors during the field study. The repeated and reiterative approach in capturing the success story of change from the same actor-Champion and corroborating it with other actors for triangulation purposes provided the necessary robustness, validity and reliability to the studies.

Based on the above, this chapter presents a framework of the attributes of an Institutional (Community) Champion which spurs him or her to take voluntary action. It is these factors which make him or her a successful volunteer-leader. These are:

25.3 Level of Embeddedness

Embeddedness is a structural variable that indicates how intimately or deeply the Institutional Champion is 'embedded' into or wedded to the overall social structure. It indicates the level at which the Champion is located in the rural socio-cultural structure. The level of embeddedness indicates the extent to which the Institutional Champion is aware of the problem at hand in terms of both its breadth and depth. Embeddedness engenders empathy, as the Champion is then better placed to appreciate the problems faced by the community in a more direct way. This proximity to the problem or the challenge itself, not only at the theoretical and mental level, but also in experiential terms, makes it more likely that the Champion will volunteer to lead the change process and also increases the ability of the Champion to 'problematise' the problem or challenge by increasing its sharedness and universality before the community. Only when a problem is defined properly will the community be able to marshal its resources to solve it.

25.3.1 Level of Involvement

Level of involvement is defined as the extent to which the Institutional Champion is seized by a desire to resolve the problem because of his/her close appreciation of the issue. Unlike the previous variable, which is conceptualised around proximity to the problem faced by the community and the Champion, the Institutional Champion's acceptability as a volunteer-leader within a rural community is dependent on the extent to which he/she is not only aware of the problems/challenges, but also how far he/she is immersed in that problem. If the institutional champion is not involved with the community in either structuring or formulating the problem or co-searching for a viable solution, then his mere awareness of the problem may not make him acceptable as a Champion to the community. Higher involvement reduces the perception gap between the community and the Champion as far as the problem is concerned, and induces him/her to volunteer to lead the change process. Hence, involvement has everything to do with being familiar with the problem rather than with the community. It reduces the physical, mental and psychological distance between the Champion and the community, which helps to increase the ability of the champion to convince and mobilise the people.

25.3.2 Level of Selflessness

Though very difficult to define precisely, selflessness can be broadly understood as counterfactual to the level of selfishness which forms the basis for the *homo economicus* in mainstream neo-classical economics. It can be loosely stated as the extent to which an individual is not bothered about him/herself, meaning thereby selfish interests, greed/desire etc. While selfishness engenders competition and the desire to win in the face of another individual's similar greed/desire, and provides the *raison d'être* for a business organisation to make profits, social entrepreneurship requires a different trait for the Champion to be successful at leading the people within a community. Until and unless the community perceives the Champion to be selfless and have no individual agenda, championship activities for convening lose credence as they become identified with the greed and selfish motives of the Champion. Selflessness is the most difficult trait for a Champion to have. But this trait alone could make volunteering easier and more sustainable as the persona of the Champion takes a back seat to the problems at hand. People realise that there is nothing for the Champion to gain from this activity; he/she is simply there to help them solve their problems. Since, as a consequence of this lack of any personal selfish interest in the solution to the problem at hand, the Champion's presence is not perceived to significantly alter the pre-existing power structure within the community, his or her acceptability increases. Through being part of the power rubric, but not perceived by the community as altering the balance by throwing his or her weight behind any single power bloc within the community, the Champion gains acceptability and credibility and is in a position to lead the people of the community. The latter shed their power-related inhibitions to listen and follow the Champion in solving the problem at hand. It is likely that an individual with a high degree of selflessness would volunteer to lead the change as it carries the hope and possibility of improving the lives of the community. Basically, selflessness engenders pro-social behaviour (Bierhoff 2005).

25.3.3 Level of Empathy

Empathy is feeling. As distinct from level of embeddedness, empathy generates a mental model which makes the champion feel the same amount of pain that the community experiences due to a problem, which induces him or her to volunteer to mitigate this pain. Even if an individual is embedded into the community and involved in the problem, there may not be empathy unless the person feels the same way about it as other people in the community. Empathy can induce a desire to convince the people to set the problem right by leading from the front rather than merely sermonising.

25.3.4 Level of Organisational Ability

Organisational ability is the ability to manage people and resources in a productive way to achieve one or more objectives. Social mobilisation literature recognises the need for a high degree of organisational ability on the part of the protagonist in order to take the movement forward. This requirement for organisational ability arises not only from the presence of a number of collaborators, but also from the need to arrange physical space, obtain infrastructural support and communicate directly with the collaborators as well as the people supporting them, in order for a successful convening activity to take place. When the level of support increases and the number of collaborators multiplies, the necessity to organise and manage their physical, logical and informational needs cannot be taken less seriously. Since communication and dissemination require very different skills, it may be necessary for the Institutional Champion to harness the resources of other collaborators in order to move forward. This requires organisational ability. Mobilisation may be possible, but it is extremely important to organise the mobilisation in such a manner that it does not degenerate into a free-for-all. An actor within the community who possesses such organisational abilities is likely to volunteer to lead such a change process.

25.3.5 Level of Education

Though education and literacy are not necessarily the same thing, for the purpose of this study they are regarded as synonymous. Education helps a person to distinguish between right and wrong. It equips an individual with more information, which in turn helps to improve decision-making. It fosters greater long-term thinking and aids analysis of data. It also helps an individual identify strengths, fish out opportunities, be aware of weaknesses and ward off threats. It would be fair to posit the hypothesis that the higher the level of education/literacy of the Champion, the greater the convening ability, because he/she is in a position to more authoritatively and reassuringly convince people belonging to different factions and with different incentives to get together for a shared goal. A set of factions inimical to each other may not have realised what opportunities are ready to be grabbed and what threats need to be tackled in a win-win manner that enables all factions to derive benefits from such opportunities instead of at the cost of one another. The Institutional Champion, being educated, will be able to convince them of this, because of his/her superior ability to explain and convince. This may also induce him/her to volunteer to lead a change since he/she is well aware of the pre-change scenario and the possible ways in which such a situation can be changed.

It would be equally necessary to assess the level of literacy of the community. It can be posited that a higher literacy level within the community should make the life of the Champion easier as less convincing would be necessary.

But while education may improve the analytical power of the individual, it may also sow the seeds of greed and interests based on selfishness, which may not necessarily be conducive to convening activities. Therefore, it may well be possible that a convening activity is easier when the literacy level of the community is low. But it could also be possible that if the literacy level is below a certain threshold, neither the community nor the Champion has the ability to identify its strengths or the available opportunities, or to ensure that weaknesses and threats are overcome.

Another concern could be the differential literacy level between the Champion and the community which might bring up the challenge of "champion capture", as a form of "elite capture". In fact, convening can succeed on a sustainable basis only if the risk of elite capture – which basically constitutes an abuse of power – is minimised.

25.3.6   Social Position of the Champion in the Society

Drawing on Bourdieu's theory of fields, Battilana (2006) has argued that the social position of the individual is a determinant for generating institutional entrepreneurship. However, social position is related to the legacy of a social structure prevalent in the community, which may also be responsible for resisting any entrepreneurial work. Battilana calls this the "paradox of embedded agency." In the typical village context of this study, this social position flows from several factors, including economic wealth, caste considerations and educational background. It can also flow out of political supremacy. While it may make the life of a Champion easy to convene, it could also make his/her life difficult because of the very social position enjoyed by him/her. For example, if, with particular reference to the context of an Indian rural ecosystem, the Champion belongs to a higher caste, it may be more difficult for him or her to convene people or factions belonging to lower castes because of the historical legacy of distrust and lack of faith. However, it can also be argued that if a higher-caste individual problematises issues relating to a lower-caste community, he/she would have a better chance of success than a lower-caste Institutional Champion because of the greater likelihood of being able to convince his or her own caste-members to cede some power or to lower-caste people for the greater good of the community.

It can also be argued that such social reputation arises due to factors like empathy, involvement, embeddedness and selflessness, which have been described earlier – the same traits that induce voluntarism in the Champion. The social position of the Champion generates a feeling of trust and respect from the community, engendering the belief that he/she can solve their problems. Hence, it is invariably the case that the Institutional Champion should be someone from within the community and not from outside. However, examples abound where an outsider, maybe an NGO, has been at the forefront of convening activities due to their pragmatic and prolonged relationship with the community based on involvement, embeddedness, selflessness and empathy.

25.3.7 Economic Position of the Champion

Defined as the position enjoyed by an individual by virtue of his/her economic wealth, it could be argued that the economic well-being of the Champion contributes to his/her social position and therefore need not be discussed separately. Economic well-being gives elbow room to the Champion to organise and convene people for discussion, as he/she has the resources to hire a meeting place and, for instance, use modern means of communication, such as a public address system, to interact with the community. It also provides the Champion with the elbow room to stick out his/her neck to volunteer a change process foreseen as good for the community. But, as argued earlier, this economic wealth may also result in loss of trust between the community and the Champion due to increased economic power asymmetry. However, as in earlier times, there is also evidence of great philanthropists, backed by a comfortable bank balance, acting as social change agents because of their involvement with the community. There is also a case for the community to think that because of the already existing wealth, a Champion is unlikely to be guided by further greed, avarice or selfish motives. This may, therefore, increase the likelihood that the community will choose to listen to and be led by that Champion.

25.3.8 Political Strength of the Champion

In the Indian village context, with democratic decentralisation and the local self-governance ethos gaining ground everywhere consequent upon the 72nd and 73rd Constitutional Amendments, it can be argued that the Institutional Champion is more likely to succeed if he/she has political strength. But, as argued earlier, it can also be argued here that political strength may contribute to the social position of an individual and vice versa. It is likely that this may spur a Champion to volunteer to lead a large-scale social change which he/she perceives as beneficial to the community. There is, however, the danger of "elite capture" because of political power.

25.4 Testing the Framework

The framework was again tested in the context provided in the four case studies.

Analysis of results revealed that of the *nine* possible attributes listed above, the attributes of *(a) high degree of involvement, (b) high to medium degree of embeddedness, (c) high level of empathy (d) high degree of commitment, and (e) a high to medium degree of selflessness* appeared to trump other attributes. The studies also revealed that the *institutional position of the Champion need not be high* though *the level of literacy of the Institutional Champion was high*. This runs counter to the argument made by Battilana that the institutional position of the Champion is of paramount importance. The relatively low significance of embeddedness which has emerged from the case studies indicates that an outsider, or external Institutional Champion, could also succeed in convening people and spearheading a social change process, provided he/she has high degrees of the other significant traits mentioned above. It was also seen that in all four cases, women took the lead as institutional champions. This raises a very pertinent question – are women the most likely or preferred candidates for social change?

25.5 Process of Volunteerism (A Dialectical Explanation)

During the early days, the Institutional Champion necessarily has to cede some conceptual space to the other collaborators to keep them interested in the larger collaboration and also to build their stake in the ultimate problem-solving or solution-seeking endeavour, since the invisible hand of "power over" would be driving each of the collaborators then. Once having volunteered to lead the change process, the Institutional Champion must strive hard, and with the utmost sincerity, to not only be guided him/herself by the "power with" conception of sharing power with the collaborators, but also try to internalise this concept by slowly convincing the collaborators of the futility of a "power over" concept compared to a scenario of empowerment through collective action. This is, however, much more complex and difficult to achieve than Ostrom's collective action theory (Ostrom 1994), as the latter assumed *homogeneity* in the group and *small size* of the group as important preconditions for collective action to take place. Ostrom also did not deal with power asymmetries in the rural context, which is at the root of "elite capture" in rural ecosystems in India. At best, Ostrom assumed power homogeneity or equity in her Institutional Analysis and Design (IAD) framework for collective action. But for the community as a whole, with heterogeneous collaborators and a relatively large size, this collective action has to take the form of social mobilisation. Therefore, the role of 'trust' as a variable is crucial because that alone can keep the collaboration going and the size and heterogeneity variables becoming less pronounced. Trust comes from power equality and equity when fairness and justice are guaranteed. It is at this stage that the Institutional Champion will not only have to internalise the "power with" concept, but will also need the moral and spiritual authority to spread such awareness. Hence the *level of moral and spiritual mettle* is an important factor in prompting an Institutional Champion to volunteer and be capable of convening both people and resources to find a solution to a nagging problem within the community. But moral strength comes from selflessness, and vice versa. Hence, for a social Institutional Champion to succeed, it may well be that, as indicated earlier, the *level of selflessness* is probably the most crucial factor in driving forward a successful change activity.

25.6 Conclusion and Way Forward

For sustainable systems, existing institutions have to change as they are based on selfishness, opportunism and interest-maximisation. This change can occur endogenously through institutional entrepreneurship by an inspired agency which volunteers to lead the change. This is a non-reciprocal relationship not based on any self-interest. Since not all individuals in a community volunteer to help effect change, the volunteer or the change agent or the champion needs to have particular characteristics. Those actors in the village community whose innate behavioural characteristics are towards a higher degree of selflessness (who love to sacrifice for others' benefits) are spurred towards volunteerism and are therefore ideally suited to be groomed as community champions. Of these, the attribute of 'selflessness' assumes importance, as it is linked to altruism. The question is how to identify such volunteers who can act as community champions. Can selflessness be imbided in an individual or a community in order to achieve sustainable systems? Can there be a community leadership development programme? These are matters for exploration and further study.

References

Acemoglu, D.; Johnson, S.; Robinson, J., 2004: "Institutions as the fundamental cause of long-run growth", National Bureau of Economic Research Working Paper No. 10481, at: https://www.nber.org/papers/w10481.

Acemoglu, D., 2005: "Unbundling Institutions", in: *Journal of Political Economy*, 113(5): 949–995.

Allen, N.J.; Rushton, J.P., 1983: "Personality characteristics of community mental health volunteers: A review", in: *Journal of Voluntary Action Research*, 12(1): 36–49.

Barley, Stephen R.; Tolbert, Pamela, 1997: "Institutionalization and Structuration: Studying the Links between Action and Institution", in: *Organisation Studies*, 18: 93–117.

Battilana, J.; Leca, B., 2008: "The role of resources in institutional entrepreneurship: Insights for an approach to strategic management combining agency and institutions", in: Costanzo, L.A.; MacKay, R.B. (Eds.), *Handbook of Research on Strategy and Foresight* (Norwell, MA: Kluwer).

Battilana, J., 2006: "Agency and institutions: the enabling role of individuals' social position", in: *Organisation*, 13(5): 653–676.

Berger. G., 1991: *Factors explaining volunteering for organizations in general and for social welfare organizations in particular* (Unpublished doctoral dissertation, Heller School of Social Welfare, Brandeis University, Waltham, MA).

Berger, P.; Luckman, T., 1966: *The Social Construction of Reality. A Treatise in the Sociology of Knowledge* (New York: Penguin).

Bierhoff, H.W., 2005: *Prosocial Behaviour* (Hove: Psychology Press).

Bussell; Forbes, 2002: "Understanding the volunteer market: The what, where, when and why of volunteering", in: *International Journal of Non-profit and Voluntary Sector Marketing*, 7(3): 244–257.

Chambers, Robert, 1998: "Beyond 'Whose Reality Counts?' New Methods we now need?", in: *Studies in Cultures, Organisations and Societies*, 4: 279–301.

Cnaan, R.A.; Handy, F.; Wadsworth, M., 1996: "Defining who is a volunteer: Conceptual and empirical considerations", in: *Nonprofit and Voluntary Sector Quarterly*, 25(3): 364–383.

Costanzo, L.A.; MacKay, R.B. (Eds.) 2009: *Handbook of Research on Strategy and Foresight* (Cheltenham, UK; Northampton, MA: Edward Elgar Publishing).

DiMaggio, P.J.; Powell, W.W., 1983: "The iron cage revisited: Institutional isomorphism and collective rationality in organizational fields", in: *American Sociological Review*, 48(2): 147–160.

DiMaggio, P.J., 1988: "Interest and agency in institutional theory", in: Zucker, L.G. (Ed.), *Institutional Patterns and Organizations* (Cambridge, MA: Ballinger): 3–22.

DiMaggio, P.J.; Powell, W.W., 1991: "Introduction", in: Powell, W.W.; DiMaggio, P.J. (Eds.), *The New Institutionalism in Organizational Analysis* (Chicago: Chicago University Press): 1–38.

Eisenhardt, K.M., 1989: "Building Theories from Case Study Research", in: *The Academy of Management Review*, 14(4): 532–550.

Friedland, R.; Alford, R.R., 1991: "Bringing society back in: Symbols, practices, and institutional contradictions", in: Powell, W.W.; DiMaggio, P.J. (Eds.), *The New Institutionalism in Organizational Analysis* (Chicago: University of Chicago Press): 232–263.

Giddens, A., 1984: *The Constitution of Society* (Berkeley, CA: University of California Press).

Holm, P., 1995: "The dynamics of institutionalization: Transformation processes in Norwegian fisheries", in: *Administrative Science Quarterly*, 40(3): 398–422.

Kark, R.; Van Dijk, D., 2007: "Motivation to lead, motivation to follow: the role of the self-regulatory focus in leadership processes", in: *Academy of Management Review*, 32(2): 500–528.

Kirk, P.; Shutte, A.M., 2004: "Community leadership development", in: *Community Development Journal*, 39(3): 234–251.

Langone, C.A., 1992: "Building community leadership", in: *Journal of Extension*, 30(4): 23–25.

Levi, Margaret, 1990: "A logic of institutional change", in: Cook, K.S.; Levi, M. (Eds.), *The Limits of Rationality* (Chicago: University of Chicago Press): 401–418.

Lounsbury, M.; Crumley, E.T., 2007: "New Practice Creation: An Institutional Perspective on Innovation", in: *Organization Studies*, 28(7): 993–1,012.

Meyer, J.; Rowan, B., 1977: "Institutionalized organizations: Formal structures as myth and ceremony", in: *American Journal of Sociology*, 83: 340–363.

Oliver, C., 1992: "The antecedents of deinstitutionalization", in: *Organization Studies*, 13: 563–588.

Ostrom, Elinor, 1990: *Governing the Commons. The Evolution of Institutions for Collective Action* (Cambridge: Cambridge University Press).

Rochester, C., 1999: "One size does not fit all: four models of involving volunteers in small voluntary organisations", in: *Voluntary Action*, 1(2) (Spring): 47–59.

Scott, W.R., 1995, 2001, 2008: *Institutions and Organizations* (Thousand Oaks, CA: Sage).

Scott, W.R., 2004: "Institutional theory", in: *Encyclopedia of Social Theory* (Thousand Oaks, CA: Sage): 408–414.

Sethi, J.D., 1978: *Gandhi Today.* New Delhi: Vikas Publishing House.

Seo, M.; Douglas Creed, W.E., 2002: "Institutional contradictions, praxis and institutional change: A dialectical perspective", in: *Academy of Management Review,* 27(2): 222–247.

Smith, D.H., 1994: "Determinants of Voluntary Association Participation and Volunteering: a Literature Review", in: *Non-profit and Voluntary Sector Quarterly,* 23: 243.

Virmani, A., 2005: "Institution, Governance and Policy Reforms: A Framework for Analysis", in: *Economic and Political Weekly,* 40(22): 2,341–50.

Wilson, J.; Musick, M., 1999: "The effects of volunteering on the volunteer", in: *Law and contemporary problems,* 62(4): 141–168.

Zucker, L.G., 1983: *Organizations as Institutions* (Greenwich, CT: JAI Press).

Zucker, L.G. (1991: "Postscript: Micro foundations of institutional thought", in: Powell, W.W.; DiMaggio, P.J. (Eds.), *The New Institutionalism in Organizational Analysis* (Chicago: University of Chicago Press): 103–107.

Chapter 26
Peoples' Power and Processes in Ushering Changes: Cases from Bihar

G. Krishnamurthi, Jaya Kritika Ojha and Amrita Dhiman

Abstract The power of people and processes in mobilising a community and facilitating community institutions can usher in changes in communities and improve the lives of people. This chapter presents a case study based on the project of the Aga Khan Rural Support Programme (India). The project focused on establishing strong processes and robust people's institutions to provide clean, safe water for drinking and regular irrigation so that people could make efforts on their own to improve their own livelihoods, agricultural production and market linkages for enhanced household income, and simultaneously, to ensure good health, nutrition and sanitation in hamlets (known locally as *tolas*).

The case study was carried out using focus group discussions (FGD) and consultative meetings with beneficiaries in selected hamlets (*tolas*) spread over four blocks of the Muzzafarpur and Samastipur districts of northern Bihar, India, the project areas of Aga Khan Rural Support Programme (India). The case study presents processes, procedural and operational interventions taken up by the programme which brought about considerable changes in rural communities and established robust institutional structures for sustainable development.

Keywords People · Processes · Systems · Interventions
Institutions · Change · Sustainable development

G. Krishnamurthi, Senior Professor and Dean, DMI, Patna, Bihar, India; Email: gk1949@gmail.com

Jaya Kritika Ojha, Associate Professor, DMI, Patna, Bihar, India; Email: jkojha@dmi.ac.in

Amrita Dhiman, Associate Professor, DMI, Patna, Bihar, India; Email: adhiman@dmi.ac.in

© Springer Nature Switzerland AG 2019
A. K. J. R. Nayak (ed.), *Transition Strategies for Sustainable Community Systems*,
The Anthropocene: Politik—Economics—Society—Science 26,
https://doi.org/10.1007/978-3-030-00356-2_26

26.1 People, Processes and Sustainable Community Systems

Change is both natural and vital. Change becomes positive when 'People' are drivers of the change and 'Processes' are the means to realise and usher in the desired 'Change'. Many development agencies perceive the state of Bihar as the ground for experimentation of such changes, be it socio-economic or political, as the indicators of socio-economic conditions are low compared with those of several other states of India. People are agents of change; people are also actors, the actual doers. The design of development interventions for sustainable community systems can be based on collective will and collective efforts through participatory development processes. Gandhi called it *"Gram Swaraj"* (village self-rule), that aims at *Sarvodaya* (the rise of all). He placed "people first" (Gandhi 1962*)*.

26.1.1 The Context

Bihar has always been considered a resource-rich state, especially endowed with abundant water resources, primarily because of the presence of the river Ganges and its tributaries. However, the reality is that agriculture in Bihar is still largely dependent on rainfall across all thirty-eight districts in the state. It is the year-to-year variation in rainfall which causes floods or droughts. In drought, the level of surface water goes down. The reports of water testing show that the surface water and water from shallow bore-wells in many villages in the districts of Muzaffarpur and Samastipur has become contaminated, toxic, with many impurities and heavy metals. It is also known that water in many parts of the state is contaminated with arsenic, though reports from different sources are not always in agreement. It is also believed among some that arsenic has entered the food chain. The contaminated water poses serious health threats. The worsening situation of water in north Bihar creates problems for people with regard to both their livelihoods and their health (AKRSP(I) 2011, 2012).

The Aga Khan Rural Support Programme (India) [AKRSP(I)] commenced a project called 'Water for Good Health and Sustainable Livelihoods' in May 2011. The project covered 40,000 women and men from weaker sections of rural society, spread across one hundred hamlets (*tolas*) in the *Muraul* and *Sakra* blocks of the Muzaffarpur district and the *Pusa* and *Tajpur* blocks of the Samastipur district of Bihar. The project could mobilise people in the communities, harness their energy and resources, and establish processes that could lead to desired changes. The case presents a clear demonstration of the rapport between the communities and a development organisation, namely AKRSP(I), genuinely concerned with the well-being of the communities, and the way it has been able to unleash the hidden energy of the people for their own good. Bihar is generally perceived as a state dominated by a populace which is divided into multiple social groups and which values individualism and individual action over any form of collective action, however profound the results of any collective endeavour might be. The results realised by AKRSP(I) show clearly its understanding of the social structures in the one hundred hamlets and its ability to work through such structures.

26.2 The Case Study

The case presents several diversified processes and procedural and operational interventions taken up by AKRSP(I) which made spectacular changes to the lives of rural communities, evidenced by an increase in household incomes and food availability; better availability of safe drinking water to community members; improved health and sanitation; increased awareness of different development programmes and schemes operating in their areas, and their convergence/integration with the programme of AKRSP(I); and the establishment of robust institutional structures for sustainable development.

The case study is based on field visits, *focus group discussions* (FGD), personal interviews and consultative meetings with beneficiaries, organised in the selected hamlets (*tolas*) and villages of the project areas of AKRSP(I) in the Samastipur and Muzaffarpur districts of Bihar as a part of the end-term evaluation of the project 'Water for Good Health and Sustainable Livelihoods' undertaken by AKRSP(I) under a contract with the European Union (EU). AKRSP(I) launched the project in May 2011 and brought it to a close in April 2016.

The project 'Water for Good Health and Sustainable Livelihoods' had the overall objective to 'Improve livelihoods and health status of poor and marginalised communities living in *flood-prone areas of north Bihar* through the management of available water resources in an integrated, sustainable and ecologically relevant manner, with appropriate support of the state'. The specific objective of the project was to 'Demonstrate a model for integrated water management in a *flood-prone area* that is evidence-based, community-led and state-supported, and suitable for wider adoption'. It was envisaged that the achievement of these objectives would be measured by an increase in household incomes, a reduction in disease, an improved habitat and better management of public services.

The project had its rationale in the fact that an estimated 37% of rural Bihar is affected by floods every year, and almost a million hectares are permanently waterlogged. This situation reduces the area available for agricultural cultivation, which leads to small land-marginal farmers becoming landless, and they are often forced to migrate to find work. Domestic health is also negatively impacted by vector-borne diseases from poor surface drainage among dense habitation patterns, and from contamination of drinking water from handpumps, a situation that worsens significantly during the seasonal rains.

The project was launched to address the waterlogging of fields, and the water and environmental health issues among habitations, through locally organised responses. These were meant to:

- Raise the incomes of poor households by creating waged employment;
- Improve food security and nutrition by bringing land into production;
- Improve health status through behaviour change and environmental and infrastructure improvements; and
- Mobilise sustainable organisational structures.

The project was designed to give emphasis to structured analysis and learning from results and approaches as the basis for engagement with other practitioners in flood-prone areas and with governance bodies whose job is to help translate relevant policies into practice.

The project was designed to cover 40,000 women and men from scheduled caste (SC), Muslim minority and other backward classes (OBC) spread over a hundred *tolas* in the Muraul and Sakra blocks of the Muzaffarpur district and the Pusa and Tajpur blocks of the Samastipur district of Bihar. The project revolved around water and the diverse roles it plays in the everyday lives of people in the area, from livelihoods and agriculture to health.

However, it came out during the study that the ground reality was quite contrary to the context for which the project was initiated. The two districts, which were assumed to be flood-prone, are in fact drought-prone, as uniformly stated by the beneficiaries in all FGDs. This has also been borne out by the rainfall data for the two districts. It is now known that the rainy days per annum, which used to number fifty-two in Bihar in the past, have been progressively coming down to between fifteen and twenty-five during the last five to seven years.

AKRSP(I) addressed this issue during the project period with a two-pronged approach. As a preventative measure for any possible waterlogging, AKRSP(I) facilitated the construction of drainage structures through community-managed interventions. Simultaneously, recognising the development of drought-like situations, the project agency encouraged community-managed group irrigation models.

This chapter, based on a study of the project, establishes the fact that the *power* of people and *processes* in bringing about change in the communities has been vital in facilitating sustainable community systems. Every village in the two blocks where the study was undertaken is divided into hamlets/*tolas*, and each hamlet/*tola* is generally inhabited by residents of a specific social group. For example, there are Paswan *tolas*, Kushawa *tolas*, Muslim *tolas,* Sahni (fisherfolk) *tolas*, Scheduled Caste (SC) *tolas*, Other Backward Classes (OBC) *tolas*, and so forth. In ushering in the change, the unit of change has been the entire community (of a hamlet/*tola*) residing in a village. As Mahatma Gandhi writes in *Village Swaraj*, "Every village has to be self-sustained and capable of managing its affairs even to the extent of defending itself against the whole world" (Gandhi 1962: 38).

The study covered a sample of eight *tolas*, and the results were similar across all of them. This chapter describes the FGDs conducted in three *tolas*, namely *Sahni Tola, Bankurva Tola and Chakhaji Tola*, and presents the conclusions of the remaining FGDs.

26.3   The Interventions in *Sahni Tola*

26.3.1   *Water, Livelihoods and Convergence*

An FGD was conducted with the representatives of *Sahni Tola*, which comes under Khaira Village in the Pusa block of Samastipur district. The issues discussed with the targeted group were water-based livelihoods, water and health and convergence.

Water: In the *Sahni Tola*, community members said, "This year is again a bad year for water. Summer is at its peak. There is no water from hand-pumps; all the water bodies have dried up." An SHG member remembering earlier days said, "A few years back the situation was more horrifying; people used to wait for hours in queues at hand-pumps for water. Lack of water used to create a threat to lives." She continued by saying:

> Now things are comparatively better. We have water because AKRSP(I) supported our community in building the community bore-well, the water tank, and the water supply points. The water project hand-held the community, built our capacity through interactions and training; now we know how to manage the available water. Even people from nearby villages are coming to our village in search of water.

AKRSP(I) started work in the village on 28 December 2013. The organisation conducted an initial baseline survey in Khaira village and, along with the community, decided to construct a water tank. Initially, the major challenge was to convince the community to co-operate and mobilise them to contribute towards the structure. With continuous efforts from the community leaders and AKRSP(I), the community was convinced to build the water tank. The community constituted a water committee named *Maa Manapurna Peyjal Upyog Samiti* (water user committee), *Khaira*. SHG groups of the village *tola* played a very important role in the process. They sanctioned loans to members for making contributions. The community spent Rs. 135,000 on the water structure by way of raw materials like sand, cement and bricks, plus about 2,200 feet of delivery pipe with associated fittings and labour charges. Fourteen water supply points have been constructed in the *tola* and one point caters for about ten families. Every household in the *tola* pays Rs. 20 for electricity and maintenance per month. Initially, they had decided to use the water for drinking purposes and food preparation. However, when no water supply was made available by the local administration the community started using the water for rearing cattle as well. The adjacent villages also take water from Khaira water supply points.

26.3.2   *Health, Sanitation and Hygiene*

A major result came out during the FGD; the village had earlier lacked sanitation facilities and there was no awareness of the need for sanitation facilities. People used to fall ill frequently. There were no platforms around hand-pumps and water points.

An important activity taken up by AKRSP(I) under the project, with the participation of the community, was the construction of toilets in every household. As reported by community representatives, initially the residents of the *tola* were not very supportive of the idea of constructing toilets. Gradually, however, they started appreciating the idea after realising the benefits of having toilets in their homes. They say that women members of the family are quite happy. Currently, there are 160 fully constructed toilets in the village of 170 households. All the toilets are in use. Community members realise the need for behavioural change in their toilet habits, which does take some time. The community is getting united and committed to improving their pattern of life.

Another important result came out during the FGD. Before the project, people used to suffer from waterborne diseases like diarrhoea, gastritis, cholera, dysentery, typhoid, scabies, trachoma, schistosomiasis, Guinea worm disease, malaria and *kalaazaar* (black fever). An elderly lady of the *tola* mentioned that she was happy that there are no more water-based diseases in her *tola* now. Previously, on average at least one person per family used to get sick due to contaminated water, and families had to pay a huge amount for treatment and medication. The incidence of the diseases has drastically come down following the project, and there have been substantial savings on health-related expenses. The project started with the collection of water samples from different water sources and hand-pumps by AKRSP (I), which had them tested. It was found that 75% of the water sources did not contain potable water. The water sources have been given permanent colour codes of green, red and yellow in accordance with the level of contamination and impurities. Green means 'safe to drink', and red shows that the water source is polluted and water from the source should not be used. Yellow indicates that the water source is under threat of turning non-potable. Suggestions were also given by AKRSP(I) to treat the water before using it.

26.3.3 Agriculture and Livelihood

When water was plentiful the community members cultivated two crops every year (*Kharif* [summer crop] and *Rabi* [winter crop]), but because of inadequate water supplies have now limited themselves to the cultivation of vegetables. With the support of AKRSP(I), they have now created kitchen gardens for the cultivation of spinach, okra, radishes, pumpkins, bitter gourds and cucumber, etc. In the *Sahni Tola*, community members also have landless vegetable gardens. Families sell vegetables if they have surplus from their kitchen gardens.

The community members have received training to make *Amritpani*, a spray mix of nutrients for crops (vegetables and wheat). Solar lights have also been given by AKRSP(I), so that they can work in their fields even when it is dark.

Bore well water is used for irrigation, but the bore wells are drying up. The village has 25–30 private tube wells for irrigation. The residents pay Rs. 120 to 130 per hour to use water for irrigating one *kattha* (a unit of land measurement, equal to 1,361 sq. ft.).

During the project implementation, AKRSP(I) conducted many training programmes on vermicomposting. The communities took an active part and nearly thirty per cent of the residents from the *tola* now practise vermicompost. They use vermicompost on their fields and sell the surplus at the rate of Rs. 600–650 per quintal. The community now believes that the use of vermicompost increases crop production, and the members now spend less on chemical fertilizers, resulting in monetary benefits. They have also found, during soil testing, that the quality of soil has improved in the process.

The community members expressed their concern over the miserable situation of aquaculture. Fishery has been on the decline as a livelihood option for the last seven years because the rainfall has been declining heavily. They now buy fish from Andhra Pradesh for personal consumption.

26.3.4 Migration

Due to lack of rain and water, people are migrating to places like Nasik for wage labour in grape, tomato and pomegranate packing units, and to Nepal for mango packing. Around 500 people migrate every season. Each migrant saves about Rs. 25,000–30,000 every season and sends it to the family.

26.3.5 Sustainability of the Interventions of the Project

The community members have undergone various capacity-building and training programmes and are aware of their enhanced capabilities; they now feel sufficiently empowered to sustain the initiatives ushered in during the project. During the FGD, they affirmed that the systems and processes have now been fully formulated and the need to follow them has also been well-established.

To conclude, the community feels satisfied with the project, and has not only accepted the changes brought about by AKRSP(I) through the intervention, but has been appreciative of the harmony and peace in the village now. The people feel the urge to retain the good practices that have been the harbinger of comfort and satisfaction in their lives.

26.4   The Interventions in *Bankurva Tola*

26.4.1   *Collective Marketing of Produce*

An FGD was held with the members of *Srimahavir Kisan Vikas Samiti* (farmers' committee/KVS) and *Sinchai Vikas Samiti* (irrigation committee) in *Bankurva Tola* of Chandauli village. The members said that KVS started functioning in the year 2009 with the support of AKRSP(I). The KVS members used to have regular monthly meetings to discuss the major issues and challenges of the *tola*. In one such meeting, it was realised that irrigation was a major challenge, and farmers concluded that if irrigation facilities could be improved in the *tola* and in the village, better economic conditions could be realised. AKRSP(I) supported the farmers throughout the project with the following activities:

First, the KVS members received vermicompost training. Initially, ten farmers were trained, constructed the vermicompost pit and started vermicomposting with the earthworms provided by AKRSP(I). They used the vermicompost on their own fields and sold the surplus in the village market. They said that because of vermicompost they had good produce. Following this, other farmers also started vermicomposting and found it a good livelihood option too, as the surplus compost could be sold for between Rs. 450 and 480 per quintal at market. Vermicompost is being used by many farmers, especially for vegetable farming; in addition, they have also gained knowledge about vermicompost from different government schemes.

A large variety of seeds, such as pumpkin, bitter gourd, okra and spinach, were distributed by AKRSP(I) at subsidised rates through the KVS; this has resulted in good yields and produce and good returns.

Gradually, the KVS members realised their inability to market the increased produce and were getting low returns. The KVS member farmers started searching for new markets to maximise their profits. It came out during the FGD that the farmers discovered the existence of three main *mandis* (markets) where farmers can market their produce during the *parwal* (pointed-gourd) season – *Samastipurmandi*, *Pusa Road mandi* (market) and the *Motipurmandi* (market) – and marketed their produce collectively. Initially, everything went well; however, over time, partly due to market fluctuations and partly because of the time needed to collect the vegetables at the *tola* level and transport them to market, the efforts started drifting. Further, the middlemen and traders of *mandis* started considering the farmers as their competitors and deliberately manipulated the prices of vegetables, leading to violent fluctuations, so that the farmers would not be able to establish a firm footing. As a result, they had to sustain some losses (about Rs. 700–800) and were aggrieved and dissatisfied with the fluctuation and losses they had to bear at *samiti* (committee) level. No solution could be found, and the farmers started doing their individual marketing in the local market. The farmers have now realised the power of being united and have formed a federation, which they are planning to register soon as a Farmer's Producer Organisation. They see it as an organised and long-term solution. They also plan to market vermicompost.

26.4.2 Management of Water for Irrigation

The farmers said that previously they used to dig shallow bores of ten feet but could irrigate only one *kattha* of land per hour. During rains, this system would not work. Under the project executed through AKRSP(I) – community collaboration – a 2,200 feet underground pipeline was laid for supplying water from the deep tube well. A generator was also installed for back-up. The bore-well covers a 22-acre area owned by eighty farmers. With the timely availability of water now, farmers are getting enough water to irrigate three *kattha* lands in one hour, with an expenditure of Rs. 90–120 (depending on whether it is powered by mains electricity or the generator). They say that they can save a substantial amount on irrigation and do not have to worry about water anymore. The water is now available year-round to all the member farmers for irrigation, the cost of which has been reduced by two-thirds. As a result, the farmers are producing good quality vegetables all year.

The farmers' contribution towards the installation of the bore-well was Rs. 117,000. AKRSP(I) supported their efforts with Rs. 345,000 for the installation of the generator and the construction of the shed. The generator was installed one year ago. The bore-well is fitted with a 5 hp motor and the generator has a 10 hp capacity. Around Rs. 500,000 was spent on constructing the community irrigation system by farmers and AKRSP(I). All the accounts are maintained well. Registers like the minutes of meetings, the farmers' ledger and the usage register are kept up-to-date. An operator for ensuring water supply to farmer members is paid Rs. 20 per hour for his services. The monthly meeting takes place regularly to discuss the *samiti* (committee) accounts and related matters. At times, the *samiti* allows the farmers to pay for their water usage after their produce has been harvested. The *samiti* also once decided to give the first water as an advance. But the norm is that before the next water demand farmers must pay for the amount of water previously taken as an advance. Efforts are being made to benefit all the farmers of the *tola*.

During the FGD the community members further said that an alternative drinking-water facility is not available in the *tola*, but water from the bore-well is used for irrigation only.

The *samiti* members said that waterlogging was not a problem anymore; in contrast, all water sources have dried up, with the drought prevailing for the last four years. AKRSP(I) has also organised many training programmes for the village on good quality seeds, diseases, crop varieties, farming inputs, pest control and vermicompost, etc.

26.4.3   Community Health

The community representatives said that previously they used to suffer from many gastro-intestinal issues. The reason was unhygienic surroundings and open defecation. AKRSP(I) has made many interventions to deal with the health and sanitation issues. Toilets have been constructed. Awareness programmes and exposure visits have also been organised. Thirty toilets have already been constructed in the *tola* and ten more are to be constructed soon. The members of the *tola* said that a normal toilet costs Rs. 18,000–20,000, but the AKRSP(I) toilet design costs only Rs. 10,000–12,000. Individual households received material valued at about Rs. 4,000 from AKRSP(I) to build toilets. Continuous efforts are being made by the community and AKRSP(I) to build toilets in *tolas*. The community believes that there is better sanitation and cleanliness in the village than before, and toilets are especially beneficial to girls and women. It is a big step that ensures their personal dignity. Gandhi wrote in *Harijan*, 1942, "If rural reconstruction were not to include rural sanitation, our villages would remain muck-heaps" (Gandhi 1959: 22).

26.5   The Interventions in Chakhaji Tola

26.5.1   Irrigation

Representatives of the *Sinchai Vikas Samiti* (irrigation committee) of *Chakhaji Tola* of Bankurwa village of Chandauli Gram Panchayat in the Pusa block of the Samastipur District participated in an FGD to discuss issues of irrigation, health, nutrition and sanitation.

Yatin, an active young leader of *Sri Mahaveer Kisan Vikas Samiti* (committee), Chandoli, and a member of *Sinchai Vikas Samiti* (committee), Bankurwa, gave a vivid account of the work of AKRSP(I), which began in his *tola* in the year 2009. He said that farmers previously had no clue about the proper management of a farm, produce or market. AKRSP(I)'s professionals distributed seeds and conducted training on topics like collective irrigation, the use of high-yield seed varieties, fertilizers, plant diseases, zero tillage, conservative agriculture, mulching, market dynamics and mineral mixture. Besides all these, they are aware of the importance of soil-testing.

The community dug a bore-well for irrigation purposes. It is 270 feet deep with an 11 × 9 feet maintenance room and a generator. Ninety-two families draw water from the community well for drinking purposes and 126 farmers use the community bore-well for irrigation purposes. The bore-well is shared by two *Kisan Vikas Samiti* (farmers' committees) and two *tolas*, spread over two villages. The *Sinchai Vikas Samiti* (irrigation committee) earned Rs. 10,000 last year. They have also built the *samiti* (committee) office and plan to build a seed store (*godown*) and another store for bulk fertilizer.

26.5.2 Collective Purchase of Inputs

Before AKRSP(I) started its work here, middlemen had been exploiting the farmers by taking their produce at low prices. The farmers gradually started getting organised for the collective buying of good seed varieties and fertilizers on favourable terms. The fertilizer companies now come directly to the village, as it is a potential bulk buyer of fertilizers. AKRSP(I) has also promoted a net house for seeds and saplings.

The community members were not aware of vermicompost prior to the entry of AKRSP(I). The farmers, who previously used to spend a lot on chemical fertilizers, now use vermicompost, especially for *parwal* (pointed gourd) crops. The switch-over to vermicompost has resulted in a reduction in farming-related expenses.

26.5.3 Community Health and Education

Until three or four years ago, the community members simply accepted the fact that there were massive water-borne diseases, especially intestine-related ones. AKRSP (I) conducted surveys and sent off test-samples of water from different water sources of the *tola*. Community members were made aware that the upper layer of water was infected and contaminated, and that water from hand-pumps was not fully fit for human consumption. According to *tola* members, the AKRSP(I) professionals educated them about contamination in water and explained the colour coding which they had done on the hand-pumps. The AKRSP(I) team informed the community that groundwater is clean and safe to drink. Now there are not many incidents of water-related diseases in the *tola*, as people are using clean water.

The intervention of the smokeless *chulha* (stove) is comforting to women and saves them from lung and eye infections. People are becoming aware of the importance of cleanliness, sanitation and hygiene for good health.

Farmers need to reach markets early to sell their vegetables. AKRSP(I) gave solar lights to community members so that farmers can work comfortably in their fields and harvest fresh vegetables early in the morning when it is still dark. The community members who use solar lights have to pay Rs. 2 per day as maintenance charges for charging of batteries and other repairs. Solar lights are also helping children to study in the evenings.

With some community contributions, AKRSP(I) supported the creation of a computer centre with the required furniture and a computer teacher. The youth in the *tola* are getting trained and many of them have started working as data entry operators outside the village and also in cities.

The community representatives were delighted that their self-confidence was growing, and they were getting stronger, made possible by the project. The community is confident that its members can collectively work for everyone's benefit and betterment. The community accepts that the members have been witnessing a paradigm shift towards development, but a lot needs to be done regularly in future as well.

To conclude, the interventions facilitated by AKRSP(I) undoubtedly prove that keeping faith in people and their capabilities can accelerate the processes of sustainable development. Margaret Mead has rightly commented, "Never doubt that a small group of thoughtful, committed citizens can change the world. Indeed, it is the only thing that ever has."

26.6   The Way Forward

26.6.1   People First …

The people of the villages of Muzzafarpur and Samastipur now have faith in their own capabilities and the practices are getting formulated into processes. People are well aware that it is only through their participation and willingness to bring change that they can do something that was beyond their imagination a few years ago. This is what Gandhi said: "Be the part of change you want to see."

Gandhi suggested, "It should be the inborn responsibility of every individual and institution to play a role in social (community) work." AKRSP(I) in Bihar has been the facilitator, continuously monitoring and extending support wherever people needed. The community understands that it is their own power and potential that they have been executing to bring about the desired changes. The most important thing has been that the people have been the actors, players, executors, and reviewers of their activities. AKRSP(I) has defined its own role by going to the people and inspiring them to become change-makers.

People have been actively participating in the step-by-step processes, watching their own errors, failures, improving upon them and then facilitating collectively to find the solutions. As Amartya Sen puts it, "…so many of our abilities to do things depend on interaction with each other." Thus, the gradual transformation of the situation has led to the betterment of people's lives, enabling them to become mini-powerhouses through useful practices which combine technology, community participation, collective decision-making and action, and governance and management systems and processes for sustainable transformation in an effective way.

If people are enabled to decide to work collectively, they can change outdated practices and initiate a series of actions of development aimed at fulfilment of their aspirations and realisation of their dreams. The individuals, institutions and interventions can bring in the desired changes to create sustainable community systems and processes, for the well-being of all – People and Planet.

References

AKRSP (I), 2011: *Study on Quality of Life: Chaur Area of Muzaffarpur and Samastipur, Bihar* (AKRSP (I)).

AKRSP (I), 2012: *Baseline Report on Water for Sustainable Livelihood: Bihar* (AKRSP(I)).

APMAS, 2014: *Quality and Sustainability of Self Help Groups in Bihar and Odisha* (Hyderabad, APMAS); at: https://www.nabard.org/Publication/QualityandSustainabilityofSHGsinBiharand Odisa (19 June 2016).

Byrne, L., "Interview: Amartya Sen on power to our citizens", at: https://liambyrne.co.uk/ research_archive/interview-with-amartya-sen-on-power-to-our-citizens-sept-2009/.

Datta, A., 2016: "Migration, remittances and changing sources of income in rural Bihar (1999– 2011)", in: *Economic and Political Weekly*, 51: 31.

Gandhi, M.K., 1959: *Panchayat Raj* (Ahmedabad: Navjivan Publishing House).

Gandhi, M.K., 1962: *Village Swaraj* (Ahmedabad: Navjivan Publishing House).

Chapter 27
Summary and Way Forward

Amar K. J. R. Nayak

This concluding chapter provides a brief summary of the various discussions and observations across different chapters of this volume and derives some sense of the future pathways towards facilitating sustainable community systems.

While the Introductory chapter outlines the challenges of sustainability of community systems on various factors, such as relationships, institutions, production, organisation, and governance, the chapters in different sections further elaborate on the issues, practices and policies that either undermine or facilitate sustainability in general and go towards building sustainable communities in particular.

The introductory chapter outlines the key dimensions towards building sustainable communities and brings forth the critical factors under each dimension that affect sustainability. The chapter clearly articulates that the sustainability of a community system is based on three axioms, namely, we are inter-connected, we are interdependent and that the strength of an inter-connected and inter-dependent system is the strength of the weakest in that system. Accordingly, these basic axioms need to be taken into consideration while designing each of the five dimensions for ensuring sustainability within each dimension and their interaction with each other at different levels.

Most importantly, this chapter not only presents the various lock-in problems that make our modern economy and society have been tied up to but also provides a framework to choose simple re-combinations of factors as a future way forward to building sustainable small communities, the foundation for making the wider society safe and sustainable.

Part I of this volume discusses the issues relating to relationships, the core of any sustainable community system. The broad outline of relationships consist of five key factors: sense of interdependence, notion of capital and wealth, mental construct, morals and values, and faith and beliefs.

Amar K. J. R. Nayak, Professor of Strategic Management Xavier Institute of Management, Xavier University Bhubaneswar, Odisha, India; Email: amar@ximb.ac.in

© Springer Nature Switzerland AG 2019
A. K. J. R. Nayak (ed.), *Transition Strategies for Sustainable Community Systems*,
The Anthropocene: Politik—Economics—Society—Science 26,
https://doi.org/10.1007/978-3-030-00356-2_27

The articles in this section discuss these factors very deeply in different contexts and from different perspectives. They argue for an alternative ontological foundation for the present man to be a force for good in the anthropocene. The empirical evidence of this is being observed in the social solidarity economy landscape that rejects the ontological position of man as *homo economicus*. It argues that moral values are fundamental to building sustainable communities. It, however, cautions that moral values need to be anchored in reality rather than simply drawn from a mere imaginary world. From a Buddhist perspective, the world today, fuelled by greed, profiteering, competition, and selfishness, is moving away from sustainability. This only leads to environmental degradation, wastage and inequality. This section argues that policies need to be designed for the well-being of society as a whole; production needs be based on the needs of people rather than artificial demand created by the market system. In other words, a new relationship needs to be established between people and nature; one of co-operation, not exploitation. The Buddhist perspective also highlights the significance of the concept of interdependence for sustainability by drawing on the interdependence among the dhammas, as in Buddhism. The idea of interdependence has also been explained from the Judeo-Christian perspective by highlighting the interconnections among the concepts of people, peace, prosperity, partnership and planet. The section goes further by expanding human development from economic, social, political, and ethical development to art and spirituality for social transformation towards sustainability.

Part II of this volume is about Institutions, the most intricate dimension in the transition process towards building sustainable community systems. The broad outline of institutions consists of five key factors: norms and conventions, rules and regulations, principles of justice, interaction intensity, and institutional loading.

Institutions needs to be understood very deeply to be able to transit out of the lock-in problems that we are in, as it is the institutions, both informal and formal, that bind us or lock us together in different ways. Breaking away from these informal and formal institutions is usually difficult, as they are usually deeply rooted, interconnected and not so clearly visible on a day to day basis. Another critical challenge to understanding and resolving institutions has been that the factors of institutions are likely to change when we take either a top-down perspective or a bottom-up perspective. However, for sustainability, a bottom-up perspective would work out well compared with a top-down perspective.

The two articles in the book highlight the significance of the convergence of various schemes and programmes for greater efficiency. The first chapter draws this argument from a top-down perspective. The second chapter argues for convergence broadly from a top-down perspective, but it qualifies its argument for contextual flexibility in the convergence method, implying the need for a bottom-up view in the convergence approach.

Part III of this volume discusses the various technical factors and related policy issues of a sustainable agricultural system, the basic production function in any community system. The broad outline of a sustainable agricultural system consists of five key factors: water (moisture), soil, seeds, diversity, and ecology.

There are various methods of sustainable agricultural practices in different parts of the world. In India alone there are over half a dozen methods with different nomenclatures, such as organic farming, natural farming, natu-eco farming, zero budget farming, homa farming, spiritual farming, low-cost integrated sustainable agriculture (LEISA), etc. The basic principles behind these different types of agriculture for sustainability include one or more of the above five factors. The articles in this section discusses in depth three of the five critical factors or principles of a sustainable agriculture system at farm level: water, diversity, ecology. The section also presents simple methods of sustainable agriculture from a nutritional and environmental safety point of view. Further, the section also discusses scientific methods to assess the performance of integrated sustainable farming systems at macro levels.

In-situ water conservation, both traditional and improvised on farms, shows the potential to resolve recharging groundwater issues at individual farm level. The section also discusses the significance of traditional ways of effectively governing and managing water resources as a common property at village level. The science behind diversity and its significance to soil health and productivity through empirical evidence is very well brought out in this section. Further, the section also highlights the practice of smallholder farmers adopting diversification as an approach to reduce risk and safeguard their livelihoods. Forest accounting and ecological sustainability has also been covered in this section. Further, some approaches and alternatives to the mainstream industrial agriculture in the domain of Moringa (drumstick) and cotton cultivation have been discussed from nutritional and environmental perspectives. Finally, two chapters in this section discuss two scientific methods to measure and assess the performance of different agricultural systems.

Part IV of this volume discusses the various factors and related policy issues of organisations, with a specific focus on Farmer Producer Organisations. The broad outline of an organisation as a critical pillar of support to a community consists of five key factors: size, scope, technology, management and ownership.

The issue of organisational design from a sustainability point of view has been a recent phenomenon. The efficiency of large corporations in the business world during the last couple of hundred years has mesmerised most in the mainstream. Accordingly, there has been an all-out effort to replicate the organisational design of large corporations for all types of organisations in different sectors, collectives and co-operatives. While efficiency is a necessary condition, it is not a sufficient condition for sustainability. Therefore organisational design based on the approach of large corporations need not be the path to sustainability. On the contrary, with the greater efficiency and growth of large corporations, overall inequity in society has historically increased across the world. Taking a cue from this observation, there is a need to review organisational design very differently from a sustainable community system perspective.

This section contains contributions from some lead researchers from India on this subject. The issues that have been discussed include the origin of legal provision for farmer producer organisations in India, the key issues that have been missing for making producer collectives sustainable, and an illustration of a very successful small-sized, multi-product, multi-service co-operative that has been in operation for over seventy years and has been highly successful in meeting the needs of over 3,000 members of the community, consisting of seventeen contiguous villages, with a very robust financial performance. Importantly, an empirical research-based article provides a theoretical foundation to determine the optimal size of a producer organisation to achieve long-term sustainability.

Part V, the last section of this volume, discusses the various factors of governance with a special focus on grass-root-level community governance that can facilitate building sustainable community systems. The broad outline of governance includes five key factors: frequency of interactions; decision-making methods; a problem-solving approach; resource dependency; and governance architecture and responsibilities.

This section focuses on the critical factors or principles of governance at grass-root level for the sustainability of a community system. This is based on the premise that the factors and principles that work at the lowest level of governance of a society are the principles that can make a larger society sustainable, and not vice versa. Interestingly, all countries and nations around the world have a basic unit of governance, such as Gram Panchayat, Ward, Council, Union, at community level, therefore the framework of community governance is applicable across the world.

The section begins with an analysis of the very idea of community and the present status of community that is alarmingly wearing down. It emphasises the critical historical role of a community for our sustainability. This is followed by a discussion on the need for community champions to help build sustainable villages and communities. It distinguishes between entrepreneurs for personal profit and champions for social causes and argues that champions need to be selfless in order to be able to resolve the problems of the community. The last two chapters in this section present two persuasive cases on the stages, procedures and processes for systematically strengthening community governance towards building sustainable community systems.

Indeed, the volume has taken only the first step in sewing together the analysis of various factors of building sustainable community systems into five broad areas or dimensions of sustainability. It presents a well-knit theoretical framework for analysis and conception towards building sustainable community systems. It is the first volume on such a multi-dimensional theme; some sections have greater and deeper contributions and some topics need to be developed further in the future.

About the Author

Professor (Dr.) Amar K. J. R. Nayak is a tenured full time Professor of Strategic Management at the Xavier Institute of Management Bhubaneswar, the flagship Business School of Xavier University Bhubaneswar. He engages in research, teaching, training and policy advice to various organisations and institutions, including United Nations Agencies, the Government of India, state governments, corporations, non-profit organisations and others on Transition Strategies for rebuilding Sustainable Community Systems.

He uses both *deductive* and *contextually rich inductive methods* of research in his analysis and adopts *action research in rural communities from the design and systems perspectives* to test his hypotheses and theories on specific ecosystems.

He has published four books, two monographs, and over a hundred case studies and research articles. He teaches Strategic Management, Research Methodologies and Transition Strategies for Sustainability. Professor Nayak is an alumnus of the leading engineering, technology and management institutions in India, namely the National Institute of Technology Rourkela, the Indian Institute of Management Bangalore and the Indian Institute of Technology Kharagpur as well as Kobe University, Japan. He has also worked in various techno-managerial positions at for-profit corporations and non-profit organisations. He has travelled widely and lived in different parts of the world, including Asia, Europe, North America and South America for studies, lectures, seminars and conferences.

Address: Professor (Dr.) Amar K. J. R. Nayak, Xavier Institute of Management, Xavier Square, Chandrasekharpur, Bhubaneswar-751013, District-Khurda, Odisha.
Email: amar@ximb.ac.in
Website: www.ximb.ac.in, http://hib.xub.edu.in/Hibiscus/Pub/faccvDet.php?client=xu&facid=XF241.

© Springer Nature Switzerland AG 2019
A. K. J. R. Nayak (ed.), *Transition Strategies for Sustainable Community Systems*,
The Anthropocene: Politik—Economics—Society—Science 26,
https://doi.org/10.1007/978-3-030-00356-2

About the Contributors

Alag, Yoginder K., Chancellor of Central University, Gujurat & Emeritus Professor; Gujurat, India; *Email*: yalagh@gmail.com

Bhushan, Surya, Associate Professor, Development Management Institute (DMI), Patna, DMI, Patna, Bihar, India; *Email*: surya.bhushan@gmail.com

Coelho, Nyla, Co-ordinator, TALEEMNET, Belagavi, Karnataka; *Email*: nylasai@gmail.com

Daniel, Joshua N., Programme Advisor, BAIF Development Research Foundation, Pune, BAIF, Pune, Maharashtra, India; *Email*: jndaniel@baif.org.in

Dash, Anup, Formerly Professor of Sociology, Utkal University, Bhubaneswar, Bhubaneswar, Odisha, India; *Email*: dashanup@hotmail.com

Datta, Sankar, Retired Professor and Head, Livelihood Initiatives, Azim Premji University, Bengaluru, Bengaluru, Karnataka, India; *Email*: dattasankar@rediffmail.com

Dhiman, Amrita, Associate Professor, Development Management Institute (DMI), Patna, DMI Patna, Bihar, India; *Email*: adhiman@dmi.ac.in

Giri, Ananta Kumar, Professor, Madras Institute of Development Studies, Chennai, 79 Second Main Road, Gandhi Nagar, Chennai, Tamil Nadu, India; *Email*: aumkrishna@gmail.com

Gupta, Abhranil, Ph.D. Scholar, Xavier Institute of Management, Bhubaneswar, XIMB, Bhubaneshwar, India; *Email*: abhranil@stu.ximb.ac.in

Hanumankar, Hemnath Rao, Senior Professor and Dean, Development Management Institute (DMI), DMI, Patna, Bihar, India; *Email*: hrhanumankar@dmi.ac.in

Kar, Bijayananda, Former Professor and Head, Post Graduate Department of Philosophy, Utkal University, Bhubaneswar, Odisha, India; *Email*: bkar.nkar@gmail.com

Khuntia, Tanmay, Ph. D. Scholar, Xavier Institute of Management, Bhubaneswar, XIMB, Bhubaneswar, Odisha, India; *Email*: tanmaykhuntia@gmail.com

© Springer Nature Switzerland AG 2019

A. K. J. R. Nayak (ed.), *Transition Strategies for Sustainable Community Systems*, The Anthropocene: Politik—Economics—Society—Science 26, https://doi.org/10.1007/978-3-030-00356-2

Krishnamurthi, G., Senior Professor and Dean, Development Management Institute (DMI), Patna, DMI, Patna, Bihar, India; *Email*: gk1949@gmail.com

Kumar, Praveen, Associate Professor, National Institute of Social Work and Social Sciences (NISWSS), Bhubaneswar, NISWASS, Bhubaneswar, Odisha, India; *Email*: Praveenkumar.kumar1976@gmail.com

Kumar, Santosh, District Project Manager, Bihar Rural Livelihoods Programme Society (BRLPS), Nalanda, BRLPS, Nalanda, Bihar, India; *Email*: dpm_nalanda@brlp.in

Mishra, Bishnu Prasad, Dean, School of Engineering and Technology (SOET), Centurion University of Technology and Management (CUTM), SOET, Paralakhemundi, Odisha, India; *Email*: bp.mishra@cutm.ac.in

Mishra, Saswat, Std-XII, DAV Public School, Unit 8, Bhubaneswar, Bhubaneswar, Odisha, India; *Email*: aamlansaswat@gmail.com

Mohapatra, Lipsa, Associate Professor, National Institute of Fashion Technology, Bhubaneswar, NIFT, Bhubaneswar, Odisha, India; *Email*: lipsa.mohapatra@nift.ac.in

Naik, Hemantbhai B., Ex. Secretary, Amalsad V. V. K. S. K. Mandali Ltd. Amalsad, Amalsad, Navsari, Gujarat, India; *Email*: hemantnaik.p@gmail.com

Nayak, Amar K. J. R., Professor of Strategy & NABARD Chair Professor, Xavier Institute of Management, Bhubaneswar, *XIMB*, Bhubaneswar, Odisha, India; *Email*: amar@ximb.ac.in

Ojha, Jaya Kritika, Associate Professor, Development Management Institute (DMI), Patna, DMI, Patna, Bihar, India, *Email*: jkojha@dmi.ac.in

Padhee, Usha, Joint Secretary, Ministry of Civil Aviation, Government of India and Doctoral Scholar, XIMB, C-14, Sector-1, Noida, Uttar Pradesh 201301; *Email*: usha.padhee@nic.in

Panda, Asish Kumar, Head-Corporate Interface, Centre for Management Studies, Nalsar University of Law, Hyerabad, Hyderabad, Telengana, India; *Email*: asish@nalsar.ac.in

Patil, Parashram J., Researcher, Institute of Natural Resources, Maharashtra, India, INR Maharashtra, India; *Email*: patilparashram9@gmail.com

Patnaik, Amar, Principal Accountant General, city, West Bengal, West Bengal, India; *Email*: amar_patnaik@yahoo.com

Prasan, Chandrakanta B., Senior Project Officer, BAIF Development Research Foundation, Pune, BAIF, Pune, Maharashtra, India; *Email*: ckbaradaprasan@gmail.com

Raj, S. Antony SJ, Lecturer, Xavier School of Human Resources Management, Xavier University, Bhubaneswar, XUB, Bhubaneswar, Odisha, India; *Email*: stonysj@xshrm.edu.in

Rath, Banashri, Divisional Head, Odisha Industrial Infrastructre Development Corporation (IDCO), IDCO, Bhubaneswar, Odisha, India; *Email*: banashreerath@gmail.com

Saha, Goutam, Associate Professor, National Institute of Fashion Technology, Bhubaneswar, NIFT, Bhubaneswar, India; *Email*: goutam.saha@nift.ac.in

Sarao, K. T. S., Professor of Buddhist Studies, Delhi University, Delhi University, New Delhi, India; *Email*: ktssarao@hotmail.com

Sen, Mou, Joint Director, Department Of Micro, Small and Medium Scale Enterprise and Textile (MSME&T), Government of West Bengal, MSME&T, West Bengal, India; *Email*: mailmousen@gmail.com

Sharma, Arun K., Senior Scientist, Central Arid Zone Research Institute, CAZRI, Jodhpur, CAZRI, Jodhpur, Rajasthan, India; *Email*: arun.k_sharma@yahoo.co.in

Singh, Piyush Kumar, Assistant Professor, Indian Institute of Technology (IIT), Kharagpur, IIT, Kharagpur, West Bengal, India; *Email*: piyushsingh. er@gmail.com

Suna, Birendra, Assistant Professor, Dept. of Social Work, National Institute of Social Work and Social Sciences (NISWASS), Bhubaneswar, NISWASS, Bhubaneswar, Odisha, India; *Email*: sona.birendra@gmail.com

Swain Laxmidhar, Network Director, Centurion University of Technology and Management (CUTM), CUTM, Bhubaneswar, Odisha, India, *Email*: laxmidhar@cutm.ac.in

Telidevara, Sridhar, Associate Professor, Great Lakes Institute of Management, Gurgaon, Gurgaon, Haryana, India; *Email*: sridhar.telidevara@gmail.com

Printed in the United States
By Bookmasters